T0258476

SELF, LOGIC, AND FIGURATIVE THINKING

SELF, LOGIC, AND FIGURATIVE THINKING

Harwood Fisher

Columbia University Press   New York

Columbia University Press
Publishers Since 1893
New York   Chichester, West Sussex
Copyright © 2009 Columbia University Press

Library of Congress Cataloging-in-Publication Data
Fisher, Harwood.
  Self, logic, and figurative thinking / Harwood Fisher.
      p.   cm.
  Includes bibliographical references and index.
  ISBN 978–0–231–14504–6 (cloth : alk. paper) — ISBN 978–0–231–51866–6 (ebook)
  1. Logic. 2. Figures of speech. 3. Self psychology. I. Title.
  BC53.F57   2008
  155.2—dc22

                                                            2008029089

Columbia University Press books are printed on permanent and durable acid-free paper.
This book is printed on paper with recycled content.
Printed in the United States of America

c 10 9 8 7 6 5 4 3 2 1

References to Internet Web sites (URLs) were accurate at the time of writing.
Neither the author nor Columbia University Press is responsible for URLs
that may have expired or changed since the manuscript was prepared.

To Helene

Contents

Preface

When Reason Is King and the Power of Imagination Is Knowledge

My father, Sam Fisher, believed, along with the legacy of Socrates and the formalizations of Aristotle, that you have no rational choice but to employ the logic inherent in your own thought. "Use your own head" was his battle cry. He'd tap his temple to punctuate his thinking. Maybe a touch of homeopathic magic would transport his weapons of thought out through his head and into mine. How could I resist caricaturing this gesture as imperious promulgating? It's the kind you love to show just can't be 100 percent right! But what else could he do? His fatherly duty was to pass on the continuing saga of his love of the lessons of law school. My mother was there to inspire us both by mythic journeys back into the time my father was a most assiduous and indefatigable pupil. I could be, too. She said so. I thought so. Sam Fisher insisted.

I was living in Brooklyn; it was during World War II. After my long day at Public School 105, grades K to 6, I did my homework. My mother shared her wonder that my father could lock himself in his room and sit for hours studying. I just wanted to go outside to role play with the kids on my block. I could be a comic book hero. I was Captain America, jumping stairs, waging battle with Captain Nazi. My Captain America identity didn't help my grades in PS 105. Homework continued its demands in the Brooklyn junior and senior high schools. It vindicated the role of logic and reason in making sense of history, math, science, civics.

I probably would not have written this book if my caricature of Sam Fisher was not trumped by the profound perplexity he helped to originate. He had another hat. When he switched from the lawyer's cap to the artist's beret, right from his head emerged a quite different theory of thought. It was in his drawings. They were another form for his conversation with me. Drawing was easy. I could do it just as he did. With a very few pencil strokes, I set sail on paper, experienced the sun's warmth, and depicted the waves of an ocean. On them was a boat. It was a sailboat, but it was not a sailboat. What was this "it is AND it isn't" logic? What kind of thought was it? If this thinking was so contradictory, then Sam Fisher would be the first to say it was not logical; it didn't rise to the status of reason. So how come, when he and I drew this image and launched this thought, I came up with this weird "sailboat category"? Who was doing the caricaturing? Was I? Was my father, who, after all, did the hat-switching routine? What was he saying about the logic of thought?

If there was logic in caricature, what was the relation of logic to rational thought? I had to know this—even right then, when sharp categories of good and evil were tragically expressed in the human costs of a war raging on four continents. Even right there, as I went from elementary school to high school, the veteran white-haired teachers were dominated by the Aristotelian view of logic and thought they had learned, just as my father did when he went to public school in the Brooklyn of his day. Mr. Kelly, Thomas J., my junior high school Latin teacher, reassuringly introduced Caesar's *Commentaries*. "All Gaul is divided in three parts," it began. When it and the beautiful logic of the language's syntax comfortingly conquered all in its path, we all knew that not only Mr. Kelly, but also we mere pupils could and should apply formal logical rules and organize knowledge of the world around us.

My father, however, who had thought in this mode for longer than I, had escaped a Romanized domination of Aristotle's version of thought by a hat trick. But when he tapped at his temple, was it an appeal to magic or just a figurative gesture? That's a question that echoes from chapter to chapter of this book.

I should stop this nostalgia for the New York public schools and how their teaching during those years came to produce clashes with the "it is and it isn't" logic of the artist's thought. I should get down to tell-ing you what I am going to tell you in my book. I claim that the same logical organization and forms reserved for reason can also apply to the contradiction-haunted "is and isn't" categories of figurative and imagina-tive thought. Can I do so coherently? If one mode of thought promotes contradiction and another seeks to eliminate it, does logic pull thought together or cut it apart? My father extolled the logic of "use your own head," yet he also showed me how to draw things that were and were not in their own category. I could choose among and switch modes of thought and their angles on logic and contradiction. So here I am—on deck to say how this is done and where logic fits in. My puzzlement morphed into this proposed resolution in the book: the self of the individual, who thinks figuratively, has the role, if not the power, to unfold the logic in his metaphors and analogies. Even as caricatures, they are his inventions. They can seamlessly become part of the thinking of artists and scientists alike. (Of lawyers, too, but I don't get into that.)

It is fortuitous, if not fated, that I look at the topic first from a child's eyes. It helps to think figuratively because my probe over a span of time reveals that the self "is and isn't" the same for me the child, for me the young person, and for the "me" whose sense of self grows and changes over the course of adult life. In a moment, the preview, but first a most immodest desideratum: you as reader and I as author may discover how the self orchestrates logical relationships within and among our own figu-rative thoughts. We may also gain insight into how we bring these rela-tionships into harmony with our many different modes of thought.

My preview spotlights the problems I address in the book: the book's purpose, which I present as main questions and a proposal for their reso-lution; relevant approaches in psychology and linguistics; the strategy of my argument; and a travel pack of clues that can point to the way out of awesome dilemmas. All this plus a note on how best to read the book and experience its journey.

The Problem of the One Logic and the Many Ways of Thinking

If I am to find the relation of logic to figurative thinking, I have to disen-tangle figurative thinking from its "look-alikes." Is the dream the same as a figure of speech? Does a child think in metaphors and ironies? Is there any difference between the "primitive" thinker's organization by families and the "civilized" person's stereotyping? Is the distinction one I can assign by summarily saying that one is literal and the other figurative thinking? It might seem that way, but I would first have to show that literal thinking is not in the same form as figurative thinking. Is this demonstration a slam dunk?

We all are familiar with satyrs—their human heads on horse or goat bodies. We also know the image of Medusa, whose locks were transformed into snakes. You might think for a moment that *you* here in the twenty-first century wouldn't think in such primitive terms—a person is a person, a goat is a goat. You don't entangle or mix forms. It's not logical thinking. But think again.

Suppose you didn't expect to win your frivolous lawsuit, but your attorney fights it mightily, and you win. You recommend him to your best friend. You say, "He was the six-hundred-pound gorilla in the room." Literally? "Well," you reply, "Just rhetorically."

I want to get beyond the language—whether it is the words you uttered, a ten-thousand-year-old pictogram somebody painted on a cave in Australia, or some prehistoric sculptor's rock carving of a thirty-two-thousand-year-old feline-headed statuette in Germany. Does the thinking of these ancient artists seem to you so different from your own bright idea that your lawyer is a "six-hundred-pound gorilla"? "I didn't mean it literally," you protest; "I simply used a figure of speech." (The term *trope* covers metaphors, ironies, and many other fantastic ways of using language. I use the term to refer to ways of thinking that can bring meaning to a pinnacle of insight or to the depths of chaos.)

The satyrs, Medusa, the six-hundred-pound gorilla—they all offer the same equation. A human being has animal features, is part animal, or, succinctly, *is* "the animal." On the face of it, these proposals have the same logic. That's remarkable, considering how much you and I actually agree that you did not intend a thought that is the same as the mythic thought of Medusa's snake head. We both disclaim an equating of your "Gorilla thought" with the ancients' logical mindset.

Underline my disclaimer. I can and do propose in this book that the logic behind the different kinds of thoughts is not the same. Even so, I can also see how it is the same. Whoops, something's amiss; that is a serious contradiction I just admitted. I need a bright idea, so it's best to go on the attack. I'll give you the problem. If you are still working to justify your six-hundred-pound gorilla thought, are you taking into account how you view your own personal logic of thought? If not, your relation to it may simply be missing. But don't take this personally. In most approaches to thought and its various modes, that *is* the missing piece of the puzzle. It is time for the change my father pulled out of a hat! Tap your temple. I propose that we revamp the very conception we have of how the logic of the individual's thought works. The new picture I draw portrays how the logic of a person's thinking relates to her and how she relates it to her thinking.

The Book's Purpose

The Questions Leading to the Disturbing Problem

I identify several modes of thought. Does each way of thinking use the same logic as that of the trope? What are the differences in the logic of schizophrenic art, primitive art, and the figurative perspectives of a trope? Does the apparent illogic of modern art justify the pernicious label of "degenerate art"? Does the label characterize the art as primitivized to reveal meanings, or is the art primitive? Is it a figure of speech when Wassily Kandinsky creates a bioform, or is that form merely a throwback to primitive imaging or conceptualizing? Is "regression in service of the ego" simply "regression"? The disturbing problem common to these questions is the extraordinary degree of similarity of logical form across obviously different and magnificently varied modes of thought.

The Proposal

I propose to recast the logic of thought so that it is linked to the individual thinker's self. But I also aim to specify the logical forms of thought and show how they are different in the trope when they are compared to the thoughts we call primitive, immature, and pathological. The self can oversee and coordinate all sorts of transformations from one to another of these different ways of thinking. The self, yours included, composes and conducts, bringing multiple modes of thought as counterpoint to an integrated resolution—just as it happens by way of the transformations and transmigrations from one form to another in a Bach fantasy and fugue or in a Mozart symphony. You can, as Freud showed, dig into a dream and find a meaning, which you can convert to a less primitive figure of speech—and even further transform into the more lofty logic of an understanding of yourself.

I set the groundwork for this proposal by reviewing a number of approaches to the self and thinking. Many are from the traditional view of thought displayed as the individual's language. Most emphasize that a person's thought is shaped by her experiences—encounters that synchronize her inner needs with environmental realities and other persons' demands. The captivating concept of a person's need for "cognitive consistency" has generated ideas and empirical studies to show how a person balances her decisions and conclusions with her attitudes and beliefs. Explanations of this sort depend on concepts of how a person combines her values, feelings, and logic with the logic of her thought. The colorful term "psycho-logic" should give you a feel for these combination concepts.

A full-dress review reveals both empirical and theoretical negatives. There are devastating gaps between theory and evidence. And a killer blow is the absence of a coherent explanation of a person's access to the variegated forms of her own thinking. The pileup of negatives strongly points to the need for a revamped idea. The way I look at it, logic is a "living thing"—a brainchild you foster; it responds by transforming itself, yourself, and your way of looking at things. That's the thumbnail of my proposal and the way I introduce the issues, the background, and the main themes. On to how I unfold the book's primary purpose and a little about how I avoid "either-or" thinking.

How I Unfold and Develop the Book's Purpose

The book's primary purpose is to show the self face to face with the logic that can describe and explain how tropes arise in thought. It would be nice if we knew more directly how Einstein thought of his pictures of clocks flying through space. However, I want to avoid making an impossible leap into the subjective world—his, yours, or anyone's. It is impossible unless you and I have the language to conceive, let alone directly penetrate, the meaning of a person's thought or the moment of its origin.

My major strategy is to separate the logic from the linguistics of the trope. But I am mindful that our glimpse into thought is after the fact. In front of our eyes is the rhetorical act. Einstein wrote his thoughts down. We reveal thoughts to each other and ourselves, but we do so through the screen of language or some other form of representation. So my major strategy is not without a bow to the facts of life: I cannot look directly at your thoughts, but I can view how they appear in the language and images you convey. Acknowledging rather than wishing away this fact allows us as fellow journeymen to keep faith with the best of the two worlds.

Here is my support line of defense. When I consider your display of thought in language or images, it reflects your contact with others and the world around us. You might say, "I will get to the theater by train because the cabs are busy this time of year." As many have shown, it is easy to map out your statement in an image space that shows specific routes and movements—even where you are the agent with a target objective and where the outcome is specific in relation to your role and to other events in the "space." If this diagrammed space for your linguistic display is reflective of the "real" movements when you actually do make the train trip, then I would have a picture that is a "good" representation—and, in some degree, a picture of how you picture these things in thought. The language you and I use appears as a series of condensed maps, which I call *schemas* in the book. This term helps to link up my ideas with the descriptions of classical rhetoric, traditional linguistics, and contemporary cognitive linguistics. Schemas chart our forays into the world to engage our own needs and others' demands. Not a bad thing to do—except for the many pretensions that it is a sufficient explanation. In the course of the book, I find a place for what is too often presented as the only show in town—schematic organization of thought as language. But I assure you, this time, in this book, the logic of thought sits right up front with self as the driver.

The task of showing the logical organization is going to involve redrawing maps that divide mind from self. Do I show the self as a logical operator? A category? There are also the awesomely varied modes of thought—the dream, primitive totemic thinking, and the art some classify as primitivized. Can I meet the challenge to depict a coherent view of their logic and show its access to the growth of the person's thought and expression? There are seminal ideas I can bring to the task, although they have to be culled from cognitive anthropology and the semiotics of Jean Piaget and Charles Peirce. Finding those points of contact is a high, but in the end a coherent big picture hinges on my thesis that the self is part and parcel of any sensible and robust logical organization of thought.

That's some burden to put on the self! It socks it to the individual to orchestrate her thoughts. The self's task is to embrace logically the differences in each of the modes of thought with which I compare tropes.

The scope of these different modes is so wide that I don't know why I wasn't immediately scared off. Why should it all come together? Probably because the issues I raise in the book actually work as feedback to resolve the childhood puzzlements that led me to those issues in the first place! Such are the intriguing, inherent, autoregulating reflective powers of the self my father sought to invoke.

For Your Travel Kit: Strategies, Clues, and Relief from Worries

My central strategy is to support the idea that logic is not alone. It does not control you. You create its overall structure and its subclass differences. To bring the self into the equation, battles must be fought. One worth waging is against the separation of logic from psychology—a venerable but penetrable philosophical Maginot Line. If I retreat, logic is closed off, imprisoned by excessive dialectical reasoning. Tradition would keep its form without psychological substance, a situation calling for a caricaturist. Ah, a familiar tactic. We can "recall to life" Voltaire. His invented character Dr. Pangloss shows how meanings are drained out of illustrious syllogisms and how dilemmas are empowered to promote and win wars dubiously, if not just tragically to fight them. From the rich logic of caricature and more so from the creative reach of the trope, we can pivot to open the connection between an overall logic of thought and the psychology of the self.

Prior to your read, here are three clues to follow my path to this connection. First, in order to portray a person's dynamics and rules of thought, I have to show how logic and the self interplay not only to *organize,* but also to *originate* the different forms or modes of thought. Second, I admit that awareness of the self is just not always there. At times, you are fully aware of the way you are forming a thought, so you are there. At other times, you are just beneath the thought's surface. But when you are not there, your logic leads you around like you have no mind of your own. When I describe the different thought modes, your awareness of self appears in different degrees in the different modes of thinking. So some fancy footwork—logical gyration—has to be performed in placing the self where it is not and in tuning it to the person's varying degrees of self-awareness. *But not to worry.* I have different ways of marking the footwork. I have my figures of speech. I have my dialectical format. I have my different hats. I have my dreams to plunge into.

Third, I use logical symbols many times to express the relationships of self and logic and to spell out the elements of a particular form of thinking. You can, if you want, skip forward and just read past these symbols. If the spirit moves you, you can come back to them later. Readers who may not love logical symbols or who like to breeze ahead will certainly get the ideas—which for me are actually more exciting when symbolically specified as operations of thought. I find this symbolic form of expression helps to create a language that lights up transformation possibilities.

The trope stands midway between the self and its access to the thought modes. The self grows richer, and so does the person's ability to mine all sorts of thinking—primitive, dreamlike, and poetic—to enrich his own ways of thinking about ideas, their logical place in relation to his personal sense of self, and his knowledge of the world around him.

SELF, LOGIC, AND FIGURATIVE THINKING

Introduction

*Major Terms, Their Classification, and
Their Relation to the Book's Objective*

When a lover refers to his heart as my "foolish heart," is he being poetic? Is he a pathological case? Suppose he appeals to it as if it had its own mind: "Take care, my foolish heart" (from the 1949 song by Ned Washington and Victor Young)? Has his thought degenerated to that of a madman? Is he just being childish? Has he slipped into a time warp and become like a person who lived long ago, before the human brain inhibited easy exchanges between fantastic projections of wishes and reality-checking propositional thoughts? In the song, the lyricist means to say the heart is foolish *figuratively*. He does not mean it *literally*, so he puts his idea in the form of a figure of speech.

Although we usually think of a figure of speech as a rhetorical device, in this book I focus on the form's logical structure. I do not focus on its meaning—that is not problematic. A figurative statement is patently not the same as a literal statement! You easily know the difference between thinking you *actually* have a heart that is foolish and singing the lyricist's phrase to bemoan how you respond when you're in love. Logical form is another matter. If we tie it to meaning, the structure remains within the realm of rhetoric. Logicians can put aside the meanings of the words *heart* and *foolish*, yet still look at how they relate as terms. When they say things like "S is P," the terms relate, but are free from meaning. Just as logicians do when they assign abstract letters to such terms, you can try this sort of thing with *heart* and *foolish*. The figurative and the literal statements undergo a startling transformation when we reduce each to its logical form. Despite the statements' striking difference in meaning and its communication, their logical forms appear tantalizingly the same. Yet my goal is to show how *the logical structures are distinctive, just as the meanings are different.*

When you think of a figure of speech, classically you think of language expressed to have some effect on an audience. This is a standard function of *rhetoric.* There is a long history of the use and study of rhetoric as a means of persuasion, a route to attitude change and to the attendant social and political objectives (Burke 1952). It *is* vital to find out how to use words, speech, and various linguistic forms to influence a person's views, attitudes, and thinking. In that pursuit, not only can you ask about process, but you can also point to forms, such as discrete figures of speech. The danger is circularity. Seeking rhetorical forms looks like a runaround to find rhetorical terms recasting rhetorical events.

In the history of definitions of rhetoric, there has been a shifting emphasis. Sometimes the beauty and mystery of language is in the foreground; sometimes the power and function of thought are. In this book, I am *not* after the answer to the standard rhetorical question of how to use language to influence yourself or others. That's one reason I focus on the role of thought in rhetoric instead of pursuing the role of persuasive meanings and forms. The focus also helps to avoid the circularity of explaining phenomena such as figures of speech in mere rhetorical contexts.

When I pose the issue of logical form, I am after the structure of thought—as it interrelates with figures of speech and with the other linguistic phenomena. I aim to show a specific logical identity for each of the different ways of thinking. To spotlight *figurative thinking,* I use the term *trope,* which lends itself to an analysis of logical form. Although I try to avoid emphasizing the loaded term *figure of speech,* these figures lurk in the background—their shadows complicating their identity.

All this would not be so perplexing if only you could disregard the whole issue of rhetorical matters. The fact is, though, that features of meaning and language *are* part of the picture! I purposely began this introduction with wildly different meanings of the phrase "my foolish heart." Did I paint my argument into a corner? It sure looks like I did. If I focus on rhetorical contexts, I lose the meanings in circular analysis; if I avoid the meanings, I leave out their differences. Can we be logical without deferring to this "either-or"?

We can get out of the corner if we show an articulated coherent relationship between the terms *trope* and *figure of speech*. Hayden White, from the vantage point of his historical approach to literary criticism and its denouement in consciousness studies, organizes these two terms. He is literal about the definition of *trope*. It is a "'turn' or derivation from literal speech or the conventional meaning and order of words" (1999:104). He casts the trope as a catchall or supercategory of the turns or derivations from meaning and order. Look at the way he places the trope as a logical category in a ladder of categories. I depict this placement as a three-tiered classification ladder with a top rung and then subcategories at descending rungs, or levels. View the term *trope* as a category on top at level 1. Next down is a level 2 category—*modes of figuration*. In the classical view White brings forward are four modes with the recognizable names *metaphor, metonymy, synecdoche,* and *irony*. *Figures of speech* are one more level down (level 3) as specific instances of those four modes, which White calls "the class names of generic categories of figures of speech" (105).

A Three-level Hierarchy of Categories

Level 1	Trope
	∧
Level 2	The four generic categories
	∧
Level 3	The specific figures of speech

The term *trope* at the top (level 1) is the most inclusive of the terms. The supercategory "trope" subsumes the generic category "irony" and in fact subsumes all four class names for the modes of figuration (at level 2). To complete the picture of the hierarchy, each of these four generic categories includes specific figures of speech. Thus, a given figure of speech, such as a striking phrase in a lyric, is at level 3. On this level, a specific instance is a member of one or more of the four generic categories of level 2. For example, the specific figure "foolish heart" is subsumed under the level 2 genus with the class name *irony*. (In an irony, two terms such as *foolish* and *heart* that would not conventionally be put together produce a contrast). "Juliet is the Sun" is another example of a specific figure, but it is subsumed under the generic category with the class name *metaphor*.

This is a great deal of classification, but it is important to show why I focus on the term *trope*. The term *figure of speech* is more familiar, and it *does* cover much ground. However, the term *trope,* because of its logical position as the supercategory for the generic categories "metaphor," "metonymy," "synecdoche," and "irony," has much more range—and perhaps more leverage in matters of logical structure.

White's concept of the trope atop a hierarchy is elegant. Yet there are the realities of thought and of how you experience the impact of a turning and twisting trope when a figure of speech comes your way. A writer such as Edgar Allen Poe hits you fast and hard with a figure of speech in his story "The Cask of Amontillado": "For the love of God, Montresor" ([2000] 2008a). But you do a slow take. You find yourself thinking about it again and again when you look back at it. One character uses the phrase to plead for mercy; the other intones it to moralize revenge. You spin around, turning from the literal to the figurative. Does Poe mean to play on a metaphor? Is there an ironic twist from literal to figurative? Do you go to the top of White's ladder and think about the nature of a trope? Do you first decide on how the literal interplays with the figurative? Do you start at the impact of irony?

If you adopt a perspective at different levels of the hierarchy as you experience them and work things out from there, it helps you to stabilize your interpretation and to sense the whole as a hierarchy of meanings. In this sense, you can also view the question of logical structure from the different levels of the classification hierarchy. These levels vary from specific "figure" to generic "category" to "supercategory of generic categories." However, if you *can* focus from anywhere within this organization, what is the value of White's classification as a hierarchy? Answer: we are dealing with some marvelously dynamic aspect of thought that can jump from rung to rung, yet leave the ladder intact.

I have another caveat. The turning of meanings around a ladder of logical ordering is not the only vicissitude of meaning that can spin you into dizziness. Be prepared that when you think about a statement or utterance, the logical hierarchy of categories does not merely go up and down. Terms at the different levels sometimes blend into each other as if the meanings, their abstractions, and their entailments are being constantly redistributed. Sometimes the abstract category is separate from the particular; sometimes it entails the particular. We will see this phenomenon when we take up particular figures of speech and consider their generic categories—all in relation to the nature of a trope. I can abbreviate here by showing this blending in relation to something basic—the definition of the very terms distributed in the classification ladder. The *Concise Oxford Dictionary of Linguistics* (1997) defines *trope* as "a figure of speech, especially one involving a figurative extension of the meaning of a word or other expression."

With that definition, the three levels of the hierarchy overlap. To stop the vertigo as we hang on our ladder, we have to note *and* disentangle their interrelations. If we are squaring all this mix of terms and levels with the hierarchy, we note the trope (at level 1) and the figure of speech (at level 3) appear to blend. But there's another mix. To get at the specifics of the classification structure, we have to make a choice from the bevy of the four different figures of speech at the generic mode (level 2). I say "bevy" because the four figures can also combine in complex forms. By identifying just which figure is apt, we still the whirl of these figures and their variants. I can show a bit of what happens when we take a stand at level 2.

If we look at the statement "For the love of God, Montresor," and try to home in on the kind of technique or figure Poe is using, it helps to make the question specific. Ask whether a nested metonymy or an irony is basic to the generic mode. That which is basic might help unfold the logical form identifying the thinking involved. In a *metonymy*, an aspect of the meaning targeted is equated with the whole meaning. If it is a metonymy Poe is using, it might be something like

Revenge ≡ Love of God.

In an *irony*, two presumably contrary aspects of a targeted meaning are related to the whole of that meaning. So if Poe's figure is an irony; the logical equivalence (≡) appears in a chain.

(Mercy ≡ Love of God) ≡ (Revenge ≡ Love of God).

We arrive at this picture of logical relationships by starting at level 2— the middle level of the hierarchy. There we have the mix of meaning and logic that we have to wrestle. We note that within level 2, internal logical relations can be unfolded. A look at the chain of equivalences reveals the specific logic of two different figures. We see the identity of each, and we see how the one relates to the other.

Blending emerges as a mixed blessing. We are dismayed that the terms at different levels are entangled and entail each other, but pulling them apart has a great reward in revealing the identity of the specific figures. The whole picture shows the three hierarchical levels not only in relation to each other, but also as discrete structures—each with its own interrelations. Is it disconcerting to start in the middle of the hierarchy? We will see the creative flexibility. We can start almost anywhere, yet when we finish our disentangling, the ladder's logical steps remain intact. It's all "natural feeling"—just as when Poe's phrase hits you. It's not in the abstract—it's in the solar plexus, the middle of things. It's nice to feel that you can come out of the reeling dislocation and that your sense of meaning and logic is not negated, but enriched instead.

1   The Problem of Analogous Forms

How can you specify the differences in logical structure in forms embedded in meanings intended as figurative compared to those forms in which meanings are mired in the literal?

It is not only songwriters seeking to inspire lovers who appeal to figurative thinking. The "hero" in Poe's ([2000] 2008b) famous short story "The Tell-Tale Heart" murders a victim—but not before reporting he hears the sound of his victim's heart. After he murders the fellow, he hears the sound again. He believes it to be that of the same heart beating. Most readers look at this belief as pathologically "literal." We look for clues to understand the character, who fears he will be found out. We think we understand that the character has concretized his fear, if not his conscience. We want to say the "sound" he hears is the equivalent of his conscience. We can look at the tactics of his "literalizing." Is the "sound," as an equivalent of his conscience or his fear, some sort of rhetoric in his communicating to himself? Is it like a trope? You have the form of a metonymy if you say he is reducing the fear of being found out to a heart's sound (or a recapitulation of that sound). There's the part-whole formulation. The thought reflecting his fear of being caught is equated with a portion of what evokes it. But the literalization proceeds further. The "hero" also believes strongly the sound is something that *others* can hear.

Despite his protestations that he is *not* a madman, he has surely gone over the line in concretizing his "hearing" of the heartbeat and consequently in believing that the sound of the dead man's heart communicates to others! However, the logical form of the idea that the heart "communicates" appears as even more metonymy.

(It's complex. *The first part-whole relation:* A portion of the hero's thought experience represents his fear, which is the more complete [whole] thought. *The second more literal level:* If his hearing were acute enough to hear the victim's heart [when the victim was alive], the hero might form an impression of a "living heart." His idea that the "sound in his head" is equivalent to that former living heart is also a part-for-the-whole form.)

When we go back to the Washington and Young song, we easily see the difference between the lyricist's figurative use and Poe's hero's more literal experience. If the form or logical structure is the same both for Poe's madman-hero's declaration and for the songwriters' lyric, why is it intuitive simply to dismiss the comparison? Poe seems aware of just this point. As if tauntingly, Poe's "hero" ironically denies his madness, asking, "And have I not told you that what you mistake for madness is but overacuteness of the sense?" ([2000] 2008b).

Using figures of speech is not madness. On a day-to-day basis, we recognize that the nature of the beliefs and the textures of the meanings *are* different when we compare figurative with pathological thought. Yet it's curious and dismaying that the logical forms of tropes *are* difficult to disambiguate from figurative to pathological use. What happens if we do not specify the formal differences? What advantages accrue if we do? The problem of the similarity of these logical forms spreads to an even wider set of cases that I explore in this book.

The multifaceted Robert Musil—one of the legendary men of letters of fin-de-siècle Vienna—was trained in engineering and mathematics, but he wrote literature and plays and was an astute student of the arts. His concept of "ideas" was that they rise above the concrete and the unambiguous. This is precisely what a trope is supposed to do. In fact, his definition of "ideas" comes close to that of capturing figurative thought in a trope! He describes the writer's task "to discover ever new solutions, connections, constellations, variables, to set up prototypes of an order of events, appealing models of how one can be human, to *invent* the inner person" ([1918] 1990a:64).

When Musil extends his thought to art and aesthetics, he realizes that to distinguish between "imagination" and "illusion," to distinguish art from psychiatric "disturbance," and to detach it from susceptibility to political and religious power would be of monumental significance. He writes, "If one reads the brilliant descriptions of the thinking of primitive peoples that Lévy-Bruel has given ... in particular the characterization of that special attitude toward things that he calls participation, the connection with the experience of art becomes at many points so palpable that one could believe that art was a late form of development in that early world. It would be unusually important for research in the field of aesthetics to devote its attention to the clarification of those connections, which, on the other side, are profoundly bound up with psychopathology" ([1925] 1990b:196–97).

In this book, I focus on the "trope" as a form of expressing ideas. The distinction Musil sought is a long-time problem that I cast this way: *logical forms of the figures of speech we call "tropes" are similar to the logical forms of pathological, children's, and primitive thought.*

I go further if I ask what happens if we do *not* specify the differences in form. A provocative answer is this: *the similarity of the forms breeds exchanges from one thought mode to another!*

This response, paradoxically, does do justice to the problem I raise in that it reveals something profoundly incompatible with the intuition that the forms are different. Yet it is also a key to the apparent similarities of the forms. So I'll spend time on an example in which the similarity of forms encourages exchanges of the different thought modes—as if they *were* equivalent.

In Don Quixote's fight with windmills, if his thought or its expression in his behavior were just a figure of speech, then the power and movement of a windmill as a "monstrous giant" would not be a chimera. It would be a thought with symbolic meaning. Is the hero's conception of "windmill as monstrous giant" simply a metaphor telling us the machines we make wind up replicating human qualities? The story is that Don Quixote believes the idea concretely and that, to support it, he argues that the giant himself has concocted a presentation of his own form as a windmill. The belief and the supportive argument seem pathological, if not childish. Nevertheless, it is what Don Quixote, an otherwise mature man, claims. And, too, with ease and conviction he invests human qualities in an inanimate source of power—an evocation of primitive thought worthy of a comparison with any of the many reports by anthropologist James Frazer ([1922] 1950). The easy passage from a mature Western mode of thought to myth making is striking, even if you want to consider the Man of La Mancha (or Cervantes, as author) as just slyly *caricaturing* the way myths are made.

It is in the nature of the similarity of the logical forms that the transactions between irony and myth making appear so uninhibited. Cervantes tells us that Don Quixote is confused and that the source of his degenerated thinking is, at least in part, the confusing nature of overdrawn "conceits"—that is, the development of tropes in the literature *he* has read. Cervantes writes about his hero: "what with little sleep and much reading his brains got so dry that he lost his wits. His fancy grew full of what he used to read about in his books, enchantments, quarrels, battles, challenges, wounds, wooings, loves, agonies, and all sorts of impossible nonsense; and it so possessed his mind that the whole fabric of invention and fancy he read of was true, that to him no history in the world had more reality in it" ([1605] 1882).

Perhaps the story of Don Quixote's intellectual meltdown reveals some point of diminishing returns—even when it is a person with a brave new *postbicameral* brain who is attempting to symbolize. Playing with rhetoric and its logical forms is playing with fire! Even with the development of governance over what we think is merely "inside" ourselves versus what we believe is external reality, our imagination, once incited, finds so much rhetorical context to energize it that it outruns the logical development of our thinking. Have we simply overdeveloped the way we relate icons? Is the overdevelopment of tropes our common fate—a weariness of too much exposure to the mixes of logic and rhetoric pervading our human life of thought and communication? Maybe it is an inherent feature of figures of speech that as we continue to use them, they advance influence over our thought exponentially—in Orwellian cadences. The heartbeat is not a bad image to use to illustrate such a cadenced march toward diminishing returns. When figures of speech reach a certain point, they invite the confounding of logical distinctions—poetically and stereotypically appealing to *resonances,* so that what's desirable and meaningful is simply like the matching sounds of familiar cadences. That which sounds as if it matches sounds and feels right. What form of logic is it when forms and meanings fit each other by their familiar appeal to the same feelings and reminiscences? *Resonating forms, meanings, and feelings bring a sense of equivalence—one thing or idea morphing into another, equating, and, in some sense, logically transforming.*

The logic in an Orwellian phrase such as "War is Peace" ([1949] 1969: 164) has the resonance of *irony.* There is logical transform in the face of irresolution. The slogan sounds trumpets announcing a universal proposition, but one that equates semantic opposites. How *is* the opposition of meaning converted to a proposition of equivalent terms? When the trope is invoked, a logical state of affairs is presented with *syntactic* oppositions of words. The proposition "War is Peace" is in the syntactic form "A is B." One meaning is placed on one side of the "is," and an opposing meaning is placed on the other. Three syntactic slots appear—two for meanings and one for the "is." But a proposition is also a logical structure, and the "is" permits an operation of equivalence. So the logical form of the proposition sporting an "is" allows for the placement of meanings in *syntactic opposition* (O_{syn}).

What appears next is an odd kind of resonance—yet resonance it is. The O_{syn} resonates with the fact that the terms in the syntactic slots (*war* and *peace*) are themselves opposite meanings. Label this fact a *semantic opposition* (O_{sem}). Now we have a matching of oppositions within the same trope. So, like a charm of reverberation, the *resonance of the oppositions* prevails! Trope trumps contradiction. (The trope achieving this resonance of oppositions is irony.)

Thus, Orwell's slogan resolves contradiction. What we intuitively experience as an inherent assault on meaning is drowned out by resonance. This scary power of the trope and the easy descent of form to the primitive are reminiscent of Freud's observation that the call to arms can evoke primitive passions and drown out the appreciation of life:

> We recall the old saying: *Si vis pacem, para bellum.* If you want to preserve peace, arm for war.
>
> It would be in keeping with the times to alter it: *Si vis vitam, para mortem.* If you want to endure life, prepare yourself for death. ([1915] 1957a:300)

These post–World War I reflections are eerie premonitions of present-day slogans in an age of terrorism! Nevertheless, no matter what the theme permutations, *irresolution and logical transformation cohabitate the same forms in the logical structure of tropes.* Further, even if logical structures *can* handle contradictions in meaning by transformations, the logical forms can go no further than their topology. Therefore, yes, they can be bent into equivalent shapes, but the irresolution is that the symbolic, the pathological, and the primitive merge!

More than one observer has reached for this high—yet dark—note. Freud's thoughts were about the inevitability of war and death. No matter what sublimations we can fashion, the instinctual pervades at bottom. "[N]ations still obey their passions far more readily than their interests. Their interests serve them, at most, as rationalizations for their passions; they put forward their interests in order to be able to give reasons for satisfying their passions" ([1915] 1957a:288).

This fact, along with the deeply rooted instinct for violence, leads to war—no matter how noble the thoughts sound, no matter how beautiful the logic seems (Freud [1933] 1964). Nor in Freud's view does the psychological development of our intellect transcend the limits imposed on meanings by our biological form. So, as to the constraints of form, we can go more deeply into the form's limits. To what extent can the cultivation of the intellect help? In Freud's thinking is the fatal pessimism of biological limits. No matter what the expansion of life and love offers, the nature of the individual organism tends toward an organismic dynamics leading to its death (Freud [1920] 1955).

So we can go on to sublimate, to make attempts to reason about the sublime and to think beyond the primitive. However, there appear points of diminishing returns in the race to go beyond the meanings that we can constrain. When that happens, the limits of form reveal themselves in the degradation of its capacities. At the least, Don Quixote's descents into merged forms ring true. Ahead for all of us is a confluence of age, experience, and the biological and cultural inevitabilities of unremittingly resonating forms. They logically twist and turn, signaling the eminently confounding similarity of the forms and meanings of mythlike thought and symbolic representations.

It would appear that we all are headed for Don Quixote's epiphanies. But if we remain aware of this fact, we are not there yet. How far can we reach if we "loop out" of our propositions and tendencies, and become aware of them and of ourselves? What's the role of awareness relative to our tropes? Does its logic interpenetrate their forms?

Don Quixote prophylactically acts in concert with the reader, who is asked to override the Man of La Mancha's concrete interpretations—because, in the reader's awareness, these interpretations are also figuratively symbolic. To what degree are they also figuratively symbolic for Don Quixote? I submit that this is difficult to determine. However, the logic of awareness has something to do with all this. When we are aware of death and taxes, we do not vitiate inevitabilities. Just so, we can be aware of the diminishing returns of symbolic and figurative reading of our lives and their meaning. Entrapped in the topology of similar logical forms, we continue in cadence on the road to entropy! Although I am *not* going to solve this dark problem of the quixotic journey to symbolize values and facts in this book, I *do* ask, *What is this ease of exchange of the primitive, pathological, and symbolic, and how similar is the logical form of the figurative and symbolic to the concrete interpretations of tropes and to "crazy" acts like dueling with windmills?*

If the windmill's power and movement were regarded as enough evidence to call it an enemy, we would logically have some form expressing that *"less than" what constitutes an "enemy" is equivalent to it.* That paralogical way of formulating things is remarkably similar for a figure of speech and for a pathological thought—and for the other modes of thought I have listed. However, what accounts can we give to explain the similarity in logic when there *are* differences in meanings and beliefs? Forgive the plethora of questions. Here's the key one: *Do these differences reflect in the logical structures themselves?*

This is a big question, but it is at the "heart" of this book's quest! Solutions on deck so far do not answer the question adequately. In developmental and clinical psychology, solutions available to account for this similarity usually say that the parallels in these forms are *not* exact. Take one explanation—so common that it has become commonplace. In relation to heuristic figurative thought, the similarities occur because of *regression in service of the ego.* (Although nonpsychoanalysts blush to use such terms, they offer them generically, anyway—without commitment to a tripartite psychic structure). Quite simply, the terms have come to mean that when we are nonplused by some lack of understanding, we look for an analogy to the primitive or childish. If mature thought and habit fail, we permit our thinking to go temporarily to an "earlier time zone" where contradiction and impossibility flourish.

"If only we could see beyond the sun, even travel there." A fantastic thought? It can launch programs and spur technologies to do these things. Such regressions to what appears impossible are the business of artists and scientists alike. So there's some ubiquitously acceptable idea that we will return from the "earlier time zone." We can count on a mature cognitive management that allows for a set of equivalences and exchanges such that excitement and imagination can enrich understanding of and even control over thought and objects.

Lapses into another time zone are at the heart of this kind of explanation. They are legitimate and desirable when we employ them to fashion tropes that advance knowledge and values. It is one thing for Charles

Dickens to write about Scrooge visiting his future and another for neuropsychiatrist Oliver Sacks (1970) to describe a patient who continuously relives a particular day in 1945 and cannot go forward into the time that *is* going forward with the rest of us. In pathological thought and experience, temporal mode shifts are not desirable; nevertheless, they are as if lapses into another time zone. As in the case of Sacks's patient Jimmie G., to equate all time with a particular time is retrogression to what we think of as a faulty generalization, yet this form of thought would have been the usual at a prior stage of development. The form is that a specific psychological experience of the memory of time (Mem_{timeA}) is equivalent to the psychological experience of memory for any other unit of time ($Mem_{timeÃ}$)

Although this form sounds like the one I suggested for "resonance," it is also analogous to the way a child might think about time or other things. But Sacks's case study does not suggest regression to childhood or regression in service of the ego. It instead is retrogression to an earlier form. In relation to pathology, the idea is that with retrogressions there are "primitivations." The dysfunction, whether psychogenic or neurological, leads to a degradation to a prior—a more "immature"—way of thinking. The individual exhibits thinking and structures congruent with a reverse direction to the development of thought.

In their attempts to explain the similarities in logical forms, analyses of this sort usually stop at a disclaimer. The similarity of logical structure is only by way of an analogy, which is necessarily incomplete. It cannot be fit to factors in the different contexts, such as those of developmental direction. Thus, as we grow into more mature ways of thinking, parents, teachers, and others present us with tasks to help the maturing process go forward. Not only this, but we ourselves can often look backward and see how we have changed the way we think and form beliefs. These experiences of the awareness of change are not necessarily those of the pathology of retrogression in cognitive function. A loss of function is sometimes, but not always, seen as one that can lead the person to "reverse reminisce" about how he was functioning when in a more mature mode.

The person with a degraded way of thinking is quite different from a person "on the way up" from a more childish to a more mature stage of thinking. When comparing the pathological case with mature and immature stages, tasks, expectations, and modes of thought, we can ontogenetically study the changes in thought, logical structure, and belief. We can cognitively and neurologically trace the degradations in pathological cases. However, in extending an analogy to what "primitive" thought might have been, we have a reality-check problem if we define *primitive* as "prehistoric." If we do look for the primitive in societies that have been insulated from change, we do *not* achieve *real* time travel back to some prehistoric state. Yet we do have some clues to how thinking might be if certain factors were never introduced.

Nor is the developmental analysis of thought totally free from similar conundrums. Putting ourselves in the place when and where we can know the infant's thought, when conscious awareness did not exist, is like postulating unconscious thought—irreducibly something we have to know about by signs and clues. So even if we have a liberal definition of what an organism can sense and know, we can only assume points in development when conscious and reflective thought "had not been." This is an assumption on which we can agree; nevertheless, we believe it only as a rational (or folk psychological) way of looking at the early stages of a child's sentient experience. We cannot ask a fetus what it is thinking, but it is not reasonable to assume that the fetus reflects upon its own awareness.

Yet empirical assumptions creep into our ideas about nonconscious sentience. These assumptions are tenuous—just as are any attributions of mental states on the basis of behavioral or structural capacities. Despite all this, there *are* guidelines for conceptualizing the great difference in consciousness between a "never was" and a "once had been." Studies of the brain patterns of persons who for various reasons have lost certain cognitive functions offer distinctions involving the presence of neurological structures and capabilities. However, the main point here is that historical facts of development attest to the "reversal" of history in the "once had been."

In a sense, a trope moves backwards in history and unearths the logical organization that "once had been." Contrariwise, a primitive belief does not move backwards. Its organization "never was" more logically articulated . When comparing the more healthy and mature thought that is figurative with the more pathologically retrogressed thought or with the immature thought that is more primitive, you are struck by the similarity of the forms; yet they are in totally different contexts. *The challenge is to show how these contexts are formal aspects of the different logical structures.*

Songwriter Cole Porter offers, "You're the top"—a proposition in a part-for-whole form. You read his statement figuratively and are delighted. You know that whoever sings this to you says you *are* the top. Even as you follow the incantations of the song, you bask in your belief in the singer's next claim. "You're the Tower of Pisa"—well, OK, but you do know the "top" and *any* top is less than "you." In contrast, you can imagine the different ways in which a child might conceive the meaning of these claims or how a person with thought pathology might read them. Yet each reader apprehends an analogous form—the part-for-whole equivalence. What is different for each is the context—which, so far, we have conceived in terms of a developmental direction.

The child is in the midst of a cognitive sequence from immature to mature forms and their interpretation. She may be stuck in a stage like that of my granddaughter, who, when six, would argue that she's not "really" a top. The argument seems her way of making the distinction as well as half-believing that I can't do so! The child, when changing toward a mature logical distinction, is even sometimes aware that the changes are taking place. Marvin Minsky (1985–86) relates a charming story about the child who realizes the difference between herself and another child at a later "Piagetian stage" of logical development. Jean Piaget observes that in early childhood, the child's *cognitive perspective* is locked into connections with sense experiences. The child's apprehension of a meaning separate from her own body and its actions remarkably sometimes allows her to go "forward" in her ability to *see* a complex set of possibilities.

This sort of articulated "seeing" is necessary to understand tropes. They have levels of reference and meaning. A "top" in the song could be an abstract notion, applicable to spatial considerations and to rankings of values. In the trope, *top* is a term in a proposition. *Cognitive perspective,* as in the history of visual perspective, requires development of near and far seeing and of the rules for each—as well as of the rules for relating the two. By classifying levels of reference and by selecting a level of meaning to apply to a focused term (such as *top*), we understand the trope. For those individuals with pathologies, however, the opposite occurs. With a degrading of function, the selection of target referents is more diffuse, and the meanings are more syncretic (cf. Werner 1948). The rules—if any—for concrete and abstract reference and interpretation are confounded. The identity of the Mona Lisa (the next comparison to "you" in Cole Porter's song) can be intermixed with "your" identity. In a pathology such as schizophrenia, "I *am* Mona Lisa" can become a distorted belief. For a schizophrenic, there are regressions and primitivations that, compared to the child's developmental sequence, move "backwards" in an opposite direction—a sequence of mature to immature.

In this book, I view the variations of time's flow within a perspective that features rules of classification, selection of levels of reference, and rules of their interrelation. Logical forms are constraining, but I would like not to be accused of attempting to "stop the clock" of time. Time continuously flows toward the particularization and realization of a form, yet *the forms themselves, as they are made, engendered, and presented, can be distinguished logically and thereby related to their context of development or degradation. That context is our cognitive perspective; it contributes to the picture of logical form. I intend to show the formal distinctions and their relationships to context as being critically dependent on the function of the self and its originating capabilities. My goal is to describe the various logical forms in their interrelation with the developmental context and the function of the self.*

Problems of Logical Structure and Meaning in Tropes and Pathological Thought

A figure of speech is easy for most of us to hear and understand. It is piquing. We can be swayed by its conviction or enchanted or amused by its focus. But we usually do not react to it as if it did not make sense. If anything, a good figure of speech enhances the sense of what's being said. In these terms, it would be hard to see or to say that figures of speech are illogical. If something makes sense, then the feeling of harmony appears similar to our good feeling when something seems reasonable or completes a logical formula. This feeling is motivating, if not generative. It's akin to the excitement of "seeing" congruence between a solution that appears possible and a problem to be solved—an excitement you feel when something "fits."[1]

Certain syllogistic forms are almost universally recognized, and their violations are easy to spot. Certain other forms—such as those involving double negatives—produce perplexity and misjudgment. In fact, the vicissitudes of negation and its effects on antinomy and opposition are perhaps made easier to bear by a good figure of speech. Although irony is not for everyone, simple oppositional conversions like that in the saying "Break a leg," uttered as a goodwill wish for theatrical performers, produce almost no difficulty in comprehension.

Yet when you inspect this trope—which "has it both ways," literally meaning the opposite of its utterance—you easily find that the logic behind it accommodates the transition from the literal to the agreed-upon interpretation of the meaning. That logic requires a two-step translation, however:

STEP 1. ["Break a leg"] does *not* mean ["Break a leg"].

STEP 2. ["Break a leg"] means ["Do *not* break a leg" AND "Do *not* have any misfortune in your acting or actions"].

Look at this logic in symbolic terms. The symbol B stands for the phrase "Break a leg." A *class* of things I label "misfortunate actions" is symbolized by MS. This class may include, as one range of possibilities, actors' errors—anything from the forgetting of lines to the bumbling misinterpretations of the author's intent. In addition, it can refer to nonacting factors, such as the presence of a humorless audience or a loose curtain that suddenly falls and messes up a scene. With the presentation of "Break a leg," these different domains of meaning appear within the class MS. Subsumed within MS—and now functioning in the form that anthropologist Claude Lévi-Strauss (1966) would describe as a "species"—is a whole series of *subclasses* that can be semantically divided into different domains. I label these subclasses L_{1-n}, which now becomes the symbol for a species, L_{1-n}.[2]

Let's go back to the two-step translation of the trope. STEP 1 becomes

$$B \neq B.$$

STEP 2, after establishing that

$$B \equiv \sim B,$$

now also becomes

$$\sim B \equiv \sim MS_{L_{1-n}}.$$

It should also be clear that STEP 2 can read

$$\sim B > \sim MS_{L_{1-n}}.$$

In this notation, the negation of B ($B \neq B$) begins the process of an expansion of its meaning. Where the translation turns the phrase in the trope to "Do *not* break a leg," we have moved the negation to a complementary form and to its consequent expansion of meaning, the nonmishaps ($\sim MS_{L_{1-n}}$). A remarkable aspect of negation, when it marks a complementary class, is the expansion of meanings in that class. So ~MS can refer to an infinite grab bag of possible nonmishaps. But, in addition, there is one of those fantastic turnabouts that we all have grown accustomed to see in figures of speech. Where $(\sim B > \sim MS_{L_{1-n}})$, there is a situation in which the particular class $(\sim B)$ *includes* the generic $(\sim MS_{L_{1-n}})$. This fascinating logical form is very similar to what Lévi-Strauss describes as the way primitive classifications take place! It's like saying that if "Uncle Louie" becomes a totemic generic case, the whole line of his descendents can be thought of as "Louies," sometimes even if they simply live in the same town or relate to his lineage by business associations.

The selection of the specifics of MS is motivated by the specifics of the theater—acting, audiences, and so on. But once the particulars are in place, a categorial logic is applied. These negative terms $(\sim B$ and $\sim MS_{L_{1-n}})$ are class complements for their unnegated counterparts.

Does this figure of speech simply have the same structure and dynamics of the primitive totem classifications? The heuristics are different, and they *are* reflected in the structural differences. In particular, the figure of speech gives flexibility in classification of categories. Interpretations can be targeted to a given purpose, and less idiosyncratic ordering is easy to reestablish. The contrast is in formal terms. The ultimate results (denouement) of a primitive classification appear in the form of "trees." The organization becomes fixed as an historical constitutive way of organizing categories. *Where the primitive classification entails fixedness, the figure of speech within a trope organization entails flexibility!*

The subclasses of the "trees" are like branches. They paradoxically may be more general than the species from which they "spring," but their order remains historically faithful. In the contrast, the trope appears less a victim of compulsive historical repetitions. A trope is organized in a context of thought structures, which have a *lattice* form. This form, which I describe later, gives flexibility to classifications and categorical ordering. Tropes are formed within a perspective that functions to bypass any historical "fossil" of species-subspecies efficacy or tradition and to return to a genus-species organization less bound by particularities or rules of contiguity. Contrast the "Uncle Louie" totem with the figure of speech "He's a chip off the old block." For this thought, we don't have to continue to keep focusing on connotations of wood, its forms, and characteristics. We easily transfer the relationship of the terms to a *model* and a *copy*. Probably the most frequent reference would be to a child and the parent. However, the child becomes the focus from which we look backwards to generalize *from the copy to the model*. "Look at her blue eyes—just like her mother!"

The "chip" superordinates the "block." The child becomes the totem. Nevertheless, in part-whole terms, the "chip" is a subcategory, although, just like a totem, it subsumes "the block." For the "copy" to become the model *for its own model* is only a focus, and to get more of the picture we need to back up and recognize that the copy had to be temporally and causally preceded by its model. If we keep this sobering thought in the back of our minds, then we can have fun with the contradictions in an idea such as "bringing up Father."

In all this, the nature of the trope emerges more clearly when its logical turnabouts and suspensions of temporal and causal contrariness are snapped back into place. This happens when meaning is more fully extended by our cognitive perspective—one that we can have by positioning our selves outside the different temporal and causal units on which we can playfully focus. Thus, from a position allowing us to look at and reorganize such units, we can look beyond the metonymy; we can see how the child becomes a parent—how the copy becomes the model not only for its own copy, but also for its own model! The child becomes a parent to her own child, and the child also becomes the model for the parent. Shakespeare uses such tropes in the "seven ages of man" soliloquy. Figures of speech step outside the different relations of terms and references, which allows us to reassemble the whole picture and reorganize return routes to genus-species organization. Such routes are possible with a trope, but unavailable in the fixed firmaments of the primitive totemic species organization.

To elaborate requires some terminology. When $(\sim B > \sim MS_{L_{1-n}})$ is transformed to *its* complement $(B > MS_{L_{1-n}})$, logically it is a *lattice:* "A *lattice* is a structure consisting of a set L, a partial ordering $\leq$, and two dyadic operators $\cap$ and $\cup$. If a and b are elements of L, $a \cap b$ is called the *greatest lower bound* or *infimum* of a and b, and $a \cup b$ is called the *least upper bound* or *supremum* of a and b" (Sowa [1984] 2007).

Within this terminology and notation, the issue of "sets" arises. Because we have been talking about categories and classes, I should tell you that my point about flexibility of logical structure, its ordering, and its operators is subject to your agreeing to suspend differences between "class and subclass," on the one hand, and "set and subset," on the other. These differences are largely terminological, but they provide for different notation and hence for different conceptualization and ways of operating and transforming (Lowe 1995; "Set Theory" 2001). I do not say that the differences are not important. However, to picture the complexities of reclassification, it helps to shift to a set concept at this point in my presentation.

In the shift, I maintain the restrictions inherent in the definition of *set*. Thus, in the example, we convert the particular class B (all statements consisting of the phrase "Break a leg") to a set, B. This set consists of all the elements that exhaust the phrase "Break a leg." Now refer back to the generic class for the class B. I have labeled that genus class MS. After we convert this class to the set MS, we note the oddity of the metaphoric transfer: the set B, however less universal in scope when compared with the set MS (its apparent genus), *includes* it.

The inclusion goes further. It serves to create a *quasi*-hierarchized organization that is within MS (which now has a subset status). Yet we bear in mind that the subset is the generic and that it is subsumed under the particular as its "set." To explain the *quasi*-hierarchical structuring, I might say that because the particular includes the generic, the organization takes place from that vantage point. Thus, the generic—even though a subset—now includes any "level" of misstep. The actor's good fortune is *not* tripping over a couch, but it also is speaking his lines "trippingly"!

Several problems appear immediately. The logic is not one of set and null set; nor is it one that is *simply* hierarchical in its organization from general to particular. Laws of contradiction and identity *appear* violated. Yet the transformations follow rules. Some see these rule-governed transformations as *different* logical laws of unconscious logic (Freud [1915] 1957b; Matte Blanco 1988); others as pathological logic (Von Domarus 1944; Matte Blanco 1975, 1988; Frith 1992; Keen 1999; Bermúdez 2001; Mujica-Parodi et al. 2001); and yet others as phases or stages in the development of logical forms (Piaget [1975] 1977). Some have tried to make the three-way comparison of pathological, developmental, and primitive logic (Werner 1948; Frazer [1922] 1950; Schilder 1950:516).

The Additional Problems of "Primitive" States and Immature Structures

I have raised the question of meaningfulness, yet the problem I take up involves comparisons of logical forms. It is not that I am saying that meaningfulness *instead of logic* is the issue. Some do think this (Frith 1992). I am not insisting on a difference between "procedural" and "epistemic" rationality. This present-day distinction may appear like jargon, but the issue is whether to keep to the idea that *logic* is like an objective mechanical series of steps, whereas *knowledge*—with its cognitive routines, subjective

experiences, and dependencies on communication and consensual processes—can be placed in another category or discipline. My question is
whether the underlying logical steps are the same when a schizophrenic,
a primitive and a nonschizophrenic nonprimitive reason, although what
is known—and how it is known—is the matter that is different. Is the
premise (and how it's conceived) different, but the reasoning the same?
Or are the steps in reasoning actually different?

José-Luis Bermúdez (2001) makes an interesting analysis by categorizing premises as separate from the steps in reasoning and by applying his
dichotomy to the interpretation of empirical research on the difference
between schizophrenics and nonschizophrenics. This approach would
imply sameness in the logic of human beings, and it would imply that
logic is a basic given in our nature. In fact, the idea that there is such a
basic logic, as developed in the "mental logic" theory of Martin Braine
(1998c), has a great deal of empirical support. Noveck and Politzer report
that their experimental work "corroborates evidence collected elsewhere
... and argues in favor of the claim that 1) there exists a basic repertory of
... mental logic and that 2) it is a necessary foundation for reasoning theories" (1998:general discussion, para. 4).

Although Bermúdez's approach is an exciting takeoff point, I fear the
chase would be on for the identification of "normal" and "schizophrenic"
premises—conscious and unconscious—and the melee would result in
fragmentation of meaning and logical routines. The delicious tasks of in-
terpolating and extrapolating meanings at different categorial levels and
interrelating these operations and levels of reference with logical steps
would be all but impossible dynamically and historically. So instead of trying to look at logic as a unitary objective series of steps separate from the
rational choices of meaning, I choose a route William James ([1892] 1920,
[1890] 1952) would call *rationalist*.

Subjective elements have to be part of the picture of logical forms. More
specifically, I propose that a check on the fit of these forms to the nature
of thought requires a perspective on the self and its subjectivity. In this
book, I focus on dynamics of thought and its forms from what I define as
the only perspective possible. An organism such as a bird can focus on its
prey and zero in for an attack. Although the angle of approach can vary
according to circumstances, it would be extravagant to say that the bird
also focuses on its own thought about all this. Yet the selection of an angle
from which to focus is the selection of a perspective. When we human
beings select our own thinking as the target in focus, we select on another

order of consciousness. I might say that this kind of selection is a power of agency, but that statement might easily be seen as circular. The kind of focus by which we look at our own thought is not in an unconscious state or an immature stage. If we look at this phenomenon in terms of the evolution of mind and brain, it would certainly no longer appear "primitive." Here I foreshadow the point that the basic differentiation of primitive and mature is in terms of the awareness of the self as agent—that is, an awareness of the subjective experience of self as having causally selective functions. The definition of *agency* as a shuttle between the self and the awareness of self does seem a circular concept, but the way to resolve this circularity is to say that reflexivity is involved. At the least, this idea helps us to conceive of action patterns of thought that go in a loop. The idea has several advantages. One, it's easy to tie to neurological pathways and patterns and thus to see possible groundwork for the psychological experiences. Two, the simple to complex patterns of intracommunicative pathways possible can be depicted in terms of analogous differences in subjective experiential contexts. Three, if those experiential contexts of self increase in reflexive capability, each succeeding order of "awareness" becomes superordinating—like a category. This last advantage brings into play my aim to show the differences in experiential contexts in terms of logical forms and their orderings.

I can briefly telescope this picture by marking three modes of the individual organism's subjective experience of self. First, the primitive state is a syncresis of self-experience and object characteristics. What the person attributes to the self allows the experience of hunger or satiation, but whether the hand extended is "hers" or an external object, or whether an external object such as someone else's hand is "her" or something other than her is not differentiated. Second, a more mature state is one in which the experience of self is as actor in relation to objects. And third, the "conscious" state is one in which the self is aware of (has a perspective on) itself as an actor. I really don't want to get into the issue of discerning the gradations of consciousness necessary for a bird to make a direct dive toward its prey and then of comparing the quality of *that* consciousness with the consciousness of a human being who ruminates about his own driving skills. Learning to make a better dive is a function of sensory-motor input and outcome, in contrast to the kind of consciousness in which there is awareness of the self as an agent. When I say "sensory-motor," I don't mean to imply only a mechanical sequence. Neural pathways and intracommunicative patterns involve sensations and feelings that are felt experiences. Thus, when falcons make the dive, the dive is on a curve that apparently optimizes their sight without their having to turn or bend their heads (Livio 2003). Because I am talking about learning here, I leave aside the evolutionary factors producing such efficiency. However, we might interpose the idea that for the falcon's effective dive, there are choices and there are intracommunicative determinants. I simply follow Karl Popper's (1981) lead and call the first determinant (learning to make a better dive) *sentience* and the second determinant (awareness of the self as the locus of decision) *consciousness*.

Regarding the human being, we can look at the comparison as two succeeding orders of awareness. For the first order, assume that whatever is the analogue of a "self" is undifferentiated. Psychoanalyst D. W. Winnicott ([1971] 1988) reasons that the child's boundaries of what is "inside" and "outside" are at first not present and that the child needs to realize these boundaries before she can have a sense of a "me." It is an individual experience, but not an experience of individuality.

So the infant, who may learn by sensory-motor conditioning, organizes the intracommunicative patterns within its organismic form. Even if the infant learns a "better" response to the feel of objects, and even if the improvement is made without what we would call consciousness, it is made within the individual organism's communicative dynamics and morphology. But remember the koan of being able to know what it was like when you couldn't know what it was like. In these terms, *to make an analysis of the logic of an unconscious state (or an automatic way of thinking) is something that can be achieved only from a conscious perspective.*

From that perspective, we can certainly do tricks with temporal mode— saying, for example, what the structure may have been in last night's unconscious logic of a dream, and then what it is likely to be the next night. When adults, we can also look at children's logic, say what its structure is, and make some comparison to both a prior (more immature) form and a future (more mature) form.

Like the immature forms, the "primitive" forms of thought are subject to the limitations of historical analysis (cf. Porteus 1931; Lévi-Strauss 1963). Imagine the time of the ancient wars of the Greek city-states. Imagine the events of the battles of Athens and Sparta. The time and the events segmented and sequential within the time would not be easy perspectives to articulate. Further, suppose you tried to relate that time and its events to the present time or to the sense of the position of a self at that present time. (The self might be you at the present time, or it might be some figure from the past that you project—as if present at the present time.) Consider just a few factors—the missing facts of history, the corrupting of the interpretations of one time by the assumptions and values of another, and so on. Now go back further—to whatever is so prehistoric (unrecorded) or so elemental that it was never consciously conceived!

Suppose you regard a takeoff point for a concept of the "primitive" to be something like "a state of undifferentiated thinking and forms." Fantasy and reality would be mixed, and projections of internal meaning and surmises about external facts would not be disambiguated (cf. Durkheim and Mauss [1905] 1963). Well, when a person is at such a point, the analysis of what it would be like to be more articulated in thought is simply not in the picture. It is not within the reference bank, nor is a clear distinction between temporal modes.

We need a magic genie. That is, some element of consciousness has to be present in order to account minimally for and interpret the primitive state of affairs. The magic genie qualifies because it's a force with a degree of consciousness removed from the characters and forces of the myth. Therefore, it's a trick on agency! The schematic of a myth might thus give the power over time to a magic genie who can move events and who can define and transcend time. *Therefore, the self as actor is not fully an agent that can link together various sequential causes from one historical set of events to another.* A primitive force—such as Neptune—can be described in terms that resemble what we think of as a "figure of speech." In the myth, the meaning of Neptune's power over a dominion is not merely symbolic.[3] By means of a genie's magic or attributions to the powers of a Neptune, events are *not* tied to agency or to its review by conscious thought or to any perspective gained relative to temporal and causal considerations.

For a concept of the "primitive," we should *not* assume that the mythic entities denude the person of agency. Instead, the mythic forms merely correspond to a condition (state of mind and brain) that existed long ago, when agency and its conscious review were "never wases"! The disclaimers do not backtrack on the issue of agency. They are commentary on the limitations of knowing about a state of mind when agency presumably "never was." The unconscious nature of the "truly" primitive cannot be reported—unless its unconscious aspect is within some gradation into conscious thought.

Suffice it to say at this point some persuasive analogies of mythic and primitive forms have the present-day functions of tropes—especially when influenced by rhetorical and propaganda techniques. These special cases, wherein the analogies are persuasive, are *not really tropes* in the sense that the self is free to supervise their play with meanings and categories. If you listen to someone sing "You're the top," you can play with several levels of interpretation without losing either your identity or your ability to come down to earth and reassert your identity. When propaganda is involved, though, the powers of the self can be neutralized. The degradation of a people by depicting them as less than human can be so influential that their humanity—in addition to their individuality—is lost. But for that to happen, the emotional arousal of the terms of the dehumanization has to be so great that it blocks people from their power to reconsider other terms and perspectives. Another technique that can cement you from access to your self and its powers of reshuffling the categorizations is to bracket the propagandistic use of a trope, making it the function of a stereotype. In the form

Pb (X's are SU),

Pb refers to "patriots' beliefs"; X's to some group targeted by the propaganda; and SU to "subhuman."

The Pb limits the self's perspective. With such factors delimiting the powers of the self in the free play and reorganization of trope categorizations, the figure of speech forms used are, strictly speaking, not "real tropes." They are more like "quasi-tropes."

In our primitive forms, we lack development of the line between self and the causes of change. In immature states of development, the self and the nonself are not differentiated. With present-day rhetorical and propaganda-driven statements, if tropes are coupled with deemphasis of the self and its experiences, we have the form "quasi-tropes." Not only are there analogues of the absence of the self's powers to reshuffle categories and meanings, but along with this "retroprimitive" interpretation and belief, we will see people react primitively—consonant with the investment in the "fixed" proposition. So in cases of certain kinds of rhetoric, such as propaganda—the quasi-tropes—the person's interpretations can be narrowed to unreflective thought or to fixed, instead of flexible, categorization and belief. The "quasi-tropes" appear to have mythlike primitive qualities—and even powers over thinking. Such may be the case with "quasi-tropes" that assimilate stereotypes such as "All X's are SU" and that, in turn, can be assimilated to a stereotyped belief function, such as Pb.

What if we find a "true primitive" in some isolated area of the world—someone who not only has escaped any biological evolutionary change, but also has been so isolated from social influences that not one scintilla of a meaning or way of thinking has entered his information stream? Even if we did find such a person, we would examine his way of thinking and his meanings from the perspective of our awareness (Lévi-Strauss 1963). As psychologist Julian Jaynes (1976) points out, our way of thinking would be occurring after the sea change in the structure of consciousness and its analytic capabilities that has occurred over millennia.

We can neither turn back history nor extract a prehistoric perspective from the viewpoint of a time *without* that perspective. Accordingly, my strategy in this endeavor is to use "leverage points" from which we can launch understandings. We need the self-awareness to acknowledge that we are persons who live in a different time, one with a perspective to somehow illuminate that other time long ago, but also one with a perspective that "never was." A "leverage point" in our thinking or psyche would be one of a self-awareness from which we can develop angles of understanding. Can a present-day person perform this trick? Yes, if she can—from *some point of leverage*—access her own cognitive processes as they would be viewed in terms representing rules long ago transcended.

The impossibilities of viewing the prehistoric are bound by the same sort of paradox as consciously viewing the unconscious. The immediate point of resolution is that each of these problems requires self-awareness as a "leverage point." Freud saw the *dream* as a "royal road" to the unconscious. The problem is that to understand the dream, the person would have to become conscious of it. Once that's the case, whatever logic you use is a function of consciousness—even though you presume to describe a different logic that's a function of unconscious representations and rules. The person thus requires suspension of the contraries of the two logics and some transformation rules that permit transactions from one to the other. Somewhere in this juggling of time and its perspectives, there has to be a standpoint, or what I have loosely called a point of leverage. I have attributed this leverage to self-awareness, but that gambit doesn't end the paradoxes of contrary perspectives.

I propose that the basic leverage point for understanding the differences in logical structure is the *self*. There is a sense in which you cannot be a self unless you know it, but there is also the sense that you are a self without your consciously thinking about it! Therefore, I must tell you that the paradox persists. Without exercising the powers of the self to assert such leverage through originating points of perspective that recategorize (superordinate) contraries, the problem of paradox simply repeats itself. Merely asserting that "the self is the leverage point" is not enough. If the aim is to express a logic that guides the self in terms that the self cannot directly access, there appears no end to the replication of the paradox of using a logic contrary to the inaccessible. There would be no end unless we understand the self to "step outside" itself—and outside either of conscious and unconscious modes—and thereby *distinguish* the functions of these modes.

So the "points of leverage" idea requires acceptance of contrary states of affairs *at each of two junctures* of the problem of understanding. Juncture one: rules of thought that have to be understood where and when conscious logic is necessary to adopt, transform, and explain unconscious thought and logic. Juncture two: the self, who is aware of not being aware and conscious of not being conscious. Juncture one has its "points of leverage" in the individual's thought and rule transactions. Juncture two has its points of self-regulation in the various levels of the person's experience, formation of representations, and agency. Within these levels is some point at which the transcending of categorical ordering can take place.

With all these vicissitudes to the representations of times and logics so opposed to each other, and with necessary assumptions about the transcending capabilities of the self, I proceed. My strategy is to wend my way to an understanding of the immature and the primitive structures by first comparing tropes to pathological thinking and by then showing the differences in relation to the conscious and unconscious modes of the self. As you would guess, I have decided to do this via a specification of dream logic. The structure one can describe relative to dreams not only offers insights for the structures of pathology, primitivity, and tropes, but also brings into focus the *conscious* perspective on the dream's logical structure.

The dream offers dramatic and dynamic possibilities for describing a logic that is unconscious—versus one imputed to a consciously chosen "originating" mode. Dreams, in addition, can be examined in light of their intimations of subsequent thoughts and behavior. I do not advance any "foretelling" aspect of the dream, as did Joseph. Not only are myths "the stuff dreams are made on," but so are tropes. Myths are not the only things to affect our sense of perspective and direction; so do tropes and quasi-tropes, as in slogans and stereotypes. Dreams, when recalled, present possibilities for tropelike forms that provide symbols for our interpretation, if for no other reason than because the extraordinary combinations of form and content are impetus to the use of tropes for understanding meaning. (See Josef Stern 2000:chaps. 2 and 6). With the logic inherent in the dream forms, you telescope the different temporal modes and thereby achieve intimacy with the subjective and conscious perspective governing the analysis of the primitive, the immature, the pathological, and figures of speech.

2   Natural Logic, Categories, and the Individual

Why Look at Logical Form? And What Would That Form Be Like?

No one would deny the differences between the imaginary status of Don Quixote and the person stepping in front of you in Starbuck's to declare, "I *am* Napoleon." Nevertheless, the logic in the song line "I am I Don Quixote" appears the same as in "I am I Napoleon." The logic of a trope can result in a Shakespeare, who writes about a king or a prince whom you experience *as if* he is a real person—and even *as if he is you.* The immature or unconscious logic, depending on how much it captures your thought, can result in your going to Starbuck's to claim you *are* Hamlet! Shakespeare's tropes and the Starbuck's denizen's proclamation are quite different in their meaning and their effects on others and ourselves. What would differences in their logical structure tell us? This chapter asks the primary question, *Why look at the logical forms?*

I call the received view "GibsonWorld." It supports several models—all different in certain basic assumptions from those I present. Each of these different models has concepts valuable—*necessary*—for the task. When these concepts occupy the same explanatory space, they are dialectically opposed to each other and even incoherent. Is this bad news? No. The antimagnetic elements, such as opposing yet essential concepts of inner and outer determinants, *can* be contributing forces if the view of "natural logic" is robust. Therefore, I assess the value of the received view, which has influenced my own—even if by its omissions.

Physicist Alan Lightman (2005) argues that introducing a new way of looking at things should proceed from the author's thinking and metaphors, but those terms have to undergo transformation before useful formulations arise. Agreed. I begin this chapter by citing my terms and offering brief notes on their meaning. To be transformed, do my terms have to be shrunk to size? As I conceive the "received view," its provisions from the GibsonWorld collide with contrary aspects of "autoregulation." I examine whether the provisions of a two-sided received view, despite their oppositions, work well to accommodate my terms. It is critical to the fate of my terms to explore how the oppositions not merely coexist, but produce synergy.

The Terms *Natural Logic, Constitutive Form,* and *E & E Self*

To depict the logical differences in the big picture, I take *natural logic* as my basic viewing point. The logical differences show up as a function of sign and symbol relationships. The role of iconic relations—relations between icons and referents can be literal versus figurative and concrete versus abstract. These bipolar swings are semiotic vagaries in the different forms of thought, but they do become stable. They take on the logical form of categories and the logical dynamics of ordering.

Categories are important to mature and immature thought alike. It's just as important to see whether "I" am subsumed under the category of "Napoleon" or "Napoleon" is a character subsumed under the "I" with *its* due powers of "reality testing." With the "I" in charge, these powers would take on the job of evaluating "my" relation to Napoleon, and they would *not* be subservient to the way "Napoleon" oversees the reality testing! Within this viewing point, *categories have an iconic status,* wherein their natural-logic organization is cojoint with a semiotic framework.

A *natural logic* is a pattern and set of rules for reasoning and for transforming terms and propositions so that they can stabilize yet change within the organism. This logic should enable the individual to originate categories and to change classificatory order. The *self,* as I describe it so far, is a leverage point from which to view—and initiate—this combination of originating and changing categories and their classificatory ordering. The self becomes a point of perspective in terms of and by means of natural logic. Its functions notably include selection of focus, classification of referents, and exchanges of categorical ordering.

By this time, you are wondering about the admixture of categorical considerations and assumptions of causality. If "All swans are birds" is a category set, it does not mean that the more-inclusive category "causes" the less-inclusive one. Nor does there appear any causal proposition if a poet were to portray a swan as a way of symbolizing all birds. Yet if the "I" were a category and its subset were "the acts of originating categories," then we do have another category set. However, in this case, the question of causality is not decoupled from the natural-categorical arrangement.

The main reason this mix of natural category and causal relations is *not* a mish mash is that *the natural "forms" have the twin features of logical structure and dynamics.* Structure shows up as categories and their subcategories—in short, as an ordering. Dynamics inhere in inner tensions that produce exchanges and transformations within and between the logical structures. Hence, the dynamics point to inner-inspired changes that rebound from one direction to another: from swans to birds, from birds to swans. The individual can be motivated to think out a genus-species taxonomy, but also to shift priorities and think symbolically, using particulars as prototypes. With this bouncing around as inner-inspired change, we can zero in on a causal process.

The self as a subsuming category can change a subcategory such as "Napoleon" and include the "self" within it. Then the "Napoleon complex" can change the features of the self. This causal bidirectionality is congruent with the reflexivity characteristic of the self and the "I." A version of *formal cause* appears inherent in the categorical analysis of such a natural phenomenon as the self.[1] When functions of natural forms not only result from, but also give rise to internal dynamics, such forms are *constitutive,* and so the self, as a constitutive form, reflexively can account for self-consciousness. When the "I" is a subsuming category, it can have itself as a subclass. In short, "I" can make a judgment about "I." *The reflexive relation between "I" the subject and itself as an object is a key feature of the constitutive form.* This feature also appears in the very nature of natural logic, if only because of the relation of natural logic to its embodiment and embeddedness within the self.

To go further with the idea of constitutive form, I specify the interrelation of natural logic and the forms of the individual and the self. A natural logic is inherent in and depicts the form of an individual. I don't want to be tedious, but the term *form* comes with a great deal of baggage. I want to clarify that which I do *not* pack into the term. I do not mean the logic of form is *the* determinant of the individual's chemical and physical nature. Formal cause is not material cause! Nor do I think it necessary to tie up a concept of "form" in pursuit of the chemical and physical sources of logic, mind, and consciousness.

To depict the individual form—a thought, a mind, a person—in terms of natural logic is not to account for the physics and the chemistry and the material reality of the organism's organs, body, neurological system, and so on. It is instead simply to depict the dynamics of the self-regulatory powers, capabilities, and rules of governance. At this juncture, I take no position about dualism in phenomenal and material terms other than to say that the logic that depicts a natural form does not depict subjective phenomena or its products without assuming they are part and parcel of the organism. Thus, *natural logic is a series of metarules that apply to cognitive and psychological functioning as embodied within the organism and as embedded in capabilities for and experiences of subjectivity.* In particular, I view the individual's "self" and its identity as *embodied and embedded within subjective experiences.* I refer to this combination as the *E & E self.*

A *semiotics* comes into the picture by way of providing a description of the self's organization of representations and self-representations. A semiotic framework consists of the "organism's organization, ordering, and interrelation of icons and their referents" (Fisher, 2003a:243).

Two points show how I place semiotic perspective in relation to natural logic:

1. Natural logic, as it relates to classificatory structuring and categorical ordering, is only one part of the picture. The total organism's representation system includes internal biological signals and signs as well as considerations of environmental sources of information (Queiroz, Emmeche, and El-Hani 2005).

2. Distinct from *logical form,* form in semiotic terms is an event within a communication matrix. It is an interaction of sign, object, and *interpretant.* An object is the source of the sign; the interpretant constitutes the effects of the object on the sign. These interactions are with the "mind" of an interpreter, and they affect semiotic acts. "The communication of

a form amounts to the transference of a habit embodied in the object to the interpretant, so as to constrain (in general) the interpretant as a sign or (in biological systems) the interpreter's behavior" (Queiroz, Emmeche, and El-Hani 2005:7).

As we shall see, form in semiotic terms is a matter of representations of action and interaction and objects—hence, the term *schema* is more appropriate.

Contraries in Current Ways of Viewing Cognition

With this set of terms and ideas about logical form and the semiotic framework of an E & E self, more conventional categorizations, such as those inherent in a cognitive information-processing model, seem not to fit. A *procedure/knowledge split* would *not* apply to the individual's categories or to their organization. Nevertheless, to show my conventional side, I punctuate the received view of thought as *processing* and *knowing*. My account does not dialectically justify setting aside the received view. Rather than disregard a multifaceted epistemological story, I show where it does and does not relate to the viewpoint I present.

Where Can Cognitive Processing Fit Into an Account of the E & E Self and the Trope?

It's not too hard to find attempts to generate metaphors, sayings, and, in short, the kind of tropes you look at on the little tape enclosed within Chinese fortune cookies. A fun exercise is to take the usual two clauses or sets of terms on that thin slip of paper and reverse them. When the causal direction is turned around, you find your mind scrambling to make sense of the new proposition. A few simple rules can result in new products, causal patterns, and ordering of categories. "An ounce of prevention is a pound of cure" can be turned into "A pound of prevention is an ounce of cure" or "A pound of cure is an ounce of prevention." I like the original, but I can argue the case for the variants. Think of all the variants you can generate by this simple reversal rule and then multiply by a more complex rule used to select other terms such as *prevention* and *cure*. Thus, if the essentials of their relation are terms such that "prevention" is categorized ~A(A) and "cure" ~(A), you can also select and assign categorized terms for "savings" and "cost reduction" and turn simple reversing rules on a whole set of propositions about them.

One can specify the term *procedure* as steps, routines, and rules of cognitive processing. To split "procedure" from the cognitive products of "knowledge" and cognitive acts of "knowing" would be to imply that production, organization, and transformation of the knowledge "products" occur by way of a mechanically driven sequence of rules. Perhaps this knowledge is a product of one person's cognitive procedure, and it can be processed by way of another person's. "Knowledge," off by itself, would be, at best, an interactive collection of information. The rule-guided processing machine turns out the fortune cookie adages, and you attribute your own set of meanings to whichever version is on the slip of paper. This procedure/knowledge split predicates a common dependency on rules, processing, information, and sequence of actions. It therefore presents each person as a *robot-monad* with inner programming of the rules of inference.

The split turns out to be a reduction "in sheep's clothing." A person's procedural capacities and capabilities define her on some basic and primary level. She comes equipped with thinking routines and cognitive pathways for thought, evaluation, and the like. But the reduction is relentless. Our person-robot's "knowing" is also reduced to procedures and routines. Her "knowledge" would simply be information processed according to rules. Try the fortune cookie variation on yourself. When you are presented with "A slice of cost-cutting is worth a loaf of savings," you are guided by your computations of the ~A(A) in relation to ~A rules. Maybe you attach some unique associations, but it's the rules that enable you to compute. Even if knowledge, as a product of the organism's actions and operations, were to add to the uniqueness of the person who processes it, this change in nature and identity would be no more than secondary to basic structure and its simple and complex processing rules.

Like one of Rossum's Universal Robots, the "monad-as-person" *does* learn. Is the learning *by* classifying experiences, or does it involve how *to* classify experiences and information? Armed with its regularizing rules and routines, learning of one sort or another can take place, and thereby our monad can improve its chances of adaptation to its environment. With my terms—*natural logic, constitutive form,* and the *E & E self*—the person is an independent agent with autonomous rules assimilated to the self and the self-interest of continuing its form and functions. The procedure/knowledge split leads to many aspects of the individual-as-ro-bot. Even so, these aspects of the epistemological view described so far *can* be integrated with this book's viewpoints. The payoff for proceeding with this mechanistic-type framework is that this description—as robotic as it can get in its implications—offers potentially valuable pathways for approaching the relation of form and thought. The following principles appear to apply, but watch for my questions.

Inner Procedures, Autoregulations, and the Affordances

The person's learning or collecting information entails that she—as monad—interrelates with an ecological environment. Yes, inner program-ming is a function of the organism's given, but there is also the information taken in. It is dependent on what is presented by others—in more general terms, on what is provided by factors and features of the monad's external context. Typically, what others do to affect us is labeled as *social* or *sociologi-cal input and determinants*. What physical objects we see and how they influ-ence us involve factors and features that are labeled *affordances*—psychol-ogist James J. Gibson's (1979) term. If we refer to social and sociological determinants as part of an ecological input system, we can mark Gibson's concept as an expansion of the "received" epistemological viewpoint. It is one that ranges over both perception and thought.

Psychologist Egon Brunswick's ideas ([1950] 1952; see also Earle 2000) set the stage for contemporary views that perception, thought, and rea-soning are founded in the organism's effective functioning in the envi-ronment. Gibson's (1950, 1966, 1979) contributions describe the inter-play of cognitive representation with environmental patterns. He (1979) theorizes that the organism adapts to and perceives in terms of "affor-dances"—determinative features of an ecological surround—by develop-ing representations that enable adaptive movements in space.

This relationship of reactive patterning, adaptive perception, and thought has become a strongly held ecological view, and it affects the way we conceive cognitive processes. How robust can the received view be? If understanding cognition as procedure and knowledge requires the twin assumptions of *wired-in procedures* and *self-regulatory capabilities*, do Gib-son's terms round out such a picture? Do the terms I introduced—*natu-ral logic* and *constitutive forms*—gum up the works? I define and clarify

Gibson's terms relative to my argument and terms, but in getting to how each fits into the received view of cognition, we run into a major sticking point raised from the twin assumptions.

Gibson's view accommodates perception and thought as inner patterns, but these patterns accommodate and subserve two opposing directions to the causal interaction with ecological determinants. With all these excursions into inner and outer workings, we have the problem of adaptive, if not accurate, representations. The inner experience ought to be veridically related to the affordances or at least related such that the person can fulfill needs and achieve objectives. One way to optimize these outcomes is to check your representations with those of others. So we're going to have to come back to the person and the logic and meaning by which she makes forays into meanings shared by others. Some of these forays are brilliant figures of speech, some are immature or pathological thoughts; some are primitive ideas. Some are sensitive to others; some disregard or deny other persons altogether. So, right off the bat, *How do you keep a coherent perspective on the logic of inner programming in relation to disruptive or influential variations called to life by sociological factors and social input?*

Internal Logical Needs and Sociological Press

Classifications are set in motion by internal rules—yours. You and other persons "need" to do the classifying as a way of organizing perceptions, thoughts, and all sorts of sentience. But you don't just classify and grind away if there's nothing to classify. You need information. Moreover, as well as you might overchew a field of events that has no new information, certain "wired-in" rules simply do not get used or engaged because information does not reach them. Some see rules of logic as basic and automatic (Braine and O'Brien 1998a, 1998b; O'Brien 1998; Hanna 2006). Maintaining a view of the automaticity of inference procedures does not block your assuming the impact of sociological influences. Many have shown how values and emotions are swept into rational inference processes (Abelson and Rosenberg 1958; Heider 1958; MacGuire 1960, 1969; Hanna 2006; Thagard 2006). These values and emotions, in the lingo of psychologist Henry Murray, constitute a *press*, "a directional tendency in an object or situation" (1938:118).

People are exposed to events that engage and elicit—if you want, construct—a series of alternate routines to make conclusions. Although the term *affordances* does a good job of referring to the characteristics of external events, Gibson is describing *perception*. When we factor in that a person also thinks, feels, and organizes meanings, there is more to the question of the influence of external events. I turn to Murray again because he looks at perception from the person's perspective. (His focus is "personology"!) He may help us to flesh out how Gibson's affordances concept fits into a GibsonWorld in which a person perceives and thinks as a function of inner and outer determinants.

Murray's concept of "pressive perception" valuably suggests that much of the environment is there, but inert. Pressive perception refers to "[t]he process in the subject which recognizes what is being done to him" (1938:119). We do not have to go to the ironic satires of Voltaire and Swift to note odd premises presented to and recognized by a subgroup of people—scientists, mathematicians, and logicians! Not too long ago, during the first half of the twentieth century, B. F. Skinner's radical behaviorist view of mind held cognitive research at bay. The banishment of constructs for internal variables such as "mind" had great appeal and mainstream adherents within the subgroup "psychologists." If you presented the proscription to ordinary folks, they would reject the idea of tossing out the concept of "mind." So an "odd" belief *can* determine the scientist's belief system. A scientific reduction, such as "thought is behavior," is alluring for many reasons. There's the resolution by simplification. There's also the delicious intrigue of the underground sense of contradiction.

Voltaire's and Swift's satires appeal to that intrigue. There's pressive perception for Candide. Premises are simplified and can guide his thought. Yet they heighten his sense of the premises as odd. The followers of radical behaviorism were aware of the premises they found alluring. They could cite them, and they could show these premises' fecundity for generating observables. They were aware that they were guided by them. But the oddity factor was not always in view; they regarded the guiding premises as self-evident. Views to the contrary were odd to them. You realize that you too—like myopic scientists, philosophers, and mathematicians—can miss the perceiving of categorizing premises as odd because to do so requires extraordinary measures. You have to open the storehouse of rules to find those governing an attack on odd but "politically correct" premises. Perhaps one such rule for radical behaviorists was to risk appearing to go back to former and outmoded ideas.

What about that store of rules requiring you to look for the presence (press) of extraordinary circumstances to access? Sometimes it is almost as if that store is not there. Internal turmoil results if you perceive external interference with politically correct and socially acceptable beliefs. But, in addition, the conflict stopping you from looking for and from finding that store may exist over the disorder you may bring about from within to threaten what is already orderly in your thought. An inner propensity to maintain order may operate to limit the chances of finding it. *The needs of the inner systems blind one to the pressing facts.*

Aside from Freud's metaphors of repression, many studies of cognitive balance, cognitive dissonance, and the effects of wishful thinking attest to our own avoidance of unlocking the stored powers of reason. In the great literary satires of Voltaire and Swift, the limits of syllogistic logic relative to responding to the press of meanings are the target of the authors' ironies. However, the authors not only present the reader with the ironies, but increase their severity for the characters they are satirizing. For you the reader—as for the characters—the more severe the irony, the greater the chance you can open the storeroom, get out your X-ray vision glasses, and see through rose-colored premises to the upsetting oddities.

It may strike you that an irony is a trope, but there is a deep relation between tropes and logical forms. An irony is a logical theme in variation. When you apprehend its logical construction of a set of premises that lead to a contradiction, you are energized to reframe the premises so that a logical form leading to a resolution *would* ensue. When you are presented with an irony such as "Running from trouble only opens you to more trouble," you reframe. Terms such as *trouble* and *running away* are marked with negations. So you might consider what happens if you counterbalance negations by proposing, "Not running from trouble allows you to resolve it." The doubling of irony permits you to face a fuller set of categories and complements so that the logical conclusions for your premises and terms can be better articulated.

We are not yet ready to specify what happens to the logical form of a trope. What *is* at stake in this analysis is the integrity of the person's autoregulating in relation to his dependency on information. Needs versus press? Yes, inner need for integrity versus the press of external influence of information. That information has all sorts of social, external, and logical potential to invade autoregulation. Particularly of concern are those rules that serve to keep internal balance—sometimes at the cost of perceiving external threat. Is autoregulation a kind of inner power that utilizes natural logic and its adaptable structures?

My purpose is to "X-ray" the individual's available logical structures, their particularities, and their relation to each other. I propose a way to look through to the logical structures of inadequate autoregulation versus the highly similar, but mysteriously more adaptable logic of tropes. The overarching idea I urge in this chapter is that *an analysis of logical structure can bring about coherence, which can support the processing of contraries in thought. The coherence can provide for a stabilizing of the contraries and the responsive capabilities of the autoregulating organism when faced with information about social demands and, in short, environmental press.*

Preliminary Look at Similarities in Premature and Immature Logical Forms

To get a quick and direct look at odd premises and at the chains we might put around a potential raid of the storehouse of corrections to logical routines, we visit the people of Chelm. There, because any rose coloring is done so ridiculously that humor counterbalances the fearful loss of order, we do *not* avoid X-raying reason. Chelm is a mythical town of foolish people who depend heavily on "logic" to resolve problems of living. Schor (2001) gives an example of their reasoning in this story:

A WISE MAN OF CHELM said, "What a crazy world we live in! The rich, who have lots of money, buy on credit, but the poor, who don't have a cent, must pay cash. it [*sic*] should be the other way around: The rich, having money, should pay cash; and the poor, having no money, should get credit."

"But if a storekeeper gives credit to the poor," his companion objected, "he could become poor himself." "So fine!" said the first man. "Then he'll be able to buy on credit, too!"

That the joke is so immediately apparent suggests we do have keys to access the storehouse of rules, those that can spot and probably categorize logical errors. The errors in the Chelm example of economic justice can be logically unpacked in several ways. In the main, the joke on syllogistic form is that of the undistributed middle in *transductive reasoning*. In this type of syllogistic pattern, the middle terms are identical. Nevertheless, it is invalid to allow for a conclusion to the major terms based on the distribution of their equivalence. Thus, the conclusion in the sequence

S is P

X is P

∴ S is X

is invalid. In the wise man's foolish thinking, the reasoning appears to be that

With credit you can buy things.

With money you can buy things.

∴ Credit is money.

In line with this book's main themes, there are close similarities between "pathology" of thought and immature thought. Therefore, it is worth noting that Piaget (1930) describes a period of early development during which the child might reason this way. In the Chelm example, it's easy to see that the transformation of the major terms (credit and money) into equivalent units occurs on the basis of their having equivalent functions. It is instructive to have a good look at the way Piaget depicts *transduction* in the reasoning of a child from two or three to six or seven years old. Here is his description:

> Owing to the fact that it does not reason by relations but is a simple combination of judgments, transduction does not attain to the strict generality of deduction but remains an irrational passage from particular to particular. When the child seems to be deducing,.... to be applying the universal to the particular, or to be drawing the universal from the particular, he does so in appearance only, owing to the indeterminate nature of the concepts employed.... [A] boy tells us that large-sized or "big" bodies are heavier than small ones; yet a moment later he declares that a small pebble is heavier than a large cork. But he does not, for that matter, give up his first affirmation, he only declares that the stone is heavier than the cork "because some stones are bigger than corks." Thus, the character "big" has not at all the same meaning for us. It does not define a class, it is transmitted by syncretistic communication to analogous objects: since there are big stones, little stones participate in their bigness and thus acquire weight. (1930:293–94)

I will not enter the debate about logic as "symbol" relations versus logic as propositional relations. Nor will I try to resolve matters by way of propositional schemas, as suggested by Martin Braine and David O'Brien (1998a, 1998b) and as considered separate from Piaget's "logic of classes"

(O'Brien 1998). I list my assumptions to show you how and why I navigate from form to content and from symbol to terms and propositions and from there to the evaluations of the self:

1. Categorization and predication are quite close to each other.
2. Elements of categorical form and their relation to propositional form lend themselves to symbolic representation.
3. Propositions are attitudes.
4. The considerations of truth values and formal coherence are interwoven. (All this is spelled out in Fisher 2001.)

Part of my main theme is that immature structures are strikingly similar to primitive ones, so I take this opportunity to say that insofar as transduction "moves from the particular to the particular," as Piaget (1930:294) states, it resembles Lévi-Strauss's description of primitive classification as species ordered. The Chelm example is thus fertile ground for pointing out some of the parallels in logic, and these parallels can help to identify the wired-in staples of reasoning that are common to the foolish, the pathological, the immature, and the creative.

The transductive reasoning in the Chelm example has two features worth noting. One is *its* similarity to other reasoning processes. It is reminiscent of the species-ordering tactics Lévi-Strauss describes. But there is an even more direct comparison of the procedure of seeing and reckoning with equivalents where the terms (or quasi-classes) are not identical. This comparison is evident when we consider psychoanalyst Ignacio Matte Blanco's (1988) demonstration of this very procedure as indigenous to unconscious logic and of how it appears in pathological thinking.

The second feature of Chelm's transductive reasoning is that its major terms (credit and money) are *parts* in the process (of buying things) denoted by the middle terms. Therefore, each of the two premises is in the form of a metonymy. This point is critical to specify what happens in figurative thinking—itself apparently very similar in logical form to the immature, the pathological, and the primitive. The reasoning in the Chelm economic justice example is from metonymies ("S is P" and "X is P") that appear equal to each other in form—just as two occurrences of equivalent mathematical symbols would be identical. Thus, $X = X$ even though the first X appears in a different place than the second. But in the metonymy a miscalculation occurs. The separate "parts" are deemed equal even when their only equivalence is in that they are parts. The sort of category mistake here is one of accepting synecdoche as symbol instead of as a denotative term with semantic bounds.

Voltaire's (1759) criticism of Dr. Pangloss (and Leibnizian logic) is based on the same appeal to the humor of stretching a symbol's place in a syllogism beyond the denotative bounds of the term. Thus, S and P in the sample are symbols, but the fact that they are different terms prevents an automatically equivalent relationship with each other.

Voltaire writes the following about the philosopher, Dr. Pangloss: "Master Pangloss taught the metaphysico-theologo-cosmolonigology. He could prove to admiration that there is no effect without a cause; and, that in this best of all possible worlds, the Baron's castle was the most magnificent of all castles, and My Lady the best of all possible baronesses" (1759:chap. 1). He then gives an example of the good philosopher's teachings: "Stones were made to be hewn and to construct castles, therefore My Lord has a magnificent castle; for the greatest baron in the province ought to be the best lodged."

Look at the latter statement as a syllogism (you can also look at it as a propositional schema):

Stones were made to construct castles.

The greatest baron ought to be the best lodged.

∴. My lord has a magnificent castle.

This reasoning appears different from the Chelm argument. However, for this analysis, assume that the terms of Pangloss's reasoning are symbolized as categories. Thus, "stones" is symbolized as A; "castles" as B; "baron" as C; and "my lord" as D. The reasoning chain then reads,

A ought to be B.

C ought to be B.

D is C and therefore is B.

Here, again via a middle term (B), the equivalence is between "stones" and "my lord," but it is mediated by the predicate "ought to be best lodged," which, if taken logically, would *not* be a universal quantifier! So the leap (or slip in logic) is from a "sometime" inclusion of the "greatest baron" in a castle to the conclusion that he *is* in one. The "ought," having been made identical to what "is," allows an equivalence of "my lord" with "stones." If transductive thinking, this string is a bit more complex, turning on the assumption that "is" and "ought to be" are the same. But the assumption as a logical one is some sort of category mistake and in the simplest terms asserts a synecdoche as an identity. One of the problems Dr. Pangloss has is a precommitment to an illogical belief and to its being a valid part of determining the rules for reasoning.

Please note. If you prefer, you can also look at the Pangloss problem in terms of an inference schema model, and the rules and constraints would follow those of a "schema for conditional proof." Braine and O'Brien describe that schema as follows: "To derive or evaluate *If p then...*, first suppose *p;* for any proposition, *q,* that follows from the supposition of *p* taken together with other information assumed, one may assert *If p then q*" (1998b:203).

The A, B, C, D reasoning chain given earlier can be seen in these terms. Thus, if that which "ought to be" is assumed to "be what *is,*" then we have a contradiction in a "prior assumption." Convincingly, Braine and O'Brien argue that within propositional schemas, contradictions and consistencies impose restraints on inference: "(...given that the suppositional argument leads to a contradiction, and that the premises are true, the supposition is false.) In that case, no conditional conclusion can be drawn that has the supposition as antecedent" (1998b:205).

Definitions of terms, or "lexical entries," are therefore involved, and these terms have sociological dimensions. The sociology gets wider because Dr. Pangloss teaches his reasoning to others, and the rules of synecdoche as identity fly in the face of realities. Nevertheless, they are hard to discard. I pick up later on this interaction between teaching, belief, and the detection of logicity.

To follow on the sociology issue, I come back to Chelm. As far as social and societal factors go, the citizens of Chelm—"wise" and "ordinary"—reinforce each other by settling for transductive reasoning patterns.[2] These examples should show the relation of the construction of societies and social rules to the individual and what may be wired in and what may remain dormant possibly through disuse or neutralization. This interaction of press and inner need is one aspect of the relation of information to wired-in rules and procedures. A deeper question is what happens to the wired-in rules if they are not used or are neutralized by different social rules or by social distortion of information, such as propaganda. An allied issue is whether any new information contributes to the change of rules that can result in the emergence of a new perspective and/or its engendering form.

Is there a way to describe the suspension, delimitation, or even dissolution of internal powers? The timeliest example of such suspension is of people who learn—or who are taught—to reason by association and contiguity. In the examples of transductive thinking, there are similarities in the "mature" form found in "foolish" people, unconscious logic, pathological thinking, and the tropes involving metonymy. A person can learn to reduce differences between individuals by associating them with the acts of others or with the affiliations of group membership. This book is being written during wartime, which leads to questions of the effects of stereotyping and its use in propaganda—long and short term. *One cause of vulnerability to propaganda may be the presence of logical thinking patterns and needs that are quite similar to it.*

In line with the aim to see how logical capabilities might be present, albeit neutralized by sociological influence, look at the effects of propaganda. If you are shown pictures of a specific tragic and dramatic event, you might accept it as a premise. From the one horrific event, you might allow yourself to conclude something quite stereotypical. But this procedure in your thought may not take place only during wartime; it may also be "the way things are taught" in a given society or subgroup (Atran 2003). Thus, a cranking up of your syllogistic process might be coupled with a sanctioned rule that you can proceed to a conclusion more universal than the premise. An example is the rule, "If an individual is a member of X group, you can generalize about that individual." The cranking up can go into full gear with another beauty: the truth-value of the premise does not enter into the picture. "Just have faith," "Trust me," and "You'll know it, when you see it" are also examples. Left out of any articulation with sanctioned but dubiously grounded rules may be several logical rules that are languishing in the storehouse. These logical rules may be of the type we would normally regard as wired in! They might require that you inspect the existential quality of the premise that you compute the logic relative to the quantifiers of the premise and that in the case of existential statements you take into account that a truth factor is in play. These logical rules may be *rules-about-rules;* they may be adaptive, engaging pressive perception and requiring a person to assimilate premises with the auto-regulative position of his categories of self and with his self-initiated values and intentions. But I would need more discussion to regard that rule configuration as standard equipment in the storehouse.

Suppose these "good" rules *are* wired in. What happens with disuse of such rules? Do they "not develop"? Do social strictures wall them off? Do they get sealed away because counterrules have a greater value in the organization of beliefs and the disposition of feelings that are difficult to bear? I cannot fully probe these questions here, but I can try to argue this depressing point: whatever the possible recovery from disuse, one cannot overestimate this environment's immediate effects, which are devastating either to the integrity of the individual's knowing system or to her gifts of self-regulation as they may be used to restore capabilities.

Sociological Pressure Versus Self-Awareness

By phenomena such as the acts and thoughts of "righteous persons," we know an awareness of vagaries can occur at a level of the self and that its powers of transcendence suggest learning—at least as reorganization, if in no other form.[3] This evidence is of persons' overcoming social pressures that would otherwise dampen down their reserve logical powers. It implies that *even in the face of awful propaganda or repression of logical rules of thought, repair of knowledge and self may be possible by virtue of self-awareness and the categorical latitudes it provides.*

A second implication is that *there is a strain between sociological pressure, like that of propaganda, and the needs of the self.* The battle may well be fought out on an internal level. There is the need to keep premises in a tolerable order, and there is a need to play them out consonantly, with the role of self in relation to the inner logical routines. All of this brings out the importance of the self in regard to the perspective necessary to reason logically in the first place. Piaget, in his early writing, is clear about this perspective—particularly in relation to transduction. He points out that the child works with particulars and, in fact, "sticks slavishly to given reality." Yet "the child assimilates the real to his own self and fails only to construct a universe of intellectual relations because he cannot strip the external worlds of its subjective adherences nor understand the relativity of his own point of view" (1930:296).

The child employs logic consonant with the needs of the self—at *the self's* point of differentiation from the "other." What happens when there is social deprivation or repression? The question of a mature person's having perspective on a logical procedure that "never was" or was taught wrongly or was suppressed may pivot on that person's being—or becoming—a *self.* That status is the obverse of Piaget's description of the child. The mature person achieves articulated recognition of a great deal that she is not. Her self can unscramble its attributions of thought and feeling—differentiating those to be assigned to her self from those to be assigned to the world around her. This point about the powers of the differentiated self appears basic, but I view it from many directions.

On the *inner frontier,* the differentiation of self and its role in natural logic is consonant with Piaget's thinking. Inner change in how a person indexes self and things takes place as if by a "time-release" structural determinant. By epigenesis, the individual's logical organization of categories changes to become more mature. Winnicott's ([1971] 1988) "inner need meets outer nurturing" account of change focuses the person's dynamic demand for balance and identity. His scope brings in the importance of outer determinants—especially the mother! Development and social interaction affect how the person labels and values self and others.

I view all this in terms of a combination of semiotics and self-regulation (Fisher 2003a). The focus is on the inner frontier, within which change brings forward structures as well as access to them and to levels of indexing essential to a description of natural logic. However, in this chapter, we also bring the individual's organization of icons and categories and her inner dynamics into play within an *outer frontier*—oriented to the exploring organism and its ecological field of affordances. This is the GibsonWorld. To evaluate the two frontiers, the first order of business is to show that internal sources of regulation of the self can be integrated within a framework that calls for interaction with external press and determinants. Here in a nutshell is an overview of how I picture natural logic—particularly the semiotic and the self-regulatory functions—in relation to the framework of an exploring organism and its ecological surround. This overview is an orientation to the analysis I make in the chapter's denouement. For some readers, it can operate as an organizer. Others might prefer to read the summary of its propositions at the end of the chapter.

Overview of Natural Logic, Self, and Self-Regulatory Functions

Essential to a description of natural logic is the combination of semiotics and self-regulation. The semiotic workings are as if wired in. By way of an organization of representations, they function as the self-anchored inner needs to articulate with meanings. This articulation is a structuring of particulars and instances. The structuring—a logical organization and ordering—results from a scaffolding of categories and wired-in rules of processing different orders of representations. (These representations are signals, symbols, and, in short, *icons*.) Meaning is also subject to syntactic forms—sentences, for example—that provide roles for concepts.

For the individual organism to relate representations and to make organizational decisions and transformations requires both a self and self-regulation. The self is an arbiter of the need dynamics that motivate regulations. This self-regulation and the organizational decisions are not only manifested in terms of logical operations and transformations in logical relations, but also interconnected with the semiotic considerations of the meaning of representations. The organism's inner needs interface with outer press. To achieve optimal coherence and harmony, the logical relations are consonant with—and dependent on—a differentiation of self and its actions and thoughts vis-à-vis the occurrences and objects of the world that the self encounters and intends.

Oskar Schindler, of *Schindler's List* fame (Zaillian 1993), compromises between the needs of self-regulation and the press of sociological pressures. I thus narrowly focus on his travail as if he were a monad whose inner rules and organizational dynamics can become visible. I would like to open a less narrow scope by expanding Gibsonian ideas of the exploring organism and its environmental surround. Can we amalgamate antithetical terms without their meanings blowing apart organization?

If we cast the individual's logical regulations as one set of the hardwired rules and procedures, this and other such sets are inner context. How do we expand the scope of what is inner context to relate to external context? How do we look at the environmental surround to understand its effects on the individual's mind and thought structures? The task is not only to cast a light on the dynamics and structures, but also to look at ecological affordances and pragmatic feedback about objects. In this expanded context, we would have to view patterns of action and to identify the effects on adaptation. In turn, the patterns and effects would have to be "items of information," which can be presented to (and can *be within*) the individual's mind.

We are used to seeing how a swimmer's stroke moves the waters. Can we also adopt a perspective to see how waves of water design a swimming stroke that would accommodate it and how they modify it to achieve successful swimming? This simplistic example sounds overly robotic. Yet the affordance/pragmatic feedback framework does *not* imply that "meaning" has no place in it. Within it, items of information are meaningful either as referents to external events or as pragmatic collections of information that serve as percepts, concepts, and schemata.

Monads in the GibsonWorld

The combination of the Gibsonian framework with a self-regulatory model appears odd. Gibson's notion of an "organism-located" stimulus, such as retinal patterns, is that they are in response to environmental features of objects or spatial (geographic) features. Are these geographic features stimuli? A quick sense of Gibson's concept is that *such a stimulus is also a response.* The person perceives in virtue of how stimuli such as retinal patterns produce correlates of objects. The stimuli are within the organism. Nevertheless, they respond to environmental features of objects and spatial background. The person acts on the objects in her environment by coordinating the organismic subsystems she can call on to navigate, grasp, move, and so on.

In relation to this interaction, the *self's* complex role is to coordinate, but also to provide a point of reference. If we put these tasks together, they form a big job. The referencing would be to the coordination of acting subsystems, grasping actions, orienting actions, and percepts of objects and their backgrounds. Moreover, the referencing would have to be coordinated and related to the person herself. Therefore, as a point of reference, the person herself *has to be located!* To account for a way the individual can coordinate her actions with objects—and with the demands of the environment—*Gibson proposes stimulus-correlates of the self.* All the stimulus-correlates are action oriented and interaction based.

Gibson's view (1950, 1966) is very unlike the concept of a self-regulatory organism. Self-regulation is presumably from the inside. The individual is closed off. With this idea—based on Gottfried Leibniz's view of a monad—the premises for self-determination appear more in keeping with the idea of a closed system. Leibniz describes in his principle number 7 that "[t]he Monads have no windows, through which anything could come in or go out. Accidents cannot separate themselves from substances nor go about outside of them, as the 'sensible species' of the Scholastics used to do. Thus neither substance nor accident can come into a Monad from outside" ([1698] 1999).

Contemporary neuroscientists argue that the self is locatable in the "cognitive brain." This view, even if momentarily suspending the retinal stimulus issue Gibson raises, reasserts his larger argument: the self has to be in the picture when the person articulates her perception and its various adaptations to navigating in her environment.

Can Tropes Thrive in a Combination of the GibsonWorld, Windowless Monads, and a Wired-in Processing Program?

To keep in tow the discussion of the combination of the opposing determinants—stimulus-correlates versus self-regulation—I concentrate on how they relate to tropes and their counterparts. With regard to issues of logical form as far as they can be examined within the "received" framework, here's what's on the combination plate.

Our monad is isolated only in its basic nature. It comes equipped with hard-wired procedural rules. But the individual organism survives and adapts by recognizing and navigating the characteristics of its environment, which implies that the already existing hard-wired system can enable *accurate* representations of environmental *affordances* and *effective* juggling of the representations. I pass here on a full-blown definition of *accurate*. In our terms, an accurate representation would not only conform to Gibson's criterion of organismic adaptability to the environmental features and demands, but also be coherent in internal coordination of perceptuomotor—and neurological—patterns of representation and logical structuring.

With this view of an accurate representation, the individual's choice of a relation of terms in a trope, like a metaphor, can yield a *good* or fertile one. Psychologist Keith Holyoak and philosopher Paul Thagard (1997) explain how to identify metaphors and analogies that fall short. Their references are not completely coordinated with all the relevant similar elements in the target terms.

I choose the word *juggling* to bring the discussion to bear on the way tropes handle meaning and logic. The juggler tosses his objects. They then move independently and can go in a variety of directions, such as up and down. You see them in changing relationships to each other. For a while, the objects and their movements are the primary focal point and the juggler's hands the secondary, but all objects have to come back to home base in his hands. "Juggling thoughts" is shorthand for saying we are constantly reorganizing our concepts—in other words, re-creating classificatory order in hierarchies from concrete (objects) to specific or abstract levels of reference. The order and selected level of reference reflect the juggler's intentions or intended patterns. With a trope, we may focus on a concrete object to express an abstraction. An organically impaired individual may do this, too—perhaps unwittingly.

In light of the similarities of the trope with other forms and the possible transformations from one form to another, the term *juggle* is apt. All sorts of movement patterns and trajectories appear to relate the objects that the juggler tosses and retrieves. The flying objects can appear in unison, in opposition, in equal and unequal distance from each other and from the juggler—to whom they all return. Classificatory order can be turned upside down and then restored—a trope special!

Tropes, Inner Classificatory Procedures, and Ecological Adaptation

The terms of tropes, logically assigned as elements in a set, have a *lattice* structure. *Merriam-Webster's* (2005 ed.) defines *lattice* as "a mathematical set that has some elements ordered and that is such that for any two elements there exists a greatest element in the subset of all elements less than or equal to both and a least element in the subset of all elements greater than or equal to both."

Piaget argues that the greatest and least elements of sets can be turned upside down in a classificatory ordering. The net result is that *the lattice structure provides a capability for topsy-turvy changes in classificatory ordering and its possibilities for paralogical and classically logical exchanges and transformations.* The array and organization of changes can help a person focus on the abstract when it is the most powerful thing to do or help him take advantage of the most concrete term or concept when its specifics would "pack a punch."

Take the phrase "the lion's den." We all know how risky it is to "put our heads into" it! And, we all know what it means—no matter how concrete the reference ("Be careful if you go into the boss's office to ask for a raise") or how abstract ("Watch out for the premise in a contract or proof that vitiates a solution or abnegates your intention"). The warning value of the "lion's den" phrase is great, perhaps because of its historical associations as well as the "allure" of its specific and sense relatedness. Its comparison value for any set of terms in a metaphor or metonymy is equally great. A phrase with specific images of the fearsome lion and its comfy den can subsume any terms of abstract danger. So the phrase "lion's den" *is* attributed a generic function, but it remains particularized.

In his semiotic analysis, Charles Peirce ([1885] 1993) describes that a given icon can be reevoked at various levels of abstraction. The icon can be reparticularized and also regrouped. Edmund Rostand's character Cyrano famously exploits this capability by his various descriptions of the use of the word *nose*. I use the terms *relational combinations* and *permutations* for these various reorganizations, exchanges, and transformations. They are procedural givens—"wired in." I intend this terminology to help optimize our exploration of one submodel within the combination framework. It's a cognitive theory, arguing for a procedure/knowledge split or for the reductive effect by which "procedure" is everything, or for both.

The view presently dominant *is* a combination; like the GibsonWorld, its two main "super"models are ecological psychology and various concepts of self-regulation. In addition to the problems of resolving their opposition, the combination is difficult to synchronize with the view's cognitive theory submodel. One problem is that different forms of cognitive theory have utilized Gibsonian ideas. Many of these forms then turn back to their source, becoming tributaries to the "received model." Brewer (2001) reviews some of these different forms as they are reflected in the shifting positions of Ulrich Neisser, the "father" of cognitive psychology. The forms include an information-processing orientation to cognitive psychology, models of direct perception that focus on field events, neurologically inspired models, and a brand of ecological cognitive psychology akin to cognitive anthropology (this brand looks for naturally occurring phenomena to investigate where and how it finds them).

The analysis in this book is *not* focused on differences among these forms of cognitive theory. These forms are primarily strategic as regards both methods and content of empirical research. If they show up as metaphysical classifications, they are clearly derived from research strategies. So if they are not vindicated empirically, their logical derivation and coherence are nowhere in sight.

Reduction Versus Representation

I recognize that some researchers may not wish to be included in a catch-all category that glosses position differences on process and method issues, such as the place of laboratory research relative to field events. In fairness to the different versions of Gibson-influenced cognitive theory, there *are* some theoretically conflict-provoking issues, such as claiming that "representations" are an unnecessary mediating assumption. My retort is that this claim is a methodological or methodologically derived assumption; it is reductive, albeit intriguing. Some reductions are as-ifs, and they *can* move things along, allowing interesting research and argument. But in relation to a well-rounded metaphysical classification of factors in natural cognitive life, this reduction would be inadequate.[4] From the word *go*, an interior locus from which to account for thought meaning and action calls for "representation" as a necessary concept. Thus, to incorporate patterns of internal communication—including some of the exciting studies of brain and neurological patterns, structures, and activities—without some notions of signal transmissions and *interpretations* would make the model more trouble than it is worth. Philosopher John Searle's ([1980] 1981, 1992) famous "Chinese Room" parody on communication and understanding classically argues that as a sine qua non the role of interpretation challenges self-defeating reductions by summoning the individual's sense, if not consciousness, of meanings to a model of cognition. My argument goes further, bringing in the self and asserting its pivotal relation to thought and its logical forms. *It is basic for my proposal that the self, autoregulation, and wired-in or inner routines can be coherently in the same explanatory space as the Gibson World.*

Regulation at an Exchange Station

Lattices as wired-in structural capabilities and even as routines for adaptability in organizing thought provide a platform for self-regulation to take place after information comes through the "window." This adaptation is Gibsonian. It brings inner needs, tendencies, and capabilities into synchronous form and format relative to environmental demands.

With a rich array of categorizing and ordering possibilities, we jump ahead from the idea that adaptation is merely advanced by the organism's general self-regulatory capacities to the idea that adaptation is *optimized* in terms of the individual's needs and capabilities. If the individual can direct and assess her needs and capabilities, and in the process organize categories with ultraliberal rules of ordering and classification, then self-regulation and adaptation have a greater chance of synchrony. Achieving synchrony is not merely a matter of identifying wired-in *available routines*. Nor is synchrony explained only by the logical forms, which enable a person's relevant capabilities. Instead, it is essential also to consider adaptation and self-regulation as matters of the self's assimilation of these capabilities to its own needs and sense of capabilities.

The nexus for self-regulation and Gibson's bideterminate interaction of the individual with the environment is at hand. The individual's capability for veridical perception—or, as Gibson would say, for "some correlate of it"—is *his* jump-off point for effective adaptation. A current concern is whether veridical perception is basically a signaling system with relays to effective action or, instead, an accurate (nonillusory) depiction of the parameters—and even of the presence—of objects and their extension in actual spaces. Neither possibility vitiates the need for some concept of representation; whether you visualize information as a series of icons or as an action schema, the visualization implies further transformation on another level of coordination. Paul Coates in his review writes, "According to Noë, '... our sense of the perceptual presence of the detailed world ... consists in our access now to all of the detail, and to our knowledge that we have this access" (2003:para. 7).

We have to have the information; we have to have access to it; we have to know we have access to it. If you look at the "we" as "you" and "I," and at each as discrete first-person experiencing, you hear echoes of Descartes's *cogito* in the last requirement. In most descriptions of adaptation, the focus would be on "thought" and "thought about thought." The "I" or self gets lost in the shuffle. I not only include the "I" in the form of its self-regulatory powers, but also zoom in on it. *The nexus for relays between signals and symbols and between perceptions and categorizations is where the self, as coordinative and governing, comes into play.*

Our combination keeps the Gibsonian framework in place while beefing up self-regulatory capabilities. Bermúdez (1998), too, bases his ideas of the self in Gibson's terms. The self is based in *nonconceptual* navigational acts coordinated with *affordances,* which are features of objects and events in the organism's ecological spaces. A long evolutionary history of the survival of navigational patterns may well be the outcome for organisms equipped with the internal structures that can mirror the patterns. (This might be the case for monads; see Leibniz [1698] 1999:principles 8–13).

Conceive the nexus of internal structures and navigational patterns simply as an exchange station. Actions and action patterns clear and coordinate there. The clearing may be a matter of new codification; signals are advanced in the form of operational symbols or categories for action (Rensink 2000). With liberalizing structures, there can be flexible ordering of categories. There is a pragmatic tone to this idea of an exchange station and to the idea that processing and conversions accommodate the differences between perceptual information and its categorization. Coates provides a nice summary of vision researcher Ronald Rensink's "coherence theory": "Rensink notes evidence for the formation of a 'low level map-like representation' at an early stage of visual processing. One role of this level is to guide subsequent changes in fixation and attention. It is only at higher levels of processing, where the attention is brought into play, that a 'coherence field' generates the representation (involving higher-level conceptual categories) of a physical object in the environment" (2003:para. 18).

Rensink states that "attentional effects are largely concerned with coherence. As used here, this term denotes not only consistency in a set of representational structures, ... but also logical interconnection, that is, agreement that the structures refer to parts of the same spatiotemporal entity in the world" (2000:19). In terms with implications for the formation of tropes, he accounts for the changes in an object's constitution: "Focused attention acts as a metaphorical hand that grasps a small number of proto-objects from this constantly regenerating flux. While held, these form a stable object, with a much higher degree of coherence over space and time. Because of temporal continuity, any new stimulus at that location is treated as the change of an existing structure rather than the appearance of a new one" (20).

Is this description another "juggling" conceit? It is a processing account: knowledge and its formation are described in terms of procedure! The forming of a cognitive object is likened to a focus on "lower-order" (less-stable) objects. Because this procedural description seems consonant with the idea of transitions to "more abstract" symbols from specific icons that feature particulars, it is in the same genre as the semiotic transformation that Peirce pictures when describing the reparticularization of icons at different levels.[5]

Such orientations veer toward robotology. A robot *is* an "individual"— of sorts. Visualize an individual robot's cognitive exchange station for percepts, icons, thoughts, categories, and concepts. Conceive a series of available procedures for adventures with tropes or tropelike forms that navigate different levels and sorts of representation and their alternating classificatory orderings. These goings-on at an exchange station can begin to look like an "alternating ego" in or by which objects change shapes and actions. The inner objects and action patterns are coordinated to pragmatic purposes and adaptations. This station and its exchanges can be a "nice idea." Along with a multileveled account of internal stimuli, representations, patterns, and structural loci, the station exchanges foster accurate orientation of the individual's position in the ecological field. The "nice idea" substitutes for a term such as *self*.

Does the Search for Self Go Outside the Gibson World?

"If you find my real self, tell him I want to meet him"   —Jackie Mason.

Despite the robustness of processing orientations, many researchers *do* keep on trying to account for a *self*. Some, to keep within the processing account, resort to considering the self a linguistic tradition—not a mysterious substance, not an entity requiring some sort of dualism. It is composed merely of names for events and perhaps cognitive operations. Out of habit, we have gotten used to names such as *I* and adopted ways of describing the "I." To describe the exchange station, we have phrases such as "I think" that we regard as convenient.

As the basis for self, the idea "who I think I am" may just be a cognitive coding for a cluster of the solid facts of sensory-motor signals and acts. In Rensink's terms, this coded cluster is like a category focused as a stable object. The focus is specific to what a given situation requires. A similar version of the self as a "conceptual category" is popular and influential in the thinking of the cognitive linguists, in particular George Lakoff.

In this book, we go beyond the idea of the self as a collection of cognitive operations or as a supercognitive operation or dispatcher of operations. A self, with a solid basis in the organism's signal and representational systems, may display cognitive capabilities, which reflect the evolutionary—and ontogenetic—outcome that the human being can *consciously* (conceptually) refer to the nonconceptual. It would be inviting to think that both conceptual acts and nonconceptual representations are automatic and something like wired-in procedures. But this leads to the question, Are consciousness and conceptualization the same thing? If they are reduced to the same thing, however, too many paradoxes sandbag questions of origin, intention, and agency.[6]

In the ecological view we have pursued here, our "monad's" hardwiring presumably enables accurate perceptions of environmental affordances. However, for reasons given and as a means of making an end run around various theories of perceptions as illusions, I use the term *representations* instead of the term *perceptions*.

Accurate representations are not only effective concepts and meanings. They also become patterns. Because they reflect the organism's actions and his adaptation to situations, they have the cognitive form of schematizations, such as syntactic spaces. The person's actions and adaptations also involve classificatory strategies and ordering; therefore, categorization and logical organization are involved. But in order to adapt, it's not enough for the person simply to make a fixed logical ordering and call it a day. New events, new situations, and new times may call for a relook at a framework or zeitgeist. The refocusing can be described in terms of a juggling of categories and relationships.

We need accuracy in perceiving features of objects and in determining how to signify obstacles, dangers, and problems. We also need effective juggling of the representations in relational combinations and permutations that are procedural givens. "Juggling" is critical, so let's go at it again. Consider it in an analogy with Rensink's categorical coding—a process that can focus on cognitive objects with visual stability. Why do we need coding and categorical search to achieve coherence in perceptual information, images, representation, and action outcomes? There are two possible causes of this need: attention deficits on the level of seeing or remembering specific detail, and details put on hold when they are not of immediate utility.

In either case, if automatic schemas are *not* wired in, mistakes or costly time lags might rule the day. Linguist Derek Bickerton (1990, 1995) proposes a *basic syntax* as the distillate of an evolutionary process. His idea is of an abstract—somewhat diagrammatic—schema as a superposing structure. Within it are positions for agents, objects, and outcomes. The relations among them—such as cause and effect—are similarly abstract. Even though one can place modes or nodes here and there in the schema, the meanings of the universe of events are nowhere in sight. If a schema is set in motion, then, generically, it is the syntactic feature of an object that has been acted upon. Nevertheless, that object, when specified semantically, might be a rock, a sentence, a person's goal, or, indeed, a person. So the facile causal rules between syntactic action and object can get complicated—or problematic—when seen on a more specific level. To accommodate the complexity, Bickerton pictures syntax as "not serially, but hierarchically arranged with structures nested in other structures" (1990:139). With syntactic structures as templates for thought, we may have more than one substructure of thinking and more than one alternative space for organizing information. The juggling of our thoughts and our choices comes up again. Enter a homely example. If you are confronted with a saber-toothed tiger or with the choice of going into a lion's den, what does your syntax—wired in or not—allow you to think and do?

Syntax, Survival, and the Trope Vis-à-vis Object-Oriented Schematics

Two ways of reacting mark the human capability when immediately confronted with a saber-toothed tiger. The first is survival by an efficient means of escape; call this way a default reaction. The second is to evaluate or review your best option.

For survival purposes and ensuing default actions, some automatic *script* is necessary. Computer scientist Roger Schank (Schank and Abelson 1977) championed this concept. Compared to a *schema*, a *script* is more specific to a particular set of objects and events, although the actions a person might take would differ from one occasion to the next. A "restaurant script" would have rules specifying nodal points for choices and slots for different sorts of objects and other variables so as to include most types of restaurant contingencies and their expected behavioral requirements.

When you find yourself confronted by that saber-toothed tiger, such a default mechanism may keep you alive as a human being. Psychologist Ellen Langer, describing "mindless," nonvolitional behavior, writes that "we act like automatons that have been programmed to act according to the sense that our behavior made in the past, rather than in the present" (2000:220).

That which would mark you as human would be a *secondary level of representation* by which you can think about how not to get into such situations, the values of different ways of escape or encounter, and so on. What is represented at the primary level is repeated on this secondary level in ways that are more tightly organized, but also more abstract (Bickerton 1995:101).

Both default and reevaluative schemas are necessary. Compared to our default procedure, the reevaluative schema appears more consonant with the idea of self-reflectiveness; it suggests a later evolutionary development. But the importance of the wired-in default sequence cannot be superseded or diminished by merely focusing on wonderful capabilities of abstraction and capacities to open intellectual, if not intelligent, possibilities.

Assume that you do *not* have a wired-in default syntactic schema, but you do often have the problem of facing a saber-toothed tiger. Each time the tiger approaches, the order of values for survival would be open to a variety of influences outside you. If evolution simply replaced the default with the "secondary level of representation," the syntactic role and sequence would expand the choice of relational combinations. I present the problem of external influences in the following terms.

If our friendly monad (M) *does have* a wired-in a syntactic sequence, it would allow her to think in this sequence: "I see the tiger → I run to avoid being hurt." The default action in that sequence may save M's life. But if M goes immediately to a consideration of options, she may not be so protected. Suppose M is lost in a forest with Dr. Pangloss, a person guided (or misguided) by the socially popular idea that "everything is for the best." "*Your* best way to look at this," Dr. Pangloss advises M, "is to think this way: 'The tiger sees you → Do not look at the tiger → He will properly decide the course of events. It will be for the best.'" Well, if the *agent → action → outcome* sequence—so splendidly a product of the evolution of M's syntax—were either not present or disengaged, more room would exist for the revised syntactic sequence *object → action → outcome*.

We see that the absence of or a suspension of the evolved primacy of the agent in relation to syntax and its effects on thought and action may have antagonistic effects. The person so confronted has the opportunity to find something better than an automatic default way of thinking and acting. But look at the choice Dr. Pangloss presents. The object (grammatical patient) is the cause or primary driver of action. If this is acceptable or influential, the *object → action → outcome* sequence—a more primitive syntactic form—would become instrumental. The net result for M *might* be something like "regression in service of the ego." This is fine when there *is* time for a nonautomatic response. And it is fine if the evaluation retains exchangeability for the wired-in combination. The situation of being thrust into a "lion's den" or in front of a saber-toothed tiger would not meet that requirement, however.

The example shows Voltaire's point that an overly intellectual approach can become overly dependent on the "specific nature" of the abstraction. If the abstraction is reified, you're stuck in an absurd space! The reification of the object as a "best of all possible worlds" product allows one to attribute the "best of all possible motives" to that object (the tiger). *This razzle-dazzle in which the best motives are subsumed within the specific or particular is prototypical of the logic of the trope.* Why is the logic of the trope not creative here without being either dangerous or absurd? When the trope is reified in this way, the effects find their way to mummify the syntactic schema accessible. Cervantes's complaint is similar. If the tropes become too alluring, there is a descent to primitivation. The wired-in checks and the possibilities for juggling the way back to a mature syntactic and logical organization are lost in the dazzle.

The Razzle-Dazzle of the Particular and the Ensuing Degradation of the Trope

Consider this analysis in the terms of a general theory of rhetorical "figures" (group μ [1970] 1981). Rhetorical figures linguistically decompose in planes—or, if you want, in levels of organization and meaning. Significant here is the group μ idea that forms of expression pass in their operations from "pure form to content" via a "progressive succession of fields." This progression from one field to another is from "pure and arbitrary form" to "content, or pure signified, not under any constraint of a linguistic kind" ([1970] 1981:28). The latter field—which group μ calls a Logical Field—is of particular relevance to this book and to my approach. We have been looking at the schema as an abstract pattern representation. But for group μ, the pinnacle of abstract organization is the Logical Field. When that field is regnant, its structures are far removed from the obligation to cover specific meanings. Yet the content it *does* cover, abstract and symbolic as it may be, is well articulated. This means that rules and organization are at their height, and denotative bounds are optimized, but connotations are lost. Even though you can make an analogy to Piaget's "formal operations," all that I take to be natural logic is not covered by the Logical Field. The dynamic aspect of the interrelation of different levels of meaning is missing in "formal operations." Thus, I promote the idea that *logical forms interact with meaning and with semiotic relations.*

The group μ observations show dynamics, but they go in a "reverse" direction. *When the cognitive "progression" to organization and rules by which to order, if not to hierarchize, meaning is curtailed by a dazzling distortion or a crippling overfocus at the logical level, the effects on semiotic relations and meanings "backpedal" to stultify the syntactic field.* The movement can be pictured more easily with the group μ chart of "progressive succession":

Plastic field

↓

Syntactic field

↓

Semic field

↓

Logical field

In the chart, the arrows move "downward" away from considerations of signals that have form but no meaning to a field where meaning is abstract yet articulated. With the downward progression pictured in the chart, there are more relational constraints as an "ascent" to the "pure signified" is made to more complex and articulated organization of meaning coordinated by forms. The more progression to a mature logic, the more meaning is articulated, but the more constraints to supporting a plethora of specifics. In plain talk, generalizations relate to more phenomena by limiting your consideration of their features. An example is that the law (rule) of contradiction might knock out the consideration of certain opposites and contraries, which, if reconfigured, might plausibly be seen to coexist.

The flexibility and juggling available by way of tropes allow you not only to focus on and then abandon the genus status of the more general terms, but likewise to focus on and then abandon the genus status of the more specific terms. The dip into specifics and the reorganization at a more "progressive" level look like a "regression in service of ego cure" for the rule and relational constraints. However, *if reification of meaning or content takes place within a figure of speech, then the progression reverses itself. The syntactic field gets primitivized.*

Primitivation of the Trope and the Submerging of Agency

There are all sorts of ways to hang up a trope, get stuck on its overconcretized meanings, and hold logical form in a paralyzed state. Primitivation is severe when it affects the syntactic field. Rules and roles are overconcretized, and the individual is captured by them. Voltaire makes the point with humor. George Orwell shows how tropes that are sloganized lead a death march. His "War is Peace" makes the icon *war* a species organizer! War superordinates all sorts of categories when it is made equivalent to its opposite (peace). That's quite an overevaluation of what should be an object in a syntactic field. The net result is the loss of the individual's powers to juggle. It occurs when the syntactic "object" is overevaluated and the focus on it obscures a view of agency. The implications are enormous.

If a rhetor socks home an icon as a species organizer, the icon has power over more than the meanings within the compass of the trope. Its power is magnified by the disengagement of your self and its agency to evaluate or change the course of the causal direction flowing from the totemic object. If the rhetorical icon presents war as a game for winners and losers, you're stuck in its dichotomous field of meanings; your syntax is a limited default field of objects and targets! Agency and self are attenuated by powerful pronouncements of unassailability of icons; thus, a terrorist may believe his instructions are from the Holy Book.

Consider from contemporary theory an example of descent into the concrete interpretation of thought when the primacy of agency is disregarded. Lakoff (1987, 1999), Eleanor Rosch (1978), and the cognitive linguistics school theorize about language and cognition as internalized strategies for effective action. Their idea is to amalgamate concepts and schemas at schematic nodal points from which action possibilities are leveraged and conceptual clusters are developed. Within a schematic, the "I" is just a role player and is not at any one unified position—or in any one categorization. The theory stops the parsing of its terms at this particularized classification of roles and nodes; the "I" is as relative to the specific schematic as any other "object" would be. "You" are the "you" in the syntactic field of your playing chess. But "you" are also the "you" as "you" see that I see "you." As in the general rhetorical theory, encapsulation within the particularized schematic turns focus away from the primacy of agency and pours it and its functions into a mix of thought with the syntax of objects and events. In such schematizations, the agent is merely an object; the "ascent" to the conscious "I" is attenuated. Regression, yes—but *not* in service of the ego.

Something has to be added to the ideas of wired-in syntax and conceptual representations, but *how far do you have to stray from the expanded GibsonWorld?* Focus on the logic of inference and the ordering of categories as relational combinations and permutations. They can be matters of syntactic role and rhetorical admixtures of value—such as "holy" and agent or "redemption" and outcome or object. These combinations and computations may also have a hardwired procedural basis. This focus may go along nicely with Bickerton's (1990) view of syntax in terms of "argument structure"; it casts "argument" primarily in grammatical terms. So perhaps you don't have to wander too far. You also might come to this combination of psychological, rhetorical, and semiotic aspects of logic—from a Piagetian analysis or from the kind of model of "mental logic" that Braine (1998) champions. Other recent shifts from a "classic" position to a more cognitivist view of rationality include Robert Hanna's (2006) philosophical view of a logical cognitivism and Paul Thagard's (2006) computational view of coherence in emotion and thought. With all this, do you escape from the confines of the schema? What's the place and the power of agency?

The basic model of self-regulation in cognitive organization *is* currently a popular and prevailing view. This view, in combination with the workings within the schema, encompasses the following proposition: *the individual's meanings are constructed as an ecological balancing of self-organizing rules with data and meanings provided by others and by the nature of the environment and its signals.*

All of these expansions of a GibsonWorld may be illuminated by a Leibnizian spotlight on the inner-directed monad, with its adaptive capabilities and combination of integrity, uniqueness, and changeability. I tilted the procedure/knowledge split to showcase the procedure facet. Procedure is wired in, and to the extent that knowledge has any separate existence, it is merely processed—with the possibilities of being converted to procedure. This viewpoint *has* captured a great deal of research in cognitive psychology, philosophy of mind, and the interaction between artificial intelligence (AI), robotology, and constructivist views of evolutionary biology.

I had better have good reason to propose an alternate view!

3 Shift to Individual Categories, Dynamics, and a Psychological Look at Identity

What's Missing in the Expanded GibsonWorld?

It's not enough to protest "I am not a robot" if a robot can utter the same protest. I have to show where the robot falls short, but also characterize the self in other terms. In the first portion of this chapter, I analyze the specific problems of settling for the received view as it would apply to explaining tropes as a particular mode of thought—figurative thinking. I use a thought experiment to explore the positive contributions, which an analysis of logical forms should provide.

Cognitive Processing and the Relation of Gradients of Sentience to Knowledge

The "organism in its ecological surround" model *does* appear to be in the same focus as the viewpoint of this book. Yet cognitive processing as a series of mechanical rules is a far cry from a natural logic within a semiotic framework. Semiosis not only supposes an internal set of representations and referents cutting across many layers of the organism's signal and symbolic subsystems, but also presupposes these points of transmission and transaction as subject to gradients of sentience. When an animal (such as the lion in Aesop's fable) has a thorn in its paw, we can assume sentience—at least in terms of comprehensible transactions of signals and feelings. We can assume representations necessary for intracommunication—so that the lion can feel the thorn and relay the information to some action or schema and then to acting. These representations and steps toward the lion's trying to dislodge the thorn may be different from those that communicate with other creatures, such as the mouse. Anyway, there may be necessary links between representations for internal communications and representations for social communications—so the lion can stop "calls for help" when they are not related to "feelings."

These links can be not merely of the individual's communicative subsystems, but also of gradients of the individual's sentience. It is one thing to say perception of the other person's reaction and its coordination with voice projection is a system of stimulus patterning. Conceive this system as one type of representation, which I call $Icons_S$ (icons of stimulus patterns). It is another thing to say you can be influenced by percepts of the other person's facial reaction and be aware of this—or not. Perhaps a different type of representation can code the *level of sentience/awareness*. I call the icons of this type $Icons_A$ (icons of levels of awareness). $Icons_A$ have variations, such as "icons for unconscious" versus "icons for conscious experiences." *Relational transactions* are the rules for relating representations and for relating one *type* of representation with another. (One way to picture the different sets of transaction rules is to refer to group μ's concept that meanings expand extensively in a less organized and less rule-constrained field. When the field is more symbolic, there are fewer logical constraints, such as those on countervailing meanings. The result is that an equivalence transaction does less damage to meaning at the symbolic level because the identity of units is easier to establish than when there is a myriad of unique factors and combinations of factors in the field.) In sum, procedural operations are embedded in knowledge in a variety of formats and in different stages of awareness. *Sentience and knowledge* are part and parcel of the formation and transaction of signals, icons, symbols, categories, and their interrelations and their organization.

There is still work for me to show it is valuable, if not necessary, to go beyond the received view. The target phenomena are tropes, their degraded forms, and their analogues in concrete, immature, pathological, and primitive forms. Because the natural-logic/semiotic viewpoint I describe is general, it remains for me to account structurally for the fact that the internal coordinates of meaning are obviously different for the different target cases. It's one thing to employ a figure of speech for artistic or hortatory purposes. It's another to believe a figure of speech to be a literal fact or an unassailable premise. A figure of speech may *seem* to have the same or similar logical structure compared to a concrete or pathological assertion. But, somehow, the trope's primitivations of form—if indeed they are that—are not just left at that. If the dream is a pathological episode, we somehow "recover." Although Walter Mitty goes off in fantasy to be a war hero in the middle of a humdrum workday, he recovers and shakes off the "daydream." When a lover tells his love she's like a "red, red rose," he snaps back to humdrum taxonomy of roses and lovers, as does his lover. *Usually.*

It is not merely a cognitive or metacognitive set of operations or relationships that needs to be specified. That would simply be accepting the procedure/knowledge split and working to show processing routines. Showing processing levels and nodal checkpoints can provide considerable specification of routines—and even engender startling simulations. Accounting for levels should be theoretically easy. Information about information would be a "metalevel," functioning like rules governing routines. A path of governance in the case of a primitive form may be treelike; the governance schematic for metacognitive and cognitive processing may be more like an administrative chart. (See Minsky 1985–86.)

There *is* room in group μ's "progressive succession of fields" to account for higher levels of organization and governing rules as one "ascends" (by downward progression) to the "logical field." However, the idea that an "agent" can be constructed within the procedural sets of rules is given more shape by AI concepts and their implications for robotology. The Minsky "agent" can be a set of processing rules governing other processing rules. They are reducible to information and operations (processing) at metacognitive levels, which have nodal points. The "agents" can *manage* information at various checkpoints. Let's not go too far. Suffice it to say that were we to do a thought experiment, the "administrative chart" would fall short of an "in principle" explanation. Let's do it anyway!

A Thought Experiment: Time Travel to Find Identity as a Contradiction

I need a coup de grace to do more than merely rectify the contraries in the robust model of cognitive processing that I outline as the GibsonWorld.

I suggest a thought experiment to show that when you look at what you *cannot* do with the ecological approach—even with a brave integration of self-regulation—contradiction weighs heavily. It is harder to face down full-blown contradictions than to tolerate coexisting contraries. The issue we need to take on is the relation of logical identity and psychological identity, but it takes a bit to get there.

Two Kinds of Mind and Jaynes's Analysis

I ask that you observe a pre-Homeric person, her mind, and that you view the possibilities/impossibilities of your changing the way she thinks via a set of thinking rules that you transport across time. A modest request, but let me explain the contexts of the two types of mind and the two time periods.

Pre-Homeric persons' decisions were made by dialogue with the gods instead of by dialogue with the self (Tejera 1971). Historian M. I. Finley writes, "For centuries the Greek interest in their past was only a mythical one.... [T]hey were concerned largely with individual, isolated occurrences of the past (usually involving direct participation of supernatural beings)...not with an ordered account of the past arranged systematically in time and place ([1963] 1977:24). Moreover, that which moved a person from within was more a matter of "spirited" and emotive forces than of rational forces. Changes appear in Homer and the dramatists and then in Plato and Aristotle. The direction of change is in the differentiation of *nous* (mind) such that "psyche and the power to perceive, or feel, or think" counterbalance "emotive phases of both action and perception" (Tejera 1971:34–35). Compare this emerging power of the mind with destiny *(moira)*, which was a force more powerful than the gods and could be explained as something like cosmic order (Tejera 1971:55–56).

Helpful here is Julian Jaynes's (1976) analysis of the changes from a syncretic conception of persons, forces, gods, and events to that of the person with a mind that is conscious, can focus internal stimuli, and is therefore responsible for its actions and decisions. Jaynes documents these changes by spotlighting the use of terms in literary works. He identifies "phases" of that change:

Phase I: Objective:...terms referred to simple external observations.
Phase II: Internal:...these terms have come to mean things inside
    the body, particularly internal sensations.

Phase III: Subjective: ... terms [that] refer to processes that we would call mental[;] they have moved from internal stimuli supposedly causing actions to internal spaces where metaphorical actions may occur.

Phase IV: Synthetic: ... the various hypostases unite into one conscious self, capable of introspection. (1976, 260)[1]

Cognitive Space, Schematics, and Categories

I cite one other item from Jaynes's analysis. It helps to contrast the earlier undifferentiated person/environment and person/internal sensations from the conscious self. This is a state of mind *(noos)* with an articulated sense of different internal mental spaces devoted to separating the "I" from forces and influences outside it—whether these forces are internal emotive stimuli or external environmental stimuli. How can the experience of emotion, if it is internal, fit into a space "outside the 'I'" if the "I" is in an internal space too? You can answer this question by saying that the separate cognitive spaces are not merely slots in schematics or syntactic sequences; they are also discrete categories. But you would still have to suppose emotions are not somehow subordinate to the "I." (How can you have an emotion that is not yours?) This is where the relationship between categories becomes, as Jaynes might say, "metaphorical," meaning that not all "juggling" is consciously steered by the "I" and that there may be spatial separations of the "I" and the "me"—or the protoself. According to Jaynes, the subjectively differentiated mental spaces allow for "internal stimuli supposedly causing actions to internal spaces." These internal stimuli do not have to be conceived as mysterious. The separation of self from objects that cause action can be a function of wired-in procedural rules—a necessary function, although not sufficient to explain the features and powers of the "I." Orders of consciousness may explain these features and powers.

A person with such capacities (wired-in procedural rules, at least in part) would fit descriptions of the mature mind both from a Freudian (dynamic) and Piagetian (structural) perspective. But in terms of the evolutionary approach, a syntactic articulation is also consonant with Bickerton's view, and the rational capabilities are consonant with Braine and O'Brien's view of logical capacities. With this nifty vignette of minds and time frames, *you* emerge as a contemporary person—on

track with Jaynes's picture: an individual—but not isolated. He characterizes Solon of Athens as standing "[a]t the beginning of the great sixth
century B.C. ... It is the century where, for the first time, we feel men-
tally at home among persons who think the way we do.... It is the way
he speaks about the *noos* that is the first real statement of the subjective
conscious mind" (1976:285–86).

Empathy for Someone Unlike You

Now start going back. A pre-Homeric person (call her PH) would believe
in the gods as external entities (see Jaynes 1976). PH might have believed
the gods determined how she felt and acted. Even if she felt and acted a
certain disastrous way, in her mindset the gods would have chanted about
their knowledge and power over these feelings and these sorts of events.
And these chants were to be believed.

Your thinking about this person is a journey—a time-travel trek. You
muse that things, in the terms of *your* day, would (maybe should) be otherwise for PH. You might give PH a set of cognitive routines whereby she
can "reconvert" her externalized category of "gods" to a category of "internal fantasies," daydreams, or tropes. You bet that she will thus be able to
redirect her decisions and their outcomes. However, your bet is on what
you can only suppose as a creature of your own time and its ways of thinking. From the word *go*, PH would have been a different creature. Before
going to the betting window, you decide to rework your scratch sheet.

Even if you print out a set of cognitive routines, and PH is able to
read your list; your supposition would have to be synchronized more with
what her mind can make of those rules. That would be a function of what
thinking was like in *her* day. Even with the set of "rules" you brought to
her, she would simply *not* be able to convert a category of "gods" to an
internal fantasy. In her thinking in pre-Homeric times, the "internal" versus "external" locus simply would not have been demarcated! The assignment of locus could not have been as *you* might picture it—a projection
or externalization. It would merely have been an attribution within a field
of events, which did not have the internal/external boundaries with which
you and I think nowadays.

Well, you adjust for this realization. It *would* be interesting to see what
she would do given an underlying incapacity to separate icons of imagination from those of external objects. The objective to "reconvert" might

have applied if the internal/external boundary were a "once had been." "Can she 'reconvert'?" is really *not* the question. Instead, you *might* ask whether she can use the routines to set up a boundary that "never was." It would be a boundary for separating from the external that which is to be assigned to the internal. From the catbird's seat of your own time, you decide that you *can* envision writing the rules to differentiate ideas, feelings, and objects that would be assigned to the "internal" or the "external"—two different loci of occurrence and determination.

Along with some *Star Trek* writer, you can probably conjure a way to ship those rules in a backwards-moving time machine to PH's day. Then, the rules having safely arrived, you can imagine observing her come face to face with them. You ask yourself what your own thoughts are about what those rules can account for and what they can accomplish if, by your conjuring, you can get PH to adopt them.

Nevertheless, you cannot expect those mere rules—which do not account for *her* origination of icons and ideas as they appeared in that long-ago time—to give her enough leverage to do anything like what you would wish to be a "reconversion." Your inquiry as to the possible utility of the rules would better be phrased as a question of reorigination. *She would have to go backwards in time* and reoriginate her thoughts and feelings in the context of a differentiated mind. So, without assuming she would be totally reformed as a person with a contemporary mind and the differentiation of self from that which is external to it, you would have to see her as the person she was—at that time.

The Limits of Different Identities

We're back at the only sensible premise. Your thought visit is to PH—a person of a time past. Oh dear, then there would be too much *not* present in whatever acts of origination *never took place* and were *not* present in her complex belief system and its interrelation with an undifferentiated sense of the internal and the external. You now believe you would see her sense of self as undifferentiated. The gods would be woven through the text of her identity, and her identity would be bound up with the gods. But that cameo gives primacy to the gods, whose powers supercede those of mere mortals. Therefore, the *external determinants* of the individual's identity, decisions, and behavior would be in the foreground. The focus appears as Jaynes's idea of the external. For the individual to develop a sense of differentiated self, she has to distinguish between internal occurrences she cannot control and her actual decision-making capabilities.

Jaynes recalls not only the lack of differentiated self when the external is a locus for the self's determinants, but also the undifferentiated "internal"—such as sensations and emotive considerations. As an example, he tracks changes in the term *thumos* from an *external phase* in which it means "activity as externally perceived" to a *subjective phase* in which it means something like "urges" (1976:262–63). If there are merely "urges" without a distinction between them as causes of action and as objects of the individual's thought, then we also have an undifferentiated state of self.

Origination as Logically "Someplace"

Now I can use your thought trek to show that it *is* necessary to go beyond the "received model" and the GibsonWorld. I can pinpoint three reasons that this model cannot be an account of origination by the individual. The first keeps the locus of origination outside the individual. The second does this, too, but adds that certain forces and factors within the individual do not constitute *agency*. The third limits the powers of agency by excluding self-identity.

First, in the "cognitive-processing" model, knowledge and procedure are split. This affects knowledge and its source. The individual's knowledge—whether of the gods or of the self—originates by way of information from the environment.

Second, with "procedure" as something already wired in, our PH monad is driven from the two sides (wired-in procedure and environmental information), but not from anything she or we could call agency. What is wired in is simply, by definition, not a matter of agency. In Ellen Langer's (2000) terms, the firing off of a nonvolitional schema, even if a well-coordinated specific script, is "mindless." Environmental determinants are by definition outside the procedural, so, at best, they would be "information," but unformed thoughts—hardly something we could call agency.

Within a model of this sort, each of these reasons leads "outside" the self.

Agency is the lynchpin for initiating one's own action, and it is a necessary condition for originating. *The self would, at the least, require the reflexive capacity to regard what is going on within the organism as subset to itself.* If "I" am aware that "I" have or experience emotions or urges, then "I" is the set, and the emotions and urges are the subsets. In short, differentiation has logical form as well as syntactic requirements. If to have the power to de-termine one's action requires internal/external differentiation of self, and if it requires logical differentiation of self in terms of set and subset struc-turing, then the received model and its variants do not provide adequate conceptualization.

The application of a Gibsonian or information-processing approach does not rise to the challenge of explaining regulation from within a monad. Determination from within the monad may be automatic via what is wired in. But in no way, shape, or form is automaticity the same thing as agency. Nor would automatic wired-in processing rules and rou-tines help PH or a PH monad to reform herself!

The "received" account is of the two-pronged determination of action and cognitive phenomena: *(a)* "given" cognitive routines and *(b)* external affordances. It comes close to saying that the monad is a "self-regulating" system—albeit one without a self. (This point is incomplete. The argu-ment that the received view is inadequate requires a concept of an "I" that can ascend to a categorical position beyond any point it settles at in its various operations. The pre-Homeric woman can think of herself within an action schema, but she can't move the locus of self outside the schema and into a separate internal space within the person. See "framing and "bracketing" in Fisher 2001.)

For a third theoretical reason, the received model cannot be an account of origination: an undifferentiated category of self and its environments, actions, and capacities begs the question of the self's identity.

In our thought experiment on the "pre-Homeric person," we reasoned—compatibly with Jaynes's argument—that within the logic (or paralogical system) of someone of that era, the inner-generated icon and the external representation are not coordinated to the articulated functions of the self. Instead, the situation comprising self, other, the individual's internal envi-ronment, and the meaning of her thoughts and feelings about all of these things is something of a goulash! The self, the idea generated by some inner urge such as hunger, and the appearance of a mythic (miragelike) creature showing the path to food are all members of the same catchall category!

The Who, the Self, and Identity

The catchall category is like a myth drama. In it, characters take on event-driven characteristics and events, and objects take on the characters' characteristics. *Who* exactly is it that "knows" this category? *Whoever* it may be is caught within the webs of the myth drama. As is the case with myths and heroic epics, episodes and persons get separated within all this webbing—somewhat. Nevertheless, to conceptualize how these separations take place, the question of "Who knows?" within the "received" model reduces to "What is the course of processing?"

Within the webs of the myth drama, the "process" of association and contiguity signifies the logic of the person-monad's identity. When T. S. Eliot's "claws creature" "scuttles" across sand, the specific nature of the associations and contiguities of the creature's motor actions and morphological units as they relate to the sand's environmental affordances define the individual creature. Its identity becomes more fully specific as a function of the particular history of happenings during its lifespan. Does the creature "know" it has an identity? If sentience is knowledge, yes. However, who is it that imputes the laws of association and contiguity? We would have to say the laws are focused from a third-person perspective. There is nothing wrong with this formulation—as far as it goes. The problem is that from the point of view of the "claws creature," it is inaccessible, and therefore we would have to talk about *an identity without a self.*

Self-Regulation Versus the "I's" Agency

All of this questioning about the perspicacity of a person versus the sentience of a "claws creature" has implications for the overarching issue of "self-regulation." *Regulation from within the monad does not per se mean that self is involved.* And it does not necessarily imply agency. Thus, although a myth drama can be a product of an individual mind, the recognition of the pattern of the myth and of its continued processing in the construction of its logical forms is something conceptualized from a third-person point of view. Lévi-Strauss (1966) observes that these sorts of logical products are "species" organizations (also see Pace 1983). For example, an action associated with a person and with a time contiguous with a value can cue the building of a statue or totem. It would be a logical product and serve as a species organizer. Then, third-person factors enter. You can bring homage to the totem. By doing so, you hope it will replicate its honored or revered person characteristics, the events, and the value of the events for you. Your actions appear deliberate—regulated, so to speak. Nevertheless, whatever level of sentience one may infer in relation to your bringing gifts and hopes, your accepting the object *as a totem* places you within the action schema that has been constructed— say, as a social or societal phenomenon. However, *it places you within your own schema of totems.* This schema includes the myth structure relative to that particular totem and the myth-drama structure as a metaschema. You chart out inner spaces, but Jaynes's journey goes on to mark signposts of mind and self.

All this processual regulation is a severe limit to the Gibsonian and information-processing model—despite its other significant advantages. The monad is closed tight relative to a metaphysical assumption that its physical and biological nature is part of its closed system. To the extent that knowledge is conceived in the same terms, the "I" or the self is either procedure or information. Therefore, *the "I," as an agent, cannot step outside any administrative chart or hegemony wherein the order of construction and deconstruction is at some penultimate point of cognitive-processing rules.* In short, the self is an actor caught in the web of a cognitive script. The self is *not* an agent free from any script (or wired-in procedure) to originate a new ordering—albeit, it *can* be conceived to function as a governing "agent." My two uses of the term *agent* here are like the difference between someone such as Scrooge, who comes to see he is an initiator of decisions for action, and someone such as James Bond, who is dispatched to represent and actualize someone else's interests. (Some might say that an "agent" bound by a script has no "will.")

The Historic Inaccessible Location of the Agent

In this interpretation of Jaynes's (1976) thesis about the pre-Homeric mind, the limitations of the "I" as agent present a person with a self that cannot step outside its "old rules"—in other words, is not able to recognize the originating of its own rules. Shades of Oedipus! If this is the case, it brings a halt to the applicability of the cognitive-processing rules for the reclassification of categories. This dilemma of processing *from* but not *by* a self occurs because the pre-Homeric self would be syncretic, not integral. Therefore, we simply might not be able to help the historic PH! The agency of a pre-Homeric self—if it did exist in some nuclear form— would be impossible to engage in service of a sea change in the meanings of icons, ideas, feelings, and the like, which presumably *can* be juggled around by someone from our own era.

Who might this person of our own era be? We cannot answer this question just by saying, "Once Solon appeared, the new era is Everyman." After all, there are those around today who follow the dictates of terrorist leaders. These dictates, enmeshed in mythic drama forms with religious themes incorporated, move the individual followers' selves around like objects that can act on other objects. Many cannot respond to an appeal to lift the concept of self out of the morass of such dictates. Even when the name of the self—or the "I"—is invoked, the agency, which would mark cognitive spaces for exploration and for changes of meaning and value, seems elusive. I accept Jaynes's idea: for a self to negotiate meanings, thoughts, and feelings—indeed, relative to decisions and actions—cognitive spaces have to be articulated. Moreover, these spaces have to be available for categorical attributions and semantic forays and "digs" while the self builds its categorical reorganizations of the internal and external worlds.

The "I" in Time Travel—Visitor, Not Denizen

The "person of our era" would be someone who is visited by and who can revisit myths and dreams. I focus on dreams because visiting myths produces "daytime" patterns of action and behavior such that even if one can emerge from the patterns, one has already engaged in them. This "engagement → emergence" process is something an artist can transcend. An actor can act out Odysseus in a film and go home as "himself." Conversely, Salvador Dali drags his dreams into the activity of painting their images; however, he and the painting emerge as figurative expression. The agent in the "Dali" and other such cases produces *as-if* products. The "actor case" and those from which the artist emerges with his own thought patterns in focus yield a different product: the agent himself is more accessible to self-judgment.

When a dream is involved, there are other complications. When you are "in the dream," you are "in" the web of the action. It is something that you *can* see as an as-if situation, but this observation is usually made in retrospect. (You can make a fine-grained analysis here. The actor is "in the role." His examination of himself in relation to the role can precede the performance, change it midstream, and change his self after it.)

From time to time, the "modern" individual immerses herself in a myth so much as to behave within its rules and schematics—for a time. Irrespective of her subsequent "recovery," the individual's behavior has in fact occurred. It becomes part of the person's life facts and her social environment. A trenchant example is the movie portrayal of General Patton's sense of himself as a military leader of timeless heroism (Coppola and North 1970). Living out a myth is like the acting case, but different from the dream.

Our Post-Homeric Dreams

I focus on the dream. The person Jaynes might call a current-day Solon presumably typifies a post-Homeric mindset. The term I use—a *person of our era* (PE)—sounds like the term for the *Time* magazine millennium cover selection! I mean to convince you that PE is someone like you or me. PE's self-based judgments can differentiate assessment of dream objects by dream logic from accurate assessment of affordances. In recalling a dream and what he felt and believed "during" the dream's sway, PE can retrospectively reconstruct what the dream logic was. In looking back at a dream of "flying," PE might say, "I believed that I was flying—when I dreamed it. But now that I'm awake, I realize that's impossible."

In the dream, beliefs about objects mix the noncontradiction of the image and the person with the contradiction of personal power and gravity. *It's as if the rule of identity can hold without the support of the logical rule of contradiction.*

PE can also dig into his own dream combinations to get an idea of what can be created as objects. There might have to be many modifications if PE were to mine a dream "idea" to invent something useful. According to group μ's formulation, because more abstract categorizing leaves out a great deal of meaning, it is valuable to time travel among protological forms and their contents before resettling on more mature logical boundaries. We come back to the "juggling" of tropes.[2]

Ordering Categories, Juggling Tropes, and Transforming Forms

What *does* allow us to juggle tropes? To answer this question by pursuing the route of procedural cognitive-processing rules is to come in after the fact! This mistake is comparable to focusing on a product's nature when we're after its origin. Instead, I assume logical form as an originating factor (I want to say *force,* but that term incites too many feelings left over from the vitalism wars). So I want to show how it is the very logical form of a trope that we juggle and change when we raise or convert its meaningfulness to another *order.*

By *order,* I mean a position of *super-* or *sub*ordination relative to a combination of things: categorical structure, genus-species relations, propositional entailment, and articulation of the classificatory structure relative to specificity and generality. I don't mean "order" based only on a posteriori superordinating function. The place of Babe Ruth in baseball legend is of great mythic value. He earned it based on outcomes—sixty homeruns. But if number of homeruns is what you're after, he's now at best a totem. We might count outcome numbers for Mark McGwyer or Barry Bonds. For the meaning I intend, the question of order ("top dog," here) is *not* merely an outcome of some function.

If I have a taxonomy of mental disorders, it might be based on some idea of parallels to brain dysfunction, or it might be based on social dysfunction. Either way, illnesses can be grouped in some hierarchy from more to less dysfunction and subgrouped by severity of symptomatology. However, once again, we would be placing the cart before the horse if we are trying to explain how a person (in this case, a scientist concerned with mental health) originates a logical category. One way to look at this is not to be dependent on a taken-for-granted criterion for dysfunction. We then conceive of an individual practitioner who "originates" the categorical criterion for his diagnosis.

Every time a diagnosis is made, the practitioner can still go to her chart to see what fits or can juggle known values such as social function in terms of the individual patient's needs and concepts. But if the practitioner actually did this for each patient, the taxonomic scheme might differ such that each time it would have to be coordinated with how the patient forms categories and how the practitioner, even when taking the patient's categories into account, forms her own categories. There would be lots of categories to juggle! Yet even with the downside of the profusion of different clinical judgments, what emerges as a hierarchy in that individual practitioner's thinking is a transformation of logical forms. It is dynamic rather than after the fact. The history of pragmatic functions and their evolution is not the first thing to consider in this account of change. The "received" model may in a variety of ways be complementary to the account of change and origination I propose. But there is a shortfall.

Again, what *does* allow us to juggle tropes—to unseat and reorganize logical orders and by doing so to enrich meanings? It is the dynamic state of affairs holding together and regenerating the order. Moreover, the resultant order itself involves not only a different kind of meaningfulness, but also a different logical form or format. Overall, we are speaking about *forms and their transformations.* To show the difference between the logic governing figurative thought and the logic that is either immature or unconscious, I ask: How does transformation of forms take place? What happens to the formal aspects of the logic? Do these aspects themselves transform to keep pace with the psychological events fostering the meaningfulness? What psychological elements are necessary?

The Purpose of the "Logical" Considerations

Why look at the logical forms? Despite the many sources of logical analysis and despite the allure of such analysis, this "why" question has not been adequately addressed. What function *does* it serve to bring to bear a logical analysis—particularly to symbolize it—in service of depicting these various vicissitudes in thinking? For one thing, there appear to be no easy algorithms to derive remediation. When a psychiatrist finds that in a schizophrenic's logic the law of excluded middle is violated (von Domarus 1944), the question remains, What steps can be taken to help the schizophrenic think more normally? Do you develop an instructional module with "thinking strategies"? Do you scour PET scans to find areas of the brain that do not light up the way they do when a "normal" person uses the law of excluded middle? Do you then launch a search for stem cells to repair or augment that area of the brain? Do you write an algorithm for the pathological reasoning sequence and then write a program with the missing steps included? Following up on that product, do you hand it over to a tutor (person or computer) to remediate the schizophrenic's logical routines? Each of these steps and products falls short of providing either the helping person or the patient with enough leverage to get at causes or remedies. The underlying problem is that none of these routes can confidently flow from the theoretical observation of the logical form. One can point to empirical trials of these routes and products and explain this way and that why there is some success. At the least, there is missing data. One can cap off various lines of research findings and the correlated assessment of potential clinical success and then argue that many, *many* variables remain. To search them out will not close the gap, but it might pay off pragmatically.

Specification of Patterns of Reduction

The deeper philosophical problem in deriving a treatment algorithm to relieve the schizophrenic's logic woes is that the causes of the schizophrenia are elusive. The researcher and theorist are confronted with many levels of theory and many domains of phenomena to traverse. There is no clear connection between the brain chemistry, neural activity, and brain activity pattern, on the one hand, and the logical form, on the other. This lack of connection is a problem not only for those who would try deriving remedies from theory or research findings, but also for the person theorizing—he has to think on the different levels and about the different phenomena. This issue of traversing levels of thought is everyone's problem—and that includes the schizophrenic! She has to think about her thoughts and in the process have many levels at which she organizes her meanings and the rules of processing. In fact, this very problem may exacerbate her logical struggle—and even be a major cause of it.

Fields of Organization

For the theorist who tries to derive remediation strategies, there are "connections" to make. Neural transmission patterns may or may not be clues to how meanings are configured. What connections there are may lie across different *fields of organization*.[3] To see what problems arise, take an example from which we can see the theorist's navigations, but also from which we can see that a schizophrenic herself may have trouble navigating such fields. Keep in mind that mere navigation from one "field" of organization to another is automatically a reduction. It's automatic in that as rules ascend in their stabilizing function, particulars are extracted—or left behind. A word about "meaning" is necessary here.[4] In this example, as in the group μ progression, we first decompose, moving downward toward a symbolic form.

A progression reducing specifics can be from specific algorithms to a general rule of equivalence. Suppose a theorist observes a person who stereotypes others. The person presents the pattern

John is cheap.

Monks are cheap.

∴ John is a monk.

as a thought sequence. The theorist sees this pattern as a violation of the law of excluded middle, which flags a thinking problem. So an algorithm expressing the law might be

J < C.

M < C.

∴ ~(J = M).

But suppose the theorist "progresses" to another formulation. In it, J, C, and M become equivalent elements.

All elements (J, C, and M) are equivalent as to a value—say, arousal of dislike.

∴ The denotation for each symbol can be reduced to that value.

Such symbolization may be diagnostically valuable. The reduction can tell you something about an individual's emotional readiness. Anyway, because this progression is to the more general rule, *the use of the law of excluded middle appears a casualty*. This casualty might *not* be a problem within a field of logical symbols or with some other form of reduction or extraction of specifics. But the problem of theorizing remains—the inevitable traversing from one "field" to another. Theorists somehow have to be able to negotiate "back and forth." And, as I say, we all have to do this.

Further, *this problem of understanding, then losing sight of, and then reconstituting the aptness of the law of excluded middle can be a signal problem for the schizophrenic. And it may well arise as the problem of traversing from one "field of organization" to another.* If the law of excluded middle is not carefully woven through the many changes one might make from one field of organization to another, its suspension can be a serious problem. For a scientific theorist, such suspension typically results in conflating levels—for example, thinking that the rules of behavior are the same as the rules of thought. For a schizophrenic, it might result in reasoning from prototypes or by way of something like emotional transduction! Fear can level distinctions between persons or phenomena.

Reductions can be legitimate. If you *are* working with equivalent units and logical identity or meaning is not a problem, then exchanges of icons or symbols can be made. However, suppose that for a theorist or researcher such a situation gets more complex, and traversing continues but goes from abstract symbols to those representations designating particulars. Say that an abstract logical formulation has to be transformed into a routine a computer can process. We move to a "field" with more specifics—coordinated to the nature of computer logic, the programming language, and to some extent the hardware of the computing device. The initial reduction is in part *reversed* in favor of more relevance to specifics. (The latter are specifics of computational cognitive processing and not of an embodied intellect and mind.) So, again, there are different sorts of fields to traverse. To take into account the complexity that requires a "reverse" of reduction, I focus on two factors: *(a)* level of abstraction from meaning and *(b)* processing rules.

Thought can move not only as a progression of fields of meaning, but also as a progression from abstract rules to specific algorithms. Picture two hierarchical progressions. One concerns human meaning. It might focus on an issue of borrowing from a sibling. As the issue progresses, concepts such as "reciprocity" become more abstract in their scope. A second progression concerning "processing" moves in a reverse direction—such as from computer logic *to* the specifics of program routines. The progressions can miss each other—both for the theorist and for the schizophrenic! If you plot the progression for the theorist, there are points at which meaning outstrips the algorithms and the general rules that can be coordinated. If you plot the progression for the schizophrenic, there are points at which pathology will flourish if the progression is uncoordinated, moves from field to field at different rates, or gets stuck at a particular point of reduction.

Therefore, to add to the group μ picture of the "progressive succession of fields," the distance from a productive or effective originating of thought in the schizophrenic would have to be plotted as two "field of organization" progressions. The picture would involve degenerative reductions and compensatory constructions, which themselves would be "progressive" reductions from the viewpoint of interacting continua. As a pressure point for leverage in navigating and coordinating the various fields of organization, I will show in due course that *the individual's originating dynamic thinking act would itself take place with whatever formation of self-boundaries and differentiation the person is working from as a launch pad for reasoning and beliefs and their organization.*

From the Theory of Logical Reduction to Formal Cause

Where *does* the theory of logical form link up with the solution to people's problems? If it is shown that the ability to think musically follows the logic of analogy (Rothstein 1995), how does this help a young composer to write better music? If I master the rules of rhetoric, will I win political debates? We wistfully observe that academics survived the 1960s style of answering questions of "relevance" for their abstract ideas mainly due to the ensuing practical successes of computer technology—the "revenge of the nerds"—and also to the success of Sputnik and the subsequent submerging of the "relevance" question in favor of a rush to competence. What goes around comes around.

Is it radical to think that theory has a value in itself? It's not really a novel view. Is it so conventional that the view offers nothing new? It does not have to be anarchic to suggest that "theory is as theory does." (Apologies to Forrest Gump!) Theory, as live and natural, is embodied and itself part of the dynamic structuring of the theorizer's thinking. If, however, you view this theory of natural forms as grounded in the theorist's logical transformations, you may very well think that a category mistake is being made. After all, some hard-nosed psychologists, in the patrimony of William James's call to pragmatism, may offer alluring arguments that the "thinker" *is* the "thinking." So is the idea of agency—that the theorist's logical transformations are "behind" the theorist's view of logical forms— overstated? Is the theorist-thinker vis-à-vis his transformations simply a tautology? Or is the nature of logical forms more akin to the recursive structuring of rhetorical figures?

As group µ ([1970] 1981) describes, there is not only a decomposability of higher-order structures in a hierarchy of forms; there also are rules and regulations within each ordering as well as for the entire structure and for transformation from level to level.

In attempts to redefine the nature of logical form and processes, many have argued for the contributing role of emotion and context. I agree; the context of meanings and the interpersonal and intrapersonal interrelations issues have to be examined and specified. Nevertheless, the prior question "Why look at the logical forms?" *goes elementally to the issue of the causal properties of the forms.* If pathology inspired by emotion and semantic context causes the denial of logical contradiction and the inability to act when confronted with facts—this is important to know. Yet the nature of the pathology remains mysterious. Its connection to analysis for purposes of remediation is tenuous. *But suppose the form itself is the cause.* This proposal sounds like a substitution of formal for efficient cause. Consider an ecumenical solution.

Imagine a category of contraries, oppositions, and outright contradictions. Call that category a *logical form*. To say that this logical form is "in a causal slot" is to place it within a schema—say, a schema for cognitive actions or operations. With this description, *both formal and efficient causes are in play.* The proposal is not an "either-or" reduction of psychological causality. What we have put into play is that the "logical form" can take the role of a "tendency" not to act normally in the face of contradictory thoughts. This position does not play favorites relative to any notion of a cause as *either* "wired in" *or,* as Gregory Bateson (1972, 1979) thought, "learned." I seek a strategic viewpoint from which the pathology would be the effect. Picture the difference between thinking of schizophrenia as the cause of the logical form and supposing that the logical form is a causal factor for the psychopathology. With the form in a causal slot, I propose we would have more of a chance to put together a metatheory with easier-to-see implications for remediation.

I say all this in the face of the many attempts made without taking on the relation of logical form to fundamental questions about particularly pertinent areas or domains. What *are* the areas fundamentally related to a person's logical dynamic structuring? Descartes asked how you can tell the difference between reality and a dream. If the theory of logical form and its transformation is enmeshed and embodied in the individual, a version of Descartes's question becomes *the individual's inquiry,* How can you tell what mode and meaning you are grasping? How can you tell if you are subordinated to your dream logic or to another person's propaganda or primitivism? Although these questions are just my derivations, many *have* attempted to use if not symbolic logic, then logical forms to shed light on the thinking that characterizes pathology, prejudice, propaganda, primitivism, and poetics.

Taxonomy and Disposition

Pathology can be described in terms of the absence of the law of excluded middle, the analogue of transductive thinking, and the confounding of rules of equivalence for rules of identity. Primitivism can be described in terms of the formation of classes that follow a "species" organization. And so on for prejudice, propaganda, and poetics. In short, logical forms are "variant" forms that abandon a rule, suspend it, or confound it with another. Once this analysis is achieved, the transformations of the variant forms can be traced out to see if they look like the understructure of the type of thinking they presumably represent. Thus, one use of this exploration of forms and their transformations is to highlight the differences between classical logic and whatever forms any of these variant modes of thinking might take. In this comparison, the purpose of the symbolic logic or schematization of logical forms is *taxonomic,* and its value is in terms of *disposition.* The taxonomy can be based on the presence or absence of the logical violations and/or fallacies. The simplest taxonomy can serve to differentiate in terms of values, such as that of pragmatically effective evaluation of information. If you show the violations of classical logic in contrast with their absence in a more reasoned argument, you may easily reveal the bigotry behind a propagandistic ad. "Disposition" may be quick and sure. "Disposition" here would mean that a value can be placed on the bigoted ads, and the ads can be consigned to some heap of actions on the basis of logic violations. Obviously, "disposition" would be a different matter for other variants, such as pathology, which involve psychiatric taxonomy and "treatment modes."

Consider the taxonomy and disposition issues in the case of an influential present-day focus: the constant attempt to find the areas of the brain associated with, if not responsible for, inspiring specific pathologies of thought (Mujica-Parodi et al. 2001). Disposition becomes a complex business of finding three-way interrelations among brain structures, logical structures, and thinking products and their amelioration. The brand of taxonomy becomes a side-by-side list of brain and logico-cognitive phenomena, which line up as a prolegomena to classification, disposition, and treatment.

The problem with taxonomies based on values is the ultimate point at which the commitment to the value is transitory. In regard to psychiatric classification of logic patterns, a seer's dreamlike jumbles might be valued one way within a society devoted to technological production, but in an entirely different way within a society devoted to artisanship and the quality of life.

Models of brain-inspired accounts of pathology are very susceptible to reductions: pathology in thought is pathology in language; pathological linguistic production is no more than deviant neurological signaling; and so on. The reductions themselves are subject to commitments, which also reach a point of transitoriness. Any theory hits a brick wall, and there comes a time to take into account all the things that the selected reduction does not. Where neurological studies do not account for the qualities of consciousness, some begin to argue that the model has to be changed and that consciousness has to be studied in its own right.

Reductions pave the way to other reductions. Pathology in its psychological terms is very open to a cascade of reductions ending in physicalist terms—chemical phenomena, atomic structures, quantum events, and so on. Logical forms are reduced to cognitive patterns. Cognitive patterns are reduced to linguistic patterns and viewed as operations, which can in turn be interestingly reduced to physicalist actions and their schematics, either by analogy with computer structures and events or by definition in terms of neurological structures and events. As an example of the latter, George Lakoff and Rafael E. Núñez have developed a model on the following premise: "Human ideas are, to a large extent, grounded in sensory-motor experience. Abstract human ideas make use of precisely formulatable cognitive mechanisms such as conceptual metaphors that import modes of reasoning from sensory-motor experience" (2000:xii).

In general, though, with regard to scientific taxonomies, the value of reductionist approaches is limited. That value is in part a function of the information available at a given time, fads in theorizing, and the popularity of certain metaphors. The idea of a "conceptual" basis for metaphor is an example of the fad based on the powerful synergy of the Gibsonian and self-regulatory views. The value of reductions is in using them for leverage, but one must make sure they do not become runaway Sorcerer's Apprentices! Thus, Simon Baron-Cohen (1997) *does* want to consider brain patterns in relation to a psychological dysfunction, but he agrees that there is neither a "mind-o-scope" that can look into mental productions to find the specific underlying neural transactions nor a "brain-o-scope" that can look directly at neural signals or their patterns and divine specific cognitive-level meanings. The reductive approach can be valuable in regard to creating heuristic perspectives for information gathering. We find out a great deal about language and thought by focusing on their causal possibilities in different ways. We similarly find out a great deal about neurology and the brain by pushing the possible physicalist causal primacy as far as it can go. However, there is a level of theorizing that should be decoupled from slavish reductions in order to appreciate and apprehend complex meanings and to search for the rich interrelations and implications for remediation.

An immediate implication for the study of variants of logical forms is in the value of their taxonomic ordering. Taxonomies should be based on a psychologically functional analysis and should not be reduced to a neurological analysis. We want to know if politicians or educators stereotype and if their logical pattern is rigid and one dimensional. We want to know if the autistic person can reason her way out of her own focus. Finding the answer to these questions does not mean looking away from correlations of psychological functions with brain and neurological structure and phenomena. Useful analogies can be drawn from these domains of information, but their value is attenuated if they are not subject to the kind of reorganization of categories that typify the value of metaphorical thinking and for which a theorist is responsible.

Interdependencies of subjective and objective categorizations is a difficult topic, especially if ontology is the basis for categorical organization. In his study of visual dysfunction due to brain tissue loss, Roland Cichowski (2001) takes the view that subjective experience is certainly tied to physiological phenomena, but also that the experience is complex and may best be seen in terms of its emergent properties. He reports that "what we experience subjectively as a visual phenomena may be composed of many other attributes drawn from other sensory inputs. Subjectively there is a sense of space in an image, which suggests it may be drawn from auditory and tactile inputs as well as [from] the more dominating visual [inputs]" (2001:3).

It would be reevoking a senseless dilemma here to argue about the predominance of a first- or third-person account or about the dangers of ontological dualism. In contrast, if semiotic relations are kept in mind, the logical exchanges necessary to shift perspectives and optimize meanings become a powerful way for theorists to go back and forth from domains of information to the development of fruitful remediation.

Predication, Propositions, and Intimations of the Self's Role

The Linguistic Reduction and Reversal: Catch-22 for Logical Form

Coming away from the problems and values of taxonomies, we can note a second use of the analysis of logical forms—namely, to question the value of classical logic as reflective of human thought. Psychologist Joseph Rychlak (1977, 1997) argues that the various inabilities (and alternate proclivities) of people to reason with straight deductive logic means that "1, 0" binary logic as a model of human reasoning is merely a "Boolean dream." Proponents of "psycho-logic" also reason this way, and, from another angle, so do Braine and O'Brien in their conception of the mix of logic and meaning. Rychlak makes the case that if you judge a person to be angry, for you to know whether she fits the judgment, you have to include its opposite meanings in your thinking about it. Human beings "predicate"—in other words, they extend the meanings that surround a given "target." In that sense, a "predicate" is an implicature. People set in motion an organized extended set of meanings that relate to a target. Meanings follow something like "formal cause." They also follow a logic of intention, so the thinker can affirm meanings that fall on one side of an oppositionality. However, *if taken as close to meaning "categorize," the term predicate yields something different than when it is considered only in relation to the schematics of syntax.*

Russian psychologist Lev Vygotsky (1934, [1939] 1975) sees predication as a step in inner speech. It is a step that advances an identification of the "sense" and context of meaning. You might look at this step in Jaynes's terms by talking about either a point in the evolution of thought or a point in the epigenesis of the thought of the mature individual of our day. At such a point, a person comes to have an inner sense of a predicate, which is separated from the self as a subject. This inner sense of a predicate would be like an idea in the rough. It reminds me of gestalt notions of ideas that are *forms not yet fully associated with meanings*—like the erstwhile Carmichael, Hogan, and Walter (1932) experimental finding that forms constructed geometrically are given different meanings in accordance with a person's "frame of reference."

If "inner speech" were at work, but in a nexus with thought, the "thoughts" we carry around might be images or imageless. Either way, they would be like elemental geometric forms we apperceive in terms of the meanings of our experiences and interests. We now have a checkpoint. At it, traffic proceeds from the forms to the agglomerations of experience, but it also moves in the other way, from the experiences back to the forms. The experiences "change" the form, but the form precludes change. Checkpoint? More like a *checkmate!*

Consider the group μ rhetorical model's "progressive succession of fields" as reductions that allow iconic objects into a place for thought and for its operations and organization. Imagine each field exists at one or more than one layer in a tectonic arrangement of *plates.* Layers can be focused on as individual plates or as plates in combination. When you move your fingers to type, you focus on the way to hold your hands so that you don't get carpal tunnel effects. The finger placement or typing behavior can be a schematic—as if one plate. The hands position or physiological signals of incipient carpal tunnel distress can be another. Each of these "plates" individually or in combination are sometimes "out of view" or out of place in your "thought." Your focus instead is on the meanings of that which you are typing.

A "field," as I use the concept, is a place. In thought, you "move" or reduce to a given field—in a progression. The field can be at a point in the progression where it is a combination of "plates." Thus, both of the two plates I described—one involving skills and distress, the other involving word meanings of the typed product—may be part of a field of meanings. The group μ suggestion is that in a given thought sequence, you first go to the meanings—as in the "typing" example—and then you progress to or wind up in the logical field. There you can check whether what you have written makes sense logically. Paradoxically, however, you are at the *checkpoint* from which you have to "return" to nonlogical rules and put your words in place in a semantic field.

First, remember the sequence moving toward greater reduction as: Plastic field → Syntactic field → Semic field → Logical field. Then reconsider. How do you move from the logical field to the linguistic field? I don't want to get hung up on a chicken-egg problem, but how do we go *backwards* from a focus on the "logical field" and its forms to the ways in which the semantic and syntactic fields are configured? For each field, use the phrase *rules of the field* (RF) to describe the different rules as they relate to representations, their relations, and their organization. When we focus a logical field, we engage RF concerning contradiction and partition. When we focus a semantic and syntactic field, we engage RF related to association and contiguity, as well as RF for roles and navigation within a schematic. On the "chicken-egg" issue, I have to ask, If the logical forms *are* subject to our experiences—as they fit a semantic, syntactic, and schematic base—then, to account for reflective thought, do we start at some Kantian transcendental level and proceed to *reverse reduction?*

The "Inner-Speech" Problematic

I refer to the forms of thought—such as the gestalt "geometric" forms— as "logical." When we get too close to the "inner-speech" idea, the logical forms of thought are apparently coterminous with the nature of language and speech! But look at the group μ progression. What does it mean to be "coterminous," when the RF are different for each field? As is the case with reduction, we reach an impasse—particularly when the relation of logical forms is undefined or conflated with more than one RF operating. In a standard logical field, a dichotomy might dictate that "Elvis is either alive or not alive." The RF are clear for contradiction. But when the field is one of a combination of plates, contradiction is not what it is in a simpler field. You contemplate that "Elvis has left the building," and thereby you trot in rules of schematization and association that confound the logic. Now the RF have substituted "equivalents," such as associations of "missing Elvis," for the terms of the simpler field, and therefore contradiction by the "alive/not alive" dichotomy is not operative. The fact that plates shift in the eye of the thought and different RF come into ascendancy makes it difficult to assign very specific features to logical forms. Within the logical field, if two symbols or signs or propositions are related to each other—say by identity or equivalence or contradiction—it begs the question to say that the relation was initiated by constancy in the semantic extensions or in the operations within schematics.

Complementary Class Shuffleboard

With progression as reduction, you have a catch-22 situation. If you're stuck in a reduction to a linguistic field, there is a reverse reduction from the idea of logical forms. But now you have made that reverse move, and you are operating with the semantic RF pack. If from within that pack you think of the way meanings are restricted to the linguistic field, it is *as if* the logical forms and their rules of relation are in a *complementary class of meanings*—that is, they are complementary to those that we *have* articulated.[5] How do you access that complementary class without navigating the progression up toward logical forms? This problem of shifting from one set of RF to another and accepting these rules as being superposing when the particular field is in focus is the same problem for Rychlak's semantic extension idea as it is for Vygotsky's inner-speech idea.

It *is* of great help to have an inroad to thought through language and speech. However, the inroad takes the individual thinker through a troublesome checkpoint from which he can checkmate his own thought through a battle of RF superpositions. Yet the inroad is traversable *in the third person!* That which we have articulated by language, speech, and the extension of meanings is by and with rules that we can recognize and negotiate. By and with these rules, we presumably can be conscious of (observe) the relationships of the terms representing them. But meaning is intrusive when it comes to logical evaluation. That "complementary class of meanings" comes into question. It, too, is a bag of implicatures! Its intrusion of relational meanings signals the reduction of logic to linguistics involved; it shows the begging of the question.

How *do* the forms and meanings and the relationships of the terms that would represent them emerge from their amorphous state in this "complementary class" of meanings? To take a page from Vygotsky, this "emergence" requires that language and the influence of its forms are in the picture. It also requires an observer. Another stretch of the reverse reduction would be to say that specification and separation of self from the objects of language results in the self "as an observer." With its linguistic and socially dependent signs—"I" and "me"—the individual self can now observe itself as an object and observe other selves as objects. All this would depend on the linguistic signs a given self can regard as meaningful. But that given self, operating from a first-person position, would be acting in a third-person role by submitting her thoughts to the socially agreed upon signs and schematizations. This way of conceiving the reverse reduction is to say that the first-person self is in fact a third-person role—another trip into question begging! Nevertheless, what is that "complementary class" if not a bundle of meanings? Is it like a "rough idea" that in some way can be unfolded?[6] Can this unfolding be accomplished by laying out the idea syntactically in a sentence format or laying it into a schematic? Either process might account for meanings, but the complementary class is at bottom a logical form. To get to something that even "looks" logical, the syntactic format is not enough. Another step needed is the conversion of the "sentence" structuring and schema into that of a proposition. The proposition *is* more of a logical form. It accommodates the terms of specific relationships such as inclusion, and the terms of a proposition might be more accessible to operations based on "equivalence."

A sentence can allow not only one other sentence to be contrary and oppose it, but also yet another sentence to oppose it from a different angle on its meanings. To get to the logical status of one predicate as the *contradiction* of another, you need to see the logical relation of one predicate to another in terms that permit stable transformations and operations. In brief, the next step would be to provide the propositional format. Rychlak deals with this requirement by way of synthesizing meaning with dialectical reasoning. However, there's the rub! The predicates have affirmations of meaning extensions. The propositions for dialectic are *not* related to each other in purely formal logical terms! Their terms are much more like the "conceptual categories" of a cognitive linguistics model than terms accessible to either a classical logical analysis or a Piagetian relational and operational analysis.

Self to the Rescue in Finding or Originating the Logical Forms

Within the rhetorical-linguistic model, there is a mature form that, to some extent, accommodates the description of a logical form. Where is it? Perhaps it *is* between that speech that is an inner sense of the predicate and its syntactic articulation into the relationship of sentential terms. Is it there—at least potentially—because the sentential terms, in turn, can be made into a proposition that *is* and *is not* logical? It *is* logical because it is dialectically turned to be an affirmation of an implicature. This turning marks the implicature in opposition to contrary meanings, and it comes some of the way toward being susceptible to logical operations and transformations. It *is not* logical because it is dialectically grounded in one-sided arguments. In sum, here are the pluses and minuses.

Within a rhetorical-linguistic model are Vygotsky-type steps in inner speech and their effects on thought. These steps are congruent with Rychlak's point about predicates, which calls for an expansion of logical formulation not only to account for the origin of premises, but also to account for their grounding in the individual's thinking and value systems. However, both of these descriptions are based on the closeness of speech, language, and thought. The model of dialectical argument comes close, but does *not* get to the issue of logical form. That requires a depiction of the self as a categorical origin.

The Self and Logical Organization

How does the juggling of categories take place? Who does it? Do they do it as observer? Is it by automatic programmed rules? Is it intentional? Is it conscious? The self may simply be a place where things get started, like the starting line for an Olympic dash. But the self may also *be* the "starter"—the person who decides where the starting line can be, the person who shoots off the signaling gun to start, the person deciding to run. Whoever it is—does she know she's doing this?

As philosopher José-Luis Bermúdez (2001) conceives *egocentricity*, it is a capability to recognize the self as the origin of a thought. This factor is not to be confused with Piaget's use of the term *egocentric* to mean either *not* socially based or *not* disambiguated from self. Nor is it to be confused with Vygotsky's notion of a "syncretic" or field-dependent perspective. In fact, Bermúdez's factor is the opposite of these ideas in that *the self is separated from the perceiving and from the field of events.* We have at *this* starting line the *post*-Homeric person! Bermúdez's idea is close to Harry Stack Sullivan's sense of an "a fantastic auditor"—an internal supervisor "who" monitors grammar and logic" (1944:14).

Sullivan ties "auditor's" rules to *consensual validation,* thereby emphasizing social contexts and their role in language and thought. More focused on the issues of self-regulation, I look at the self's capability to recognize *its* originating role. To depict this capability, I view the self's structure as a key to its semantics and even to its causal powers. Its syntactic—but, more strikingly, its logical—structure is similar to that of the Cartesian *cogito*—not in the assertion of the existence of the "I," but instead in the assertion that *the "I," on a level above itself, thinks about itself.* This is the structural aspect that I relate to the self's capacities to juggle the paralogical ordering of the trope and to view the trope from both its novel and its reality-oriented perspectives. In all, I agree with this view of self—the "I" and thought—not only as Bermúdez puts it, but also in its more general and encompassing Cartesian implications.

Self, Categories, and Icons

The psychology of a sentence with its subject, object, and type of action may have a direct, indirect, or nonexistent relation to the grammatical position of subjects and objects. To talk about the specification of grammatical patterns and their relation to psychological ones is only one piece of the action. Many attempt to diagram meanings in terms of the grammatical functions that correspond to action schemes, but grammar is not the only picture from which formal representations can be made. The logic inherent in a sentence is also subject to the extraction of forms, such as those of the propositions and terms enmeshed within the sentence's language and implicatures. Based on the premise that categorical structuring requires the self as origin, *propositions, too, should be viewed from the vantage point of one or more the various orders of the "I."*

The orders can be thought of as levels of awareness and as fitting within syntactic and linguistic levels of complex construction. We fit the levels of awareness within rhetorical-linguistic "fields of organization." But the levels can also be cast as logical categorizations. The "I," in a dynamic mode, orchestrates order and can be inside that order and outside any preexisting set of categorical hierarchizations.

From this dynamic exercise of the "I's" functions and structure, it's a hop, skip, and jump to describe how the "I" represents propositions themselves as icons! The "I" focuses on a proposition as if compressing it. The "seeing I" proceeds to define and produce a perceptually schematized representation—an icon. The "I's" process and products are *its abstractions* from the self's position in a field or fields of its sensibilities or needs. This abstracting is the "I's" selection of elements and values compressed into icons and constituting a "perspective."

Abstracting the "I," Abstracting from the "I," and Traversing Organizational Fields

ABSTRACTION   The "I" lifts itself from a globalized experience that includes itself, others, objects, and any experience sensed and thought. This "lift-off" is often described as a differentiation of self and other. The differentiation occurs in stages. A self can be aware that its "mother part" is not part of it. This kind of awareness can be manifested in the infant's change from treating the mother in the same way the infant treats his own hand and in the reduction of empathic feelings as if those of the other person were his own. The self is not the other, but simply a different player in an action sequence or an encounter with needs and events. There's a second "lifting" of the "I" from a self who remains "slotted in" to a series of actions and relations with other objects. Lift-off from being enmeshed in the syntactic slots of a schema involves the "I's" recognition of itself as being outside the schematic—if in no other way than to decide if and how to participate in a given action or how to relate to its objects.

All this lifting off to an outside position culminates in *abstraction from all fields.* Kant uses the term *transcendental,* which does cover the logic resulting from the "lift-offs." Each time there is an abstraction, what results is a logical form involving superordination of a less abstract category of self by another more abstract one. If you prefer, keep the Kantian term *transcendental* in back of your thinking about these phenomena. However, the term *abstraction from all fields* implies two other processes and their related forms. One of these two processes involves traversing the different fields of organization; the second involves mixing the terms and rules of the different fields.

PROGRESSION AND OTHER TRAFFIC PATTERNS IN FIELDS OF ORGANIZATION   *Traversing fields of organization* is the process of moving toward reduction of particulars and rules of equivalence. It also entails a backwards or reverse reduction, so that after a symbolic level or order is made available for equivalence operations, the person can dip back to meanings and semantic extensions to make use of the symbolic transactions and perspectives. (The "fields of organization" range from linguistic schematic forms to the logical forms of the "logical field" that group μ describes.)

In the process of juggling, the terms and rules from organizational fields are mixed in various cognitive constructions. This mix produces species-type categorizations that get used variously—in primitive, pathological, heuristic, and figurative ways.

Transcendental levels of the self and of logic *are* present. However, the self's navigation and the effects on the cognitive and semiotic fields require further articulated structuring.

Abstraction and Icons

The term *abstraction* shows ordering relative to the "I." This ordering can be described as that of logical forms, but *abstraction* also denotes a semiotic process—similar to a person's shuffling a field organization by reduction! An "abstraction," as an icon produced by the self, is a representation. Just as the "I" abstracts "objects" from *their* fields, it also abstracts the self from *its* field. In that case, the "I's" perspective on itself results in an object that is a particular abstracted *aspect of the "I."* Picture the "I" as Faust ascending to get an aerial view and surveying some circumscribed territory—the "aspect." It entails an implicature—such as the "I's" self-worth or state of consciousness. Focusing in on that aspect reveals a specific semantic extension. We now have on hand an "organizational field." What transformations of form got us here?

Is an organizational field brought into view by reduction or by reverse reduction? How do you describe the direction from which the "I" selects a perspective? Consider matters by symbolizing the "I" at different levels of its "lift-off." First, symbolize the "I" who would be enmeshed in schematizations—like the infant defining everything in terms of objects and actions relative to a globalized undifferentiated "self"—as in the relation

$I \cup Sa,$

where Sa refers to the "I"—included within a universe of schematizations. The relation reads out, "The 'I' is part and parcel of cognition as formatted in schematizations." Now, symbolize an *aspect of the "I"* (such as self-worth) as an abstraction of the "I" in relation to a *particular* schematization or set of schematizations, thus:

$I \cap (S_1),$

which reads out, "The 'I' is outside of (abstracted) but in relation to a particular schematization or schematization set (S_1)."

This abstraction changes the "I's" view of its field—by reduction of the "I"—as an object. The "I" is abstracted from the necessity of its being schematized, but it is functioning vis-à-vis a particular field. In that sense, *as regards transformation of the fields of organization, the "I" is reverse reduced, but its schematized aspect is reduced.*

The two transformations here are reciprocals, but there is also a case of congruent change in field organization—where both the "I" and its perspectivized aspect are reduced. Thus, when the "I" symbolizes itself and another person semiotically—say, as different diagrammatic entities—its perspective is focused on self and other as a *different sort of object*—namely, as a *relationship* between objects. *This sort of object entails less an implicature of particulars as subcategories of the self and more of a shift to symbolic representations.* A relationship between symbols of the *self* and of the *Other* (see the distinction between *Other* and *other* in chapter 5, page 124) may be in the form of a geometric projection or of an algebraic summary that extends to other objects. (The "objects" may be persons or things). The iconic field change in *these* cases avoids semantic extensions. Now we have some relation like this:

$I \cap AS.$

The "I" is abstracted, but its aspect-in-view is of objects qua an abstract or *symbolized schematic* (AS). This status of the relation of "I" and some level of schematization requires two changes: the "I" becomes an object out-side its symbolized relationships to others; and the symbolized relation-ships become an object outside the particular events associated with the relationships of self and Other. Both changes in field organization are by reduction. All of these cases of the "I's" perspective define the representa-tional boundaries and relational qualities of objects. They are—in Peirce's conceptual terms—*icons* (cf. Emmeche 2002).

Abstraction, Icons, and Sources of Agency

From various levels of observation, the "I" can look at itself, but for the seeing to relate to objects other than the self, the "seeing *and* "what is seen" must be separated from the self. This is a difficult point to make when the "I" *is present to do the seeing,* but the "I" *as the object seen* is that which is separated from that object—which is the nonself. Here is an example. It is one thing for a child to think that the moon is following her (as one of Piaget's children did). With this phenomenon, the moon and the self are not separate. The movements of the moon as an object are interconnected with the child's movements. It is another thing to say the isolation of the moon as an object is contingent on there being a human being not only to "see," but also to configure that object as a nonself object. In this case, the "I" as agent is necessary to the seeing, but the object seen is not tied to the object self. Yet our "seeing" is still tied to the "I" as the perceiver, and that tie embodies the icon for the object, moon. *The moon, even though it's separate from the "I" as an object, is only the moon—as we see it.* This icon we form has its boundaries and features set by our capabilities. What the "moon" is for a creature other than a person may not depend on the particular boundaries accessible to our total abilities to sense, react to, and conceptualize. So the "seeing" of and by the self, whether perceptuomotor or conceptual, is intercon-nected with the organism's individual and various subsystem relations to its total organismic needs and structure (see Glick 1983 on Werner). Bermúdez's point about the self's recognizing its originating calls for specification of the logic of self—which I have detailed in another book (Fisher 2001). In the present argument, it is critical to specify the logical forms and their interrelationships, which accommodate nuclear mean-ings that the person puts forth. (Nuclear meanings include terms, ideas, imagos, and icons.)

Origins and Propositional Relations

Where do we see clearly the birth of a "new" nuclear meaning? I ask about a new meaning because the originating of old meanings is twofold. An individual "thinks of them," but they have already been devised by others or as part of some cultural context. To get as close as possible to the origin of a nuclear meaning, I have chosen the *trope*—particularly in that it refers to the "essence" of a thought or meaning (see Bacon 1997). If we look at the thought as a relationship of symbols, we can see that its selected form is logically that of a premise. The origin of the premise (or of a predicate that can be articulated as a proposition) is reflected in the logic of tropes. *For the trope to be viewed as origin of a thought, its logic must be displayed in some propositional form or format, accommodating what Peirce ([1902] 1950) calls "abduction"—a way of drawing conclusions from specifics to specifics.*

A scientist makes a specific observation of a relation of a specific organism's—say, a sea urchin's—specific anatomical form to the specifics of its dynamic regulations. Then, by adopting this relation as a metaphoric prototype, the scientist may go on to propose *another* specific: the human being shapes ideas in terms of *its* perceptual limits or boundaries, which determine its gestalts (Polanyi 1966). Reasoning moves from the specific propositions related to a sea urchin to the specific propositions related to a human being. *This* pattern of propositional relations differs from induction and deduction—two methods devoted to a general/specific relationship, although for each method the sequential direction is the opposite. The specific-to-specific reasoning in *abduction* resembles a series of metaphors related in *analogy formats,* creating all sorts of interesting possible equivalence relations of terms and of relations.

Behold the Trope Bearing New Meaning

Although the immediate question is the role of the trope in relation to the "birth" of a new meaning, we must acknowledge the difference between a thought, which is a manifestation of a generic "creative agency," and the more rare case of a thought that we would label "creative." A theory of evolution or a theory that differentiates the changes in an individual from change in a species may very well be a thinker's creative product. Although the theory is posited in a way to allow deductions for further hypothesizing and inductive reasoning to flow toward its verification, *the formulation of the theory requires metaphoric comparisons and analogies that reveal specific-to-specific relations.* But so does a poetic trope such as "The face that launched a thousand ships"! Nevertheless, we would surely distinguish these scientific, technological, and poetic thought patterns and their products from the metaphoric-type relations of a child's sand castles to "actual" castles or from the thoughts of a madman who reasons from his cloaking himself in Napoleon's garb that he *is* Napoleon. This reasoning from specific to specific is thus also a dramatic instance of this book's central problem—differentiating between the logic of tropes and the logic of pathology and primitivations.

How *do* you tell the difference between an abduction and a childish thought, within which the logical connection is not secure? One way to deal with this question is to ask what constitutes a "good" idea. I made some preliminary observations on this issue in earlier papers (Fisher 1976, 1985). In our present discussion, the factors seem very much related to the sense of self as an agent—separate (abstracted) from the objects of thought—the presence of a "fantastic auditor," and the accessibility to consensual validation. A childish definition of a term embeds it in specifics of his thoughts and actions. A pathological thought so easily crosses the boundaries from figurative to concrete meaning and from self to nonself that beliefs turn out to be bizarre.

Although a scientist might make a very good guess, no one, including the scientist herself, knows the thoughts and perhaps other propositions "behind" her intuitive focus on a heuristic specific. As a clue to the difference of one of her propositions from less mature logical form, consider that the scientist suspends "bizarre conclusions" because the intuitive guess is something she "regards" as a jump-off point from which induction and deduction are to make a technology out of her "art." In this regard, Michael Polanyi lists the elements for the scientist motivated to form original ideas (1966:76–92). These elements include mastering a sea of information, sensing problems, and using imagination. But they also include devotion to validation by tying meaning to external things and to observation by others. Accordingly, the "fantastic auditor" appears to allow unconscious and almost automatic submission of idea- and term-juggling products to various syntactic and propositional templates. Then there are the banks of information that afford the thinker comparisons to social demands and the routes to verification. These demands, practices, and routes should serve to square theory with what others know and think. If the thinker is to do this kind of squaring, she requires access to her own unconscious reviews of memory and to her conscious sense of the limits of self in relation to Other.

At first blush, despite all this formulation, it remains very difficult to disentangle the propositional logic format of Peirce's abduction process from the specific-to-general constructions in immature thought; from Lévi-Strauss's description of the primitive logical ordering, "individual as a species" (1966:chap. 7); from Piaget's "transductive thought" (1930:293–300); and even from the schizophrenic logic Matte-Blanco describes as an "alternating asymmetrical/symmetrical type of bi-logical structure" (1988:33).

But we should put all of this in terms of a focus on tropes. Tropes are expressed in terms of the "properties or relations" of particulars. Take the term *face* in the beautiful metaphor "the face that launched a thousand ships." The property "power to produce movement" relates *face* to the term *ships*. A focus on relations might be intriguingly illuminating if one were to relate *face* to "power." If *face* is the set, then "power," as a subset flows from a particular constellation of properties. If "power" is the set subsuming *face* as a subset, then one's focus is on a generic idea of power and the amazing relation to the face as either an instance of it or a spark for it. Well, you see the "juggling" capabilities that must be present for the poet to achieve the right categorical configuration—and to come up with a trope that can produce simulations of his own juggling in the minds of the persons who hear or read the beautiful metaphor.

With all of these complex requirements and effects, can the form by which a thinker creates a trope be logically equivalent to the specific-to-specific reasoning we have been discussing? When the trope is considered as an instance of "a property or relation," one can reason that *as an instance* it is reflective of a linguistic approach to predicates (Bacon 1997). This perspective would cause us to look at the rhetorical form and function of the trope because it affects the levels at which meaning is asserted and the organization of relations of units of meaning—in syntactic and semiotic dimensions. However, as in the case of my juggling of the set/subset relations of the terms *face* and *power,* when we view the superordinating/subordinating aspects of an instance, we see that classification and categorization are involved. In this regard, we have a different take on the trope as an "instance." If we had a photograph—or hologram, for that matter—of Helen of Troy, the face we would describe might be considered a whole particular—in relation to that specific person. But *face* as a term in the "launch" metaphor is not so simple; its features are unstably organized in a continuing figure-ground shifting of its genus-species relationship to the "power to launch." Therefore, *in relation to a trope, an instance is not merely an account of a whole particular; it is a particularization that is a part of whatever class of things or events to which it refers.* Therefore a trope is also reflective of a logical form, and the negotiations of that form pivot from the exchanges and transformations of specifics.

Paralogic and Reasoning

The situation regarding equivalent logical forms becomes even more complex if we note that Eilhard von Domarus's (1944) description of schizophrenic reasoning patterns not only very much resembles the form of transductive thinking, but also compares to the use of paralogical or protolinguistic metaphor (Keen 1999:418). Another candidate for similarity of structure is Heinz Werner's (1948) concept of the schizophrenic use of *pars pro toto* for the structuring of the self. Here, too, the equivalence of self with its parts or with different "parts" of the individual's experiences involves a mix of propositional logic and protological transformations. I come back to this issue later because the role of the self is so central to my thesis. In fact, I answer the problem of similarities among the forms by advocating that *the logical structures can be specified by way of clarifying the originating functions of the self.*

In the meantime, I can point out that disturbance in differentiating the self from nonself and others is often seen as a hallmark of pathological thinking (Vogeley et al. 1999). To the extent that meanings of tropes depend on the incorporation of shared meanings and on a concomitant distinction that differentiates one's own from others' attributions of meanings, the role of the differentiation of self and other in the interpretation of figures of speech appears self-evident. I treat this more specifically later, but there's one more parenthetical point to make.

When you think someone has reached a weird conclusion, and, indeed, when the person outlines his reasoning and the reasoning *does* seems strange, you can follow up with questions. You might find that when you ask a question such as, "How did you get from *that* premise to your conclusion?" the person will reveal an assumption. That assumption will seem odd or unjustified—to you. Even in the most grievous case, like that of justifying ethnic hatred, where you disagree with some invidious premise, there nevertheless *is* a premise. The point is that, like it or not, the other person *is* reasoning with the same procedural rules that you use.

Doubt has been cast on the use of *procedural irrationality* as an explanation for pathological thinking (Bermúdez 2001). On the basis of empirical studies, Bermúdez asks if the logical pattern of reasoning from premises is not really the same in most people—irrespective of so-called thought pathologies. In that case, *the "procedure" of reaching and assessing the logical nature of a conclusion is less the issue than the selection of the major premises and the meanings of the terms.* The latter two are not procedural issues, but instead epistemic (knowledge) issues. They involve not only what is known, but also the semantic structure and context dependence of the premise's terms. In terms of the example of ethnic hatred, the semantic structure of a stereotyped claim (proposition) about an ethnic group may reveal a term that with no clear evidence is applied to a target sociological domain. Instead, the implicatures for the term in play may be displaced from what is a possible sociological context in another. The nonbigot may be clueless to the fact that the bigot reached her conclusion by "importing" a premise from a different context. (see Hayakawa's [1941] argument). This example deals with "context dependence" in that the bigot drags feelings and ideas from an old situation to a new one. Another aspect of such dependence may simply be that the semantic structure involves an overblown conclusion—that is, a generalization from an acute dependence on one instance and a disregard for other instances.

The pendulum of our discussion has swung from an interest in structure to an interest in context and interaction. But I stop at this point in the swing to show a deemphasis—if not an attack—on the nature of the individual. In brief, at the extreme, the *rhetorical-social viewpoint* is that there is no thought without language, and there is no language without social compacts and knowledge bases. Hence, cognition is a contextual matter, and any dynamic understanding involves a system analysis—largely schematics of action patterns and cognitive analogues in operations and their organization.

Once again, the pervasive "received view" shows itself, and this time the ideas are in a synchrony with the swing toward social and semantic contexts. Ideas about interaction of the individual "explorer" with the objects and events in her environment are greatly influenced by Gibson's ecological psychology. But the social and linguistic contexts as ecological systems are combined with derivations from G. H. Mead (1934) and various social theorists of the 1930s and 1940s. Sullivan's ideas of consensual

validation fit here, too, as does Vygotsky's inner speech. A current deriva-
tive we have identified is cognitive linguistics.

In this atmosphere of contextual determinants and cognitive explorations, more current views do *not* look at the logical structure and its
procedural sequences. Instead, the concern is the knowledge that the individual focuses on—targets, for example—the premises put forth (the
rhetorical positions on arguments and hypothetical propositions for goal
achievement) and the importance of recognizing counterexamples (Mujica-Parodi et al. 2001). All are looked at as cognitive analogues of action
and interaction patterns, and their schematizations are related to specific
contexts for meaning.

However, a semiotic account is more than this, and it is a necessary part
of the analysis of a thinking pattern. The account of how representations
relate not only to an ecological surround, but also to other representations
as referents goes some distance in placing logical transformations in the
individual's thinking pattern. These logical transformations not only include the classic rules of contradiction and identity, and their standard
applications, but also, as we have seen with tropes, they are present in the
creation and reconstructions of categorical patterns. In virtue of such a
variety of forms and dynamics that enable this range of transformations
and plethora of rules for transformation, the individual's overall logic appears characterized by what I temporarily label *vicissitudes* of reasoning.
This is perhaps too facile a way of riding the swing of the pendulum back
to the argument for the primacy of logical forms. Yet from that vantage
point a deductive picture of these vicissitudes has to be made for an adequate theory of mind. Consider that an empirical view operates from
performance data and can provide only *schematizations*. Not only would
the search for self-regulatory principles then be endless, but the search
for agency and indeed for causality would be lost in ever-widening interactive fields of context and unwieldy extensions of meaning.

I won't go too far in defining the term *schematization* as I use it here. To
the extent that we buy into the *rhetorical-social viewpoint*, schematizations
are supervening on syntactic spaces, and these spaces have certain components we all have come to expect, minimally including agents, actions,
objects, and outcomes. Because from the idea of an "outcome" we assume
a spatial component and a temporal sequence, we will include *mode* as
one of the minimal components for a syntactic space. We can cognitively
arrange nodal points and scalars and vectors, which, when diagrammed,
relate to the syntactic space *within which an empirical observation is made.*

To get at this concept of schematization, I withhold the role of the "I" of the agent, who can function "outside" of the "I." But even if giving plenty of space to a pendulum swing to the *rhetorical-social viewpoint,* we need a notion of an agent as one who makes observations, walks through empirical contexts, and collects data or information. I describe this "observer" playing a role within the "schematization." Therein, the function of agency falls short.

To leave matters of the nature of mind and thought to schematizations is valuable only as these diagrams and schematics relate to specific cases. There is the Kantian view of a schema as an abstract template to which many scenarios may be fit. Because, in my terms, a schema would require instantiation, it might take either of two forms. It might be part of a system of abstracting and categorizing. (On page 95, I previously marked this schema type AS. I also marked as type S*a* the schemas that are engaged only in the course of specific contexts and relations.) When, however, the pendulum is swung to the rhetorical-social view, schematization is conceptualized as highly specifically embedded in ecological and contextualist terms.

The downside is that when weighted toward the rhetorical-social view, the concept of schematization entails a major assault on a concept of individual agency via a truncated view of the "I" (see my review, Fisher 2003b). If the "I" can only be something *within a schematic,* it can look at *that* schematic only from *within another* schematic.

What can I say now, short of a big discussion about the Aristotelian problem of the "prime mover"? I end the chapter by summarizing my complaint, but also by voicing my hope.

As a way of explaining the agency of the self, "schema to superschema and back" is a shell game! It leads to the relocation of *any* causal powers or *any* source of origination to forces and context outside the self. At the very least, this relocation blurs the lines between self and objects observed by the self. I hope to redraw these lines to separate not merely the "I" from the schematization, but also, perforce, to separate myth from trope and primitive and pathological thought from rational and creative thought.

4 Form Versus Function

Logical Form, the Dynamics of Consciousness, and Tropes

In the Piagetian model of thinking, as in the Freudian, are dynamic drivers, such as the drive, need, or intention-based nature of logic (Piaget and Inhelder [1966] 1969). Piaget, like Werner, accounts for dynamic organization of thought in terms of the overall biological regulations of the organism. For Piaget, cognitive operations achieve successive states of equilibrium through reorganization (Flavell 1963). Adaptive cognitive balancing by resolution of negations unfolds choices in logical organization and leverage in the individual's classifying. These choices, as I present them, are a function of the individual's different levels of consciousness; they are textured by the different orders of the "I's" superordination.

Even though hierarchical organization of coordinated actions and their representation is characteristic of scientific—and mature—thinking, sometimes the person's overall goals can be subordinated to a single subsystem. In this way of describing logical organization and classification, we see that sometimes the person has a classical set/subset hierarchical organization, but sometimes the organization involves a reversal of superordination wherein *particulars are the supersets*. Lou "the Toe" Grosza, unlike other football players whose functions would include running, blocking, catching, throwing, was brought into a game only to kick the ball over the goal post. A person conjuring or contemplating Lou "the Toe" is exhibiting metaphorical thinking, replete with its transferability of genus and species in category relations. In general, figurative thinking has the characteristic of focus on some particular or prototype for the construction of a classificatory ordering. (In logical terms, the genus-species transfer marks a metaphorical phenomenon. Strictly, in rhetorical terms, the exchange of part for whole is more metonymic.)

By contrast with what "should" be the case at some "mature" stage of scientific thinking, Freud's "scientific" description of the dynamics of psychic organization is itself metaphorical! Ironically, within his description, he does take up the issue of governance in the psychic struggles of different suborganizations. Yet, to express all this, he uses metaphors to depict the suborganizations, such as the id, the ego, and the superego. Well, what about this metalevel of his "scientific" description of a state of affairs that involves paralogical forms? Scientists penultimately use logic to describe nonlogical phenomena, which creates a paradox. Freud does not take the necessary step to advance a solution. He does not fully or directly articulate the issue of governance inherent in the interplay between logical form and the emotional and cognitive modes.

I don't want to get into trouble with the psychoanalysts. Many elements of their proffered interrelations of the metatheory with the phenomena *are* elucidating and *do* cope with the dilemmas I identify. Obviously, the competition of primary and secondary process thinking is described with logical elements, such as the use of negation, on one hand, and the sway of ideas that eschew contradiction, on the other.

One can deal with contradiction by changing the order of classification. But one can also deal with contradiction by negating negation, thereby allowing what would ordinarily be contraries to coexist. In this regard, Freud describes *condensation* as it works in dreams: "a selection of those elements which occur several times over in the dream-content, the formation of new unities (composite persons, mixed images), and the production of common means" ([1900] 1911:chap. 6).

With certain principles of organization, a syncretic structure can accommodate objects that appear contrary or contradictory when we consciously think about them. One might argue, as does Matte Blanco (1988), that at a less conscious level there are different principles governing the structures. The argument can be turned to claim that the governing is by way of the different subsystems as they vie for control. Moreover, it can certainly be argued psychoanalytically that "building" cognitive structures may treat an imbalance or help an individual toward a better balance of whatever psychic—and nested logical—system governs thought. According to this formulation, in the category of "first-order" determinants we have "cognitive structures," which can be concepts and/or orders of awareness. If a person argues with his wife, his friends, and then his business associates, he can learn that he is "argumentative." This category can also be reviewed. It may turn out to be contradictory to the person's stated aims of having "a good marriage," being "well liked," and "succeeding in work." The contradiction now can be up for solutions. Therefore, in the "building" formulation, logical organizing principles come last! It's like the primitive apprehension of a match of particulars. The "species" order becomes the central conceptual point, and, from there, logical relations are established. Nevertheless, may not the logical structures, when they finally are in place, govern concepts, determine preference for principles that govern categories and tolerate contraries, and even set the groundwork for unconscious or conscious mode? With all these credits, plaudits, and disclaimers, though, my point is that the question remains. *Do the forms govern or merely reflect the thinking?*

In the Piagetian model, the same question remains unaddressed, except that it's even broader. The issue is the logic vis-à-vis overall organismic growth. Is the primal urge of material growth motivating logical order and structure, or the other way around? This breadth to the origin issue widens the contexts of determinants. We take a good look not only at the internal and the external contexts prompting and delimiting the growth of an organism, but also at the interaction of these contexts. Therefore, the broadening can also lead to a systems argument: all the interrelations and interconnections are necessary to take into account what would be a nonlinear view of causation.

It might be worth a word to take Freud's and Piaget's seminal thinking to the floor, where cognitive scientists exchange bids for models and exchange the currencies of their products. They attempt to work the basic ideas we have just reviewed so that cognitive operations are compatible with computer modeling of cognitive architecture, which in turn is compatible with a possible neurological account. The phenomenon of *empathy* is a fertile one to view. Empathy involves both affective and cognitive aspects. Hence, on one hand, an account stimulates the kind of explanation that is loaded toward a cognitive processing. We understand other minds by forming a *theory* of the other mind (or theory theory). On the other hand, a competing explanation is that empathy is largely *simulation* of the emotional state of the other (simulation theory). Paul Barnes and Allison Thagard (1997) attempt to resolve the difference by suggesting that where factors that would explain how the other person's mind works are familiar to you, they are *local*. This situation can be handled by drawing on your own feelings and simulating the other's. Where such factors are complex and unfamiliar to you, they are *distant*. In that case, to empathize with the other by *analogy*, you would list and compute categories and causal patterns and their relations to those of the target understanding. For local situations, you use simulation; for distant situations, you use a more theoretical approach—namely, analogy, computation of categories, and pattern matching. Whether a cognitive or an affective mode is activated, the primacy of forms remains a question.

Lattice of Self and Sign

The "I," Self-Reflexivity, and Agency

Insofar as I include consciousness and tropes as phenomena governed by constraining structures and regulating dynamics, I propose that both cognition and affect are at work. In the semiotic view, the proposal that there are the two aspects—structure and dynamics—of representations flows from the more encompassing proposition that *an icon has originating capability*. In this ordering, icons, or signs, reflect origination at the apex of their categorical classification. Thus, if the "I" is the "master icon" at the "top," it is originating, and its subcategories include factors of structuring and regulation of dynamics. These factors, in turn, govern a variety of different sorts of cognitive operations, affective patterning, and behavioral action schemata. In this regard, it's easy to see that icons should be considered part and parcel of an organism's functioning and action, as well as of their outcomes. This is a pragmatic viewpoint, but we can go further. Claus Emmeche (2002) argues that Peirce's focus on the sign as action infers a form of "agency." The sign is a locus within the individual and entails internal relations with all sorts of individual subsystems as well as with their demands and constraints. All of this is quite in line with my emphasis on the place of agency in the self and in respect to the "I."

Although I do not intend an exegesis on the relation of Peirce's approach to the issues of agency and self, the idea of agency at *some* sort of sign locus or node requires an explanation. Otherwise, this idea would get lost simply because it sounds mystical. How can an icon, which is a sign of something that the individual perceives, "have agency"? If I say that the individual, by dint of *her* agency, produced the icon in the first place, the problem is temporarily worsened. This sounds as if I'm begging the question. However, the logical calculus of something that has more than one version of itself or exists at more than one point in time or space can be solved with markers such as subscripts. Or there is the dynamic idea that identity is a process of accepting that change and variability in instances are part of a changing genus-species ordering. Either or both of these explanations can help to support the answer on which Emmeche (2002) relies—the icon is *self-reflexive*. I briefly spell out *his* explanation in the terms of *my* search for form, I articulate the locus of agency vis-à-vis the self and its *latticelike* relation with the individual's icons.

The icon, like any signal, may well be one cause in a chain of causes, yet it functions to set other phenomena in motion. If I think Helen of Troy is more beautiful than I can say, I may dwell on the experience, and then I may set in motion a series of commands that would "launch a thousand ships." The resultant icon that combines Helen's face with a ship launch can function to produce pleasurable contemplation. It would be an effect that may not be in the chain from the original "Helen" icon to the command to the launching of ships. So I can look at the degrees of congruence between the icon as a cause, the peripheral causes and effects to which it is linked, and the line of cause-effect events to which it gives rise. I can decide that dwelling on the beautiful combination icon detracts from my income in ship building or that launching ships is an overreaction and I should learn not to let love signals go that far and be so costly! This "love and love worship" subsystem needs regulation. In the business of the self-regulation of any subsystem or suborganismic form, an icon is in a dynamic state. Something like Piaget's process of *equilibration* is constantly either engaged or on alert. I have to keep in mind that although love worship can go overboard, it would also be dangerous to scorn a woman. Because one of the major functions in equilibration is readiness to reorganize in order to resolve unresolved negations, such readiness motivates not only internal integrity, but also systemic balance in the subsystem's interrelation with other signals, signs, and symbols. I hope that any fears of the idea that the sign is a locus of agency have begun to be demystified. The "I" is, so to speak, infused in the icon! That's where the dynamic of the icon can be explained. Yet the idea that the icon itself has a "power" requires a few more remarks to dispel any taint of anthropomorphism.

Emmeche writes that "in biological systems there is an inner connection between the informational (which ... we will call here the semiotic aspect of a living system) and the functional aspect" (2002:19). If I want to know *what* information you have, one of my best clues is to know how you access, store, retrieve, code, transport, and catalog it. The *informational* side of things (the "what") is captured by the sign or the icon, the *functional* side (the "how") by relationships of the parts of the organism to the whole. The "parts" may be structures or discrete processes. At certain intersections of the "what" and the "how," causal interchanges may have nodes that are "in signs" and nodes that are junctures and junctions of functions serving the interrelatedness of the whole organism. Let me come down to cases.

My point is a corollary to Emmeche's idea of the self-reflexivity of icons: *not only the self is reflexive; agency is too.* Who you are is something transported back and forth from one equivalent version of the "I" to another. So are your powers of agency. Causal powers are thus tied in with the reflexivity of self and agency. This connection occurs through the dynamic nature of icons.

The self and its consciousness may play an essential role in producing the icon, but its dynamic state serves as a signal to readmit the self and consciousness to the readjustment of its denotation and meaning. Some people know quite early in life that they will enter into a particular endeavor, such as being an actor, doctor, artist, and so on. "Actor" is a good example here. Many actors report periods of feedback that vary from no success to superstar status. The superstar perks and the changes in life circumstances that they afford are new facts, which are fed back to the machine of identity to grind into the redefinition of that same self. Who can forget Eric Hoffer, the longshoreman with success as a writer, but with the commitment not to change his work or lifestyle? Anyway, that is one sort of solution. The changes invite the self in, and the changes are thenceforward interoceptive. Ergo, icons spark agency! In sum, *outside events and information take the self outside its agency. When that self is readmitted into the thick of its own icon, the agency within that sign locus is not what it was at first. Yet, whatever its shape, it is now coterminous with the self that is within the sign locus.*

I have argued that the way to depict these events is to transform them to visualizable forms. So I have pictured the relations of the "I," self, origins, and agency in terms of concentric helical spiraling from origin points (see the full account in Fisher 2001).

The "I," the Self-as-Whole, the Self-as-Part, and Agency

In this book, I picture agency as *necessarily* included within the category of self. To depict the logical form involved, I view the inclusion as a categorical ordering issue. There is also the task of showing a record of the changes in what is included in the self and in the subsequent variations of the self's relation to its agency. To picture these changes and variations, logical form should show change, foster changing, and accommodate reorganization of categorical positions and their ordering. Tropes have these capacities within their logical functions.

I make this picture of ordering and change in symbolic terms, organizing my concepts such that the "I" is the most inclusive, the self the next most inclusive, and agency a subset of the self. This order of phenomenological experience of awareness is consonant with linguistic form, such that a person would say, "I am aware of myself." But she would not make sense if she said it the other way around. The superordination of the "I" by the self would have to be based on something *sentient*, but not what we would think of as "awareness." Where the self would be considered a deep set of urges and sense of one's own organismic limits, the self can be said to govern and even sense the musings of the "I." This view is possible in the Freudian picture of the psyche's vicissitudes. It is also a fair picture in relation to models of brain inspiration of language and consciousness. But that an urging or a material base comes first in a causal chain is less the issue than the phenomenological order of awareness of sensing or feeling. I make the order one of subjective experience because the potential of the "I" would be missed unless we read the hierarchy in terms of awareness at the highest point of superordination. In short, for the picture I offer, where S represents the self and A represents agency,

"I" > S > A.

Now I'll flesh this out. The self may undergo changes and/or be viewed at different points in time or space or both. I note this series of possible variations by the symbol $S_{1 \dots n}$. Assume at some origin point that the "I" refers to a self, $S_{1 \dots n}$, which can be engaged at any point in its denouement, even though, for instance, S differs from point 1 to point 2. We can show this symbolically as

$$\text{"I"} \equiv S_{1 \dots n} \vee S_1 \vee S_2 \vee S_3 \vee S \dots {}_n.$$

Thus, the self is represented by $S_{1 \dots n}$, but at another point (say S_2) the self (S_1) has absorbed additional features imposed by outside events($_{ao}$). At that point the self (S_1) is marked S_{1ao}.

$$S_{1ao} \equiv S_2.$$

In this manner, there are changes not only to the self, but also to what outside factors can bring to the self's agency. But let's begin the look at agency at some origin point prior to the accommodation to these outside factors. I'll choose, arbitrarily, S_1.

$$S_1 > A$$

is a class/subclass arrangement in a specific categorical order. S_1 is the set, and "agency" is the subset. However, for the self when it has been affected by external causes and has then incorporated these causes into itself, the meaning of agency inherits the "tags" of that former outside determination. Hence, "agency" is a subclass still, but its implicature is now more complex. It simply has semantically expanded. I put that expansion this way:

$$S_2 \geq A.$$

Now at different points in these transactions, there's a shuttling from self as a "whole" to self as a variation in different and discrete occurrences.[1] This is a change that involves the ascending of the self-as-a-whole to the *set* order—at one point in time. At that point, the "parts" (or several instances) of the self are *subsets*. However, the focus can shift, when the self-as-a-subset (or part) becomes the *set,* and the self-as-a-whole becomes the *subset*. In terms of the symbols I have adopted here, I would say that when considering the "I," we keep in mind that even though "I" can define the self as any of the S points in

$$S_1 \vee S_2 \vee S_3 \vee S \ldots {}_n,$$

at a given point in time it is also the case that

$$\text{``I''} \equiv S_{1 \ldots n} \vee S_1 \vee S_2 \vee S_3 \vee S \ldots {}_n,$$

and then the self-as-a-part can "cash out" its semantic extension in its equivalencies.

What happens with these exchanges of set and subset status is very similar to and dependent on the dynamic of the metaphor in its transfers of categorical superordination. In these variations of the self's status is a challenge to the awareness/sentience dichotomy. When the perspective on this dichotomy is "awareness > sentience," the metaphor-like exchanges of the self's set and subset status determine the "'I' > S > A" ordering. However, the metaphoric structure of self is one thing, whereas the metaphor that a self can think out and produce is another. Thus, there is a difference between a self with the power of change at any time and the phenomenon of a trope, which is stably formed. A metaphor such as "March goes out like a lion" has stability—a point at which the hierarchical order has changed and the order appears permanent. The change, however, is evidence that the figurative thought, which has taken place, is like either a "regression in service of the ego" or, for some moment or degree of departure, a "regression *from* the ego"! This moment appears to be an opportunity to suspend governance by the "I" and, by this means, to focus on a reversed perspective—which would otherwise be primitive or partial. So in saying "I'm a bundle of nerves," you get a message across. You have redefined yourself within the "awareness < sentience" perspective, but the presence of the "I" in control of the trope is suspended—it is not lost![2]

Therefore, with all this juggling of logical order or the awareness of the "I" relative to its product, the trope, it's important to see how the metaphorical structuring of the self interacts with the presumed fundamental—categorical organization. It's almost an appeal to take on the question of whether the dynamics of metaphorical structuring or the formal aspects of categorical structure is the more fundamental. Well, both forms are subject to dynamic changes. A clue to the role of self in this transaction of dynamics and form is to regard the categorical as closer to the ontological facts and the being of the organism, while correspondingly regarding the metaphorical protologic as closer to the "I's" apprehension of the metaphysical nature of the organism. This is like saying that the "I" is aware that "wholes" and "forms" are what beings are, and that the cataloging of beings, which we do by *employing* logic, always subserves perception and sentience. As such, the cataloging is perforce protologically synecdochic. We are trapped in a formal perspective that is our way of relating to the "parts"—or more properly, to the constrained "perspectives"—to which we can assign icons. This appears Platonistic in relation to the focus on form! However, the welding of "being" to categorical form is more Aristotelian. In any case, the question of what's real is something that straddles the ontological and metaphysical aspects—which is a massive topic, so I hope a brief comment here may suffice.

The being of the organism is what we might call the fact of its life—a concept which, if named, either takes you back to Aristotle's view or begins to attract objections to some sort of "vitalism." Nevertheless, this "fact" is what I called a "whole," just as the "fact" of the "I" becomes the superordinating category for the various instances of and vicissitudes of the self. You and I observe ourselves. This observation is of the "self" as the "I" views it. But in making this observation, we also make a necessary metaphysical posit of the self's logical structure. In sum, life makes a statement; structure is what we observe—irrespective of its time of origin as a thing-in-itself. This metaphysical structuring, which is the product of our representations, is captive. It is governed by all levels of forms (physical, logical, linguistic) and by their internal and interactive rules, order, and organization. At a given time or instance, this governance can be variously focused on one or another level or on several of these levels; however, it is constrained by all. The constraints by a given level, though, can be offset by constraints by another level.

Interplay of Category and Metaphor in the Self as a Logical Form

Now, what is the relation of the categorical and the metaphorical features and aspects of self? The self is not only a *term*, but also a *category*. As a term, it has meaning and a place or slot in a proposition—which, in the model I present, is inseparable from an attitude with *its* multilayered ties to affective meanings and values. However, as I have argued, each of these units—term, proposition, and attitude—despite its complexity, is a logical form; that is, it is generically a category. The self in particular is one of the major superordinating *categories* for the individual's sentience. The self can be fleshed out in terms of each of the several "units." So the category of self can be described by unfolding it as subsets, terms, propositions, and attitudes.

This view of the self as a natural category avoids the sense in which the self may be cast as an empty set. As a category (S), it can include a subcategory, such as "self-esteem" or "quiet self" or "agency." I have formulated this category organization as flowing from the "I" to its subcategories:

$$\text{"I"} > S > A,$$

where S represents the self and A represents agency. However, let us now stay with the subcategory "agency" (A) as subclass of the self. For the present purposes, I put this as

$$S > A.$$

The category "self," however, is also always related to another category. Here's where the metaphoric structure is at work. "Agency" (A) can be in a metaphoric relation to the self (S). The relation is expressed as ":" and hence the relationship as

$$S : A.$$

The ":" denotes an unstable categorical order as if it implies

$$S \lesssim A.$$

The inclusion relations are dynamic and therefore changeable and transferable. Accordingly, where the self is equivalent to (S > A), the "I's" invocation of a metaphorical relation can be a variable function. Thus,

$$(S : A) \leq (S > A).$$

That is, the categorical relations of the self and its agency are subject to the function of the metaphorical relation and hence are unstable as to degree.

You recall that as a person expands her experiences through growth and encounters, new information about the self can add extrinsic factors to agency, but it remains bound into the self. Nevertheless, this instability is computable—marked here by the "$\leq$" in the symbolic expression. The computability is within whatever degree to which the self (generically) is the superset for its agency (S > A).

What I mean by referring to the "generic" self (S) here refers back to the $S_{1 \dots n}$ defined in the previous section. Thus, on different levels of the person's apprehension of self, the "I" can view the whole to be subsumed by a given part or instance. For example, the adolescent version of the ideal self can be an insistent presentation of the self to the self! The "I's" perspective on the self, from its catbird's seat, can reverse that order, so that you and I know that the adolescent's self is just a "phase"! But does the adolescent know? As a saving grace for all concerned, the adolescent *can* sometimes say that he is "in a phase"! (Such recognition often does not occur, although we can hope, along with an orthodox psychoanalytic way of thinking, that the recognition is there, but temporarily suspended.)

To sum up:

> *The dynamic relationship of self and agency* (S : A) *is metaphorical in form.*
>
> *The relationship of self and agency in their categorical inclusion relations* (S > A) *is one of set and subset.*
>
> *The relation of these two relationships,* (S : A) $\leq$ (S > A), *is metonymic!*

The computation of either side of the relationship (S : A) $\leq$ (S > A) is as a part of the other. I call the "$\leq$" relation *metonymic* because it's not merely a matter of either side's exchanging superordinating order. That exchangeability is a feature of the relationship, and that feature *is* metaphorical in form. However, the dynamics (S : A) are stably included within the self—here marked as S in the (S > A) relationship. The question of the degree of this ">" inclusion is the only thing that is unstable. The important point is that whatever shifting characterizes that instability, it takes place *within a self that is varying in terms of the whole-part shifts* intrinsic to the (S : A) relationship.

I can also state this point in terms of the function of self (fS) in relation to elements and operations. In this way of putting things, the logical function of the self (fS) is unfolded in terms of its operations. These operations would include the various transformations in, of, and by relations, such as those within a metaphorical structure and those within a categorical structure. What I mean by *elements* requires characterizing the

metaphorical and categorical relationships of terms and categories themselves as subsets of the unfolded functions—or operations—of the self.
Thus, relative to the two relationships I have discussed, I might put the
function of S in this way:

fS ([S : A] ≤ [S > A]).

So if both of these relationships, (S : A) and (S > A), *are* elements of the
set S, then *both* of the following are the case:

[S : A] ∩ [S > A]

and

[S : A] ∪ [S > A].

If the self functions in a metaphorical mode, (S : A), this can affect the
interpretation of the self vis-à-vis its own agency, as would be the case for
[S : A] ∩ [S > A]. If the self "takes control" and begins a causal event or
relates to its own agency in terms of its determining powers, (S > A), then
an alteration in the self's metaphorical view may be an outcome, but not
necessarily. (Hence, [S : A] ∪ [S > A] also remains the case.) For example, if
I discover I can change my view of myself by learning the piano, I may not
change my overall view of my control over my agency. It is possible to conclude, "Oh for heaven's sake; the piano is now ruling my life!" but this degree of reversing the relationship of agency and self does not have to take
place. Either the variation in the particular learning may not be something
I raise to change the nature of my concept of my agency, or I may note the
learning, yet not think it significant enough to change or reload the agency
concept. In fact, as I have reasoned, where there are metaphoric transfers,
they can go just so far. I made this point to say that there is stability to the
self's subsuming its own agency. But another way to view the same point
is to consider that the functioning of the self can work from the "other
side." There can be changes in the view of the self relative to vicissitudes
affecting the degree of agency the person feels. You might get something
that sounds like this: "I have always been a person who is very much influenced by the power of music. Now that I play the piano, it is just more so."
Yet even in this case, the self remains stable as the subsuming category.
Look at this point in terms of the symbolic analysis given earlier.

We have a lattice structure here; nevertheless, the integrity of the self as a cat-egory is not impugned by the operations that permit metaphoric transfers. This point may be further specified by saying that fS itself should be marked to include not only a metaphorical relation among the relationships it governs and that are overseen in their denouements, but also a categori-cal relation. I use the subscript [:] to mark the metaphorical relation and the subscript [>] to indicate inclusion as the categorical mode of relation. Thus,

$$fS \geq fS_{[:] \cap [>] \cdot [:] \cup [>]}$$

I realize that these statements and relations are tied to the assumption that there is adequate functioning of the self—say, in comparison to psy-chopathological functioning. However, as you know, my general purpose is to distinguish the trope from a psychopathological equation in a logi-cal form. The integrity of self as I have spelled it out in the earlier formu-lations is a critical description in this project. This stability or insepara-bility of logical and psychological identity is the reason the appearance of agency at a node within a sign is *not the same* as the giving over of agency from the self to an object that might occur in pathological thinking. It is also the reason that the conception of agency as a sign locus is not anthropomorphism but instead represents a departure from primitive or primitivized forms.

So I would say that the " · " I have assigned to dynamic possibilities of fS indicates a transaction involving *self-reflexivity*. In short, the fS defines agency, which is found in a dialectical dance between the self and the individual's signs—or icons—of self and its relations. Thus, *the logical structure of agency can also be seen as a lattice, L, with elements a and b (self and icon) in a partial ordering ≤ and with two dyadic operators, ∩ and ∪.*

Lattice Understructures of Perception and Boundedness

It is not too often that Peirce's idea of an icon is compared to ideas in or points made in Freudian and Piagetian analyses. (Jørgen Dines Johansen's [1993] analysis of Piaget's semiotic views is one exception. Carol Schreier Rupprecht's [1999] analysis of the Freudian ideas on language and trans-lation vis-à-vis dream interpretation is another.) However, these thinkers share a major assumption that should allow us to draw points from all three of their disciplines. To view the common assumption clearly, I work backwards from the model I am presenting to the position held by these illustrious "ancestors." The purpose is to find the points that would penetrate the presently understood categories and draw out the essential relationships between logic and the dynamics of thought. The model I have been presenting assumes a *product* status for the ideas and the terms, as well as for whatever meaning and/or elements turn out to be members of sets, features of categories, and implicatures of concepts. All are products of thought and feeling and of their originated representations.

Now go backwards. Both the Freudian and the Piagetian traditions include *perceptual-driven and object-driven formation* of the ideas, terms, and "members of classificatory sets" utilized in the individual's logical operations and relationships (Piaget [1966] 1969; Johansen 1993).

Perceptual-Driven Formation

In some way, *any* sign is connected with a perceptual fact. The fact may be a referent that involves information about the organism's external context, or it may be an internal signal, which, again, serves as a referent. (Or both the external context and the internal signal may be referents.) In addition, any sign or icon is assimilated to a drive or intention. I won't quibble here about whether these drives or intentions are needs, purposes, or the like. Instead, I prefer to refer the dynamics of signs to the organism's creation of objects and its relations to these objects. Thus, the sign—and whatever form or order of icon—that can be assigned to percepts is an object in spatial *and* motivational terms.

Kurt Lewin (1935) would have called this combination of space and motivation a "field." His "classical" concept of a psychological field is based on the values, positions, and directions of personal tensions, needs, and goals, expressible in vectors and scalars. In the latter half of the twentieth century, the concept as applied to psychological understandings seems to have given way to various models of evolutionary and ecological constructionism. In recent days, though, there is an awakened interest in consciousness and self-regulation. As a candidate for a viable explanation of dynamic change motivated by the very nature of an organism and its inner experiences, the "classical" idea is not out of the picture simply because there are newer concepts of quantum fields. Brian Flanagan defines a classical field as "a kind of tension or stress which can exist in empty space in the absence of matter. It reveals itself by producing forces, which act on any material objects which happen to lie in the space the field occupies ... the characteristic mathematical property of a classical field: it is an undefined something which exists throughout a volume of space and which is described by sets of numbers, each set denoting the field strength and direction at a single point in the space" (2002).

A quantum field—which I am *not* prepared to discuss in this book— has these features, but included within it are additional compactified (in-finitesimally) small dimensions.

Hence, the space or field is bounded *and* dynamic.

Object-Driven Formation

Now for a difficult point. It is difficult because it is rarely discussed. We think of a sign in relation to other signs, and this relationship exists within a semiotic or cognitive space. We think of a perceptual object, then, also as something in relation to its referents and to the constraints of a psycho-logical or phenomenological space. So which is the "field"? Is it the space within which we locate the percept or the sign? Is it the sign itself because the sign inherently would have to have referents and relationships? If so, then an "object," such as a perceptual object, would have its own "field" relations. If the answer to my query is that the "field" is *both* the object and its surround, then we have something like "fields" within "fields." As cumbersome as that sounds, it's less mystical than baldly assuming that a sign or an object has its own potential solely because of its dynamic and structural combination of properties. Shades of Plato's focus on "forms" and Aristotle's idea of "formal causes"! The fields-within-fields dynamic and structural combination and its intimations of potential are precisely what I attribute to a natural logic, a categorical system, and the nature of the "I" and the "self." Such a conception would appear to be present in Peirce's intended meaning in his argument against severing the icon from its perceptual ties to the organism's consciousness.

One quick way to view Peirce's intention is to consider his require-ment that a sign or icon have a relation to an *interpretant* in addition to its relation to an object. This requirement can be sorted out to re-fer to the way in which another person's mind becomes conscious of the sign or representation you use (Jorna and van Heusden 2003). With

Peirce's concept of the interpretant, the icon is part and parcel of a natural thought process. It is not merely a representation *of* an object or a classification or categorization, but is also actively part of the individual's thinking (Jorna and van Heusden 2003). When the icon is the categorization we are considering within the person's thinking or "cognitive field," the vicissitudes of the "I" are in the vanguard of the play. Peirce (1931–58) pictures signs or icons in relation to a consciousness, which can well be the consciousness of the other. But, as I argue, it can also be consciousness of the self if you consider that the "I" becomes conscious of the self—and of the icons by which to refer to the other's self. I now return to the "field" issue.

If the *object* in the field is a term that we are probing or viewing, the meaning of that term has a value within its form. In sum, this Peircean assumption that terms and ideas—as signs—are percept and object driven (in other words, that they have referents, on the one hand, and bounds and dynamics, on the other) is congruent with the semiotic notion that *"no symbolization is not an icon."* This major Peircean tenet provides the basis for a dynamic view of the use of symbolic logic to see the similarities and differences in the thinking that leads to variants such as pathological and defensive thinking, on one hand, and to forms such as tropes and poetics, on the other.

Well, that's a bit of "back to the future"! The common assumptions of Freud, Piaget, Lewin, and Peirce show the point of embarkation whereby the psychologically dynamic, the logically computational, and the self-regulatory aspects of logical form can be pursued to chart out the journeys of the "I" and to find the depths and distances of natural forms of thought.

5 What Is the Difference Between the Logic Governing a Figure of Speech and the Logic That Is Immature or Unconscious?

Self, Sentience, Perspective, Logical Structure, and Semiotic System

The self by grace of the "I" is agile—creating and leaping over categories. To think figuratively, the self's categorical nature has to mirror the same flexible lattice structures and structuring I have attributed to tropes. But I have also characterized unconscious and immature thought with the same descriptions of logical and paralogical forms! Can I unfold the relationships between the self and the trope and the functions of the self vis-à-vis the "I" as they may inform our understanding of conscious versus unconscious modes?

A trope can be "ready made" and presented to a person. A says to B, "Ah, that experience is like a breath of fresh air." A does not *consciously* "make" the trope, and B doesn't consciously parse it. B says, "Oh, you're so right! You always say something that captures the experience. What made you use the phrase 'breath of fresh air'?" And A answers, "I don't know. That's what I felt. The phrase was there; it just came up!" Although A has some sort of perspective on his experience that he knows consciously, the formation of the trope appears less than conscious. Yet there is a "fit" of the phrase to the experience and to B's understanding, and I would describe that "fit" as a logical match of icons—albeit icons at different levels of categorical order. What I have to say relates unconscious logic to perspective on the trope and its application.

An unconscious perspective would be one with which the person is still "sentient." Person A says he sniffs something pleasant about the air. Although B is unaware that she empathizes or has specific associations with pleasant air, she examine the air's freshness when she hears A's statement. The word *unconscious* refers to the logical structure and the semiotics rather than indicating that B's state of sentience is without a specific feeling, sensing, and perspective. I focus on a semiotic system of reference that has its own logical structure. This system also has the functional capacity to characterize the differences in the logical structure of the forms and relations to which it refers. I focus on the self as that system of reference. From this viewpoint, its logical structure can be viewed by the "I."

The "I," the Self, and Conscious and Unconscious Modes

Here, I clarify the relation of the self and the "I," although this clarification is probably best made by your trundling through the various applications and relationships that I describe in the book. An organism can have a self totally without an "I." To picture that relation, focus on a mature person. There are degrees to which the "I" is not *in* the picture. It's like saying you cannot view your "whole" self at any one time. "I see that I am angry" is a statement by which you can refer to some things about your self. But the "I" that "sees" that anger may not see other things within the self to which the anger actually refers. There is an order of sentience. You can say, "I feel angry." But you can also say, "I see that I feel angry," which may be a higher order of the sentience of self. (I refer to this "higher order of sentience" as *awareness*. Dependent on its order, the mode in which the "I" is functioning is a conscious one.) The "I" can view the self merely in terms of an order of sentience that implies ownership and boundedness. "It is *my* anger!" Nevertheless, the whole of the individual's system of reference to itself is *not* within the "I's" view and/or purview.

Where there is little, if any, awareness, you would still have an "I," but one in a less than conscious state. The self includes sentience; in feeling a toe stub, you can certainly refer to *your* toe—it and its pain are "owned" by you. You might take this to say that you *refer* the toe to you. You see why I cast the self as a "system of reference." It also has a logical structure. The simple homely one here is

$$S > (< S_{1 \dots n}) \lor (\sim S).$$

The self as a whole (S) is superordinate to any experience, phenomenon, organ, or part of the self ($< S_{1 \dots n}$), and it (S) is disjoined from (or not bounded within) anything other than itself ($\sim S$).

You view your own self within a perspective of awareness that is a function of the "I." You might say, "I'm very sensitive about the pain I feel when I stub my toe. I know I react more than most others." The degree to which the "I" has an aware perspective on the "toe" phenomenon as one element of $S_{1 \dots n}$ can be specified. At the least, for this phrase, the "I" knows that there's pain and that it is in "his" toe. I would represent this as

$$\text{"I"} \cdot (S_1).$$

S_1 represents the "toe" experience. But the phrases "sensitive about pain" and "react more than others" have different degrees of the aware perspective. I represent these phrases with subscripts, thus: the degree to which the "I" has an aware perspective can be expressed as a range within the $S_{1 \dots n}$. Where S_1 represents a "toe" experience, S_2 represents "sensitive about pain," and S_3 represents "react more than others," the degree of the awareness perspective ranging over the self can be cast:

$$f\text{"I"}(S_{1-3}).$$

Moreover, the statement, "*I'm* very sensitive about the pain that *I* feel when *I* stub my toe. *I* know that *I* react more than most others" can be seen to have different orders of the "I," which can be cast this way symbolically:

$$\text{"I"} > \text{"I}_{1 \dots n}.\text{"}$$

Thus, there are the levels of the "I's" observation of itself. In the phrase "*I'm* very sensitive about the pain that *I* feel" are two orders of reference—the first "I" in the phrase appearing to observe and "know about" the second. *This succeeding order of reflexive reference is the hallmark of consciousness.* The first "I" in ("I" > "$I_{1 \dots n}$") is continuously in a superseding mode;

no matter how wide, the range of "$I_{1 \ldots n}$" is within the aware perspective. Thus, the n term should really be expressed as $(n + 1)$, and the first (or superordinating) "I" has the built-in function and capacity to supersede whatever expansion of the $(n + 1)$ takes place. The number of degrees of the self in the aware spotlight is still always less than is included in the self. Therefore, in relation to the way the "I," the self," and consciousness interact, the specific distribution of the "I's" perspective within the $S_{1 \ldots n}$ is an additional factor in describing conscious and unconscious modes. For example, some persons are more aware of their social role than of their task competency.

Characteristics of the Self Critical to Logical Form

The central issue is the trope's logical form and its resemblances to the immature and unconscious logical forms. I should show that the self is an integral part of the dynamics of logical operations and thereby that its structural character differs from one mode to another (conscious to unconscious) and from one linguistic and psychological use and its patterns to another (e.g., from pathological formation to well-formed trope).

Note the two ways to contrast the self with "other than the self." One is "the other" as another person or self. A second, capitalized here, is the "Other," the nonself. With this category, the complement of the self is a grab bag of anything else—internal or external to the person—that is *not the self*. This *nonself* grab bag can include the stars and galaxies, other people, and even the person's internal parts, such as blood or organs.

The Three Principles for Differentiating the Logical Forms of Tropes from Unconscious, Less Mature, and Pathological Forms

Three principles about the self are so critical that they are *necessary* to make the distinction between a trope and an immature, unconscious, or pathological construction:

1. The differences in logical form can determine whether the self is distinct from the other. A number of important functions, including the use of communication to interpret and isolate a signal or icon are dependent on a view differentiating self from other. Moreover, a syncretic view of self

as intermixed with the Other (nonself) augurs less ability to distinguish figurative from literal. There would be less capacity to articulate oppositional and reverse-oppositional meanings, or to engender and manage latticelike exchanges of superordination of particulars and universals. There would also be less capacity to manage, disentangle, and organize mixtures of rules of equivalence and identity in categorial transformations.

2. The view of the self as originating provides decision making and evaluation relative to any interpretation of an already existing trope. In addition, the idea that the self is originating illuminates the nature of the trope as an invention to change perspective on meaning and to shuffle categories and the exchanges of genus order. The self is not only responsible for interpretations of the trope, but also for comparisons of its meaning to the self and subself (as needs, wishes, feelings). Last, not only can the self make its own tropes, but also the self's relation to its part-selves is trope-like.

3. In concert with the second principle, the self can look at itself looking at the trope. This reflexive reference would be by the self. It is a function of the "I," which can be expressed $[f"I"(S_{1 \ldots n+1})]$. In this formulation, the subscript $(1 \ldots n+1)$ refers to the self looking at a given trope that either refers to the non-self or to a version of the self. Thus, the capacity of the self to *look at itself looking at the trope* not only provides for levels of perspective and evaluation, but also LOGICALLY PROVIDES DIFFERENT ORDERS FOR CATEGORIZATION. Therefore, the logical structure can be differentiated from one in which the perspectives and orders do *not* appear and/or can*not* be part of a process of genus exchange or an articulated premise shuffling and do *not* accommodate sensible (intelligible) alternations of equivalence and identity rules.

How Do the Logical Forms Differ?

The differences of trope from less mature and unconscious forms affect and are reflected in the "I" as a category. Keep in mind that the "self looking at the self" is a function of the "I." I focus on the "I" instead of on the self. This focus heightens the sense that we are talking about awareness. Also, it helps to use language and syntactic forms to specify meanings of the awareness. Thus, "I see that I'm doing or feeling such-and-such" is a linguistic display of two orders of awareness.

As an example of different orders of awareness in a tropelike form, the "I" may be compared to the "I" at some other time—without conscious intent. You reach for a cup in a cupboard that no longer exists, but it did a long time ago, when "you-now" were "you-then." You realize that by your act with the phantom cupboard, you were comparing some aspect of your self as you think you are now to an aspect of the self at that long ago time. When you consciously think about it, the comparison

$$\text{“}I_1\text{” : “}I_2\text{”}$$

is in a metaphoric form. You realize that you are making this comparison. Your realization codifying an "I" looking at "I_1" and "I_2" and at their ":" relation can be cast

$$f\text{“}I\text{” (“}I_1\text{” : “}I_2\text{”).}$$

Formal differences attend the unrealized and the realized comparison. You can look at these differences as they are given earlier—complexities in the structuring of a metaphor. You can also view how the different forms affect—and are reflected in—the vicissitudes of recategorizations, which include the "I." The adolescent, looking back, can recategorize the "I" and re-view different perspectives of the self, but when that adolescent becomes a much older individual, very different values and versions of the self may dominate the metaphorical juggles. Take this point as a page from Erik Erikson's (1950) view of changes in the person's conflicts over identity as a function of her stage of development. When you're an adolescent, nicknames given to you by others can affect your self-confidence. They may be very much in focus to define the self. When you are an older person, your own sense of what you are able to produce and do might equal and displace your concern over the nicknames. In terms of my analysis, very different subscripts of the ($S_{1 \ldots n+1}$) are spotlighted.

A person can attend a trope and know what it means when he is "on automatic pilot." Similarly, the individual can talk about his "I" without being immediately aware of himself. He may say "I 'hear' what you're saying" and mean "I understand and empathize with your statement and the feelings behind it." When he says, "I hear you," he is not necessarily focused on his "I"; in fact, we guess that he feels his "I" at some less than conscious level. When we look at the "I" as a category that affects the interpretation of a trope, we have to think about consciousness. There are a couple of ways to do this.

The first way is to view the consciousness as a *potential*. The person would have to have the capacity for it, although the time of its active appearance may not be the present. It may have appeared in the past or will appear in the future, or both. In the past, the person would have to have been aware of the "I" if he were making a trope or varying one. There's no way to make up the figure of speech involving "I hear you" if *hear* and *understand* are not related to an articulated difference between the "I" and the other. So, even though you don't consciously label that difference, there is some point at which, in some form, that difference is consciously known. The disclaimer on this "conscious" state is that the knowledge does not have to be abstract. The upshot is that a "particularized 'I'" may come along with baggage. The comparison of self and other can be limited to a species-type classification, as in "a member of a social or ethnic group," it may be tied to a place within a schematization, such as "when confronted with a challenge of a particular sort." Thus, "'I' hear 'you'" may be limited to a role: "As a fellow bus driver, I fully understand how you feel." Or it may require a specific scenario: "Since we both face root canal work, I know exactly what you are experiencing."

A second way of looking at the point of view of consciousness is for the "I" to look at itself. This comparison is of a more abstract order than, say, a comparison of self and other in terms of "my possession" and "her possession" or "my sense of humor" versus "her sense of humor." If you compare your possessions to those of your wealthier neighbor, the sense of self is conscious, but specific instead of abstract, *and metonymic instead of based on the self-as-a-whole*. In this sort of case, a future appearance of conscious consideration of the "I" in "I hear you" would occur when you are either asked or self-prompted to consider your capacity *indeed* to empathize.

"I hear you," you say to your neighbor.

She retorts, "You often seem oblivious to my feelings. Are you *sure* you really get it now?"

You answer your neighbor, "I know there are times when I avoid understanding, but I feel it now."

Empathy can have passive elements, but it can also become a proactive tool. So for this second way of indicating consciousness ("I know...when I..."), the logical ordering of an "I" looking at itself should be fleshed out in terms of the implications for levels of conscious experience and their relation to causality (agency).

The Logic of Unconscious Thinking

We process categories mostly in terms of situations to which they fit or do not fit. In Otto Selz's (Schilder 1950:559; van Strien and Faas 2004:180)) terms, they are fit to a schematization. If we go to the image of a creature foraging in a field, "its decision" to advance on some plant to eat it is a logical one. I don't mean "logical" in the somewhat trivial sense of a logical class existing simply because any bounded action or object is what it is and is *not* what it is not. For an immature or relatively nonconscious organism, the object "plant" is syncretically embedded in the need for food, yet this object is distinct from one that would be harmful. So the logic of the decision is based on the ontological existence of both the class of eatable plants and the complementary class of noneatable plants. All this is still primarily logic stuck in its schematizations. The class distinction may be especially dependent on a decision, which, in turn, can only be reliably made for plants in *that* field. The field is syncretic to the determination of the eatable plant classification in that the classification holds only whenever the creature is present and hungry, and we have that plant in a category with others not in a category of the harmful ones. Further, though, the plant may be embedded not only in the need for food, but also with other objects and experiences within the "scenario"—for example, short grass. If any of these scenario contingencies is not present, the category can break down. We have a schematization that "fits," but only with a series of "ifs"—*if* these elements are present, *if* the method of approach is not dangerous, and *if* there's no trap (or sign of a trap) that should signal a different schematization, such as going around the other way to grab the plant.

Well, that takes care of obvious schematizations, even if by now it's clear that the term *unconscious* is a very difficult one. We may not have an unalloyed term for it; it may be a state of affairs in a constant relation with a series of opposites that we would call *conscious*. The term *unconscious* may be well embedded in sentience, so it is always a concept and never a category! Any time you consider a state of sentience "unconscious," you find that you have a range of conscious phenomena to think about. Thus, you assemble varied features; you do not identify elements.

It's inviting to think that a clear line of consciousness begins at a certain point of neurologically reflexive exchanges of information (see Melloni et al. 2007). However, that sort of definition of consciousness is by an observer of something other than the subjective experience of consciousness. In this book, I focus on a logical structuring. Although you can argue that the role of logical structuring *also* requires an observer, the feel of your control over logical transformations and alterations is quite different from the notion of a juggling done by your own neurological events and pathways.

The observer of a neurological event views it as an object—even if it "belongs to her." You can imagine a neurologist looking at a monitor that shows the variations in her own neural events, which she can influence by just viewing the neurological events of the moment as compared to those of some other moment or of some other person. But looking at that monitor is a different matter compared to the neurologist's "look" at her own feeling of excitement as a phenomenological (subjective) event. I reject the idea that simply because the event is subjective, her look at it has neither existential nor determinative status. Her look inward is instead an aspect of a reflexive act that has a logical structure—the "viewing" subsumes the "viewed." *Hence, subjective is not mystical or extraontological; it is related to an action function: the subject—as an action mode—is determinative. In the form of the "I," the subject is an origin. The object is a product.*

The focus on logical structure leaves the neurological conditions and prerequisites intact. Yet the problem for those who would deny the determinative role of the self and the "I" is the resultant status of the neurological and "material" conditions and prerequisites. They would be neither sufficient nor relevant explanatory concepts in regard to subjective and self origins, the individual's conscious recognition of these origins, and the distinction of self and other as loci of meanings and organization of thought. This is a bitter pill. Some researchers are currently attempting to gloss over the subject/object dialogue and division of action modes. Their quantum concepts and fuzzy sets approaches are worthwhile to further articulate relationships, such as those of the material conditions of action and the phenomenon of action, which are no more reducible to each other than are the qualitatively different functions of matter and energy. However, the difference I focus on is not the impossibility of conversion, but the reduction of form to matter. Alfredo Pereira (2007) makes a refreshing distinction between the qualities and the structure of conscious and unconscious thinking. *I have argued and will argue more fully later that these subjective phenomena and their representations are part and parcel of logical structuring. The specification of such logical structuring is the way I choose to go for this analysis of consciousness.*

Do Schematizations Tell the Story for the Establishment of Categories?

A script is a schematization that provides a pattern for the inclusion of instances—and, thus, something very like the determinant of a category. Are there schematizations that allow us to recognize categorical patterns of inclusion? Would this be the case for the self? Can the various role contingencies and decision nodes for the self be made to follow the rules that would underwrite a general schematization for the self-in-any-time-and-any-situation? What about the "I"? Can we specify the different logical orders of the "I" by way of including the action, objects, physiological states, and as many things that can be distributed to the nodes, slots, and modes within a general schema? If so, are they sufficient to account for the role of both conscious and unconscious factors in regard to logical structure?

Categorical implications of schematizations are recognizable either by way of the inductive defining of a category (fitting it to previous experiences) or by deductive formatting (fitting the category to previous structures). When I enter a restaurant, a whole series of actions that would fit into my restaurant script are very dependent on the "chair I can sit on" category. If it doesn't compute, I might simply leave the restaurant. However, there are variables that on the basis of past experience call for chair testing and the like. So much for inductive defining. The deductive formatting would be a matter of "procedural" aspects that determine structure. The position of the category (or object) in a proposition-like sequence allows an automatic sense that the sequence of propositions is "valid." So it may be that a "wired-in" protologic of syllogistic thinking is

X is Y.

Z is X.

Therefore, Z is Y.

If this wired-in "valid" procedure refers to some terms and propositions, as in

All men are mortal.

Socrates is a man.

Therefore, Socrates is mortal.

then these propositions are obviously valid in form. But if they are inherent or entailed in a determinative schematization pattern, then the terms used have semantic extensions and additional logical considerations that require variations in what can be coded as an equivalence. The equivalence of X and Y can be predicated on an exact identity of the two, or it can also be established by a rule of inclusion such that the subclass is always a member of the class. In our example, the validity of the proposition depends on the term *man* being included in "men." So the format here is that the X terms have to be such that the specific case is included in the more general. Moreover, the general case has to be exhaustive. Thus, "all German men" as a premise would be insufficient for this reasoning about Socrates. Also, "all Greek men" might open the problem to researching Socrates' proper ethnicity and running the risk of historical turns in his ancestry as torpedoes for any clear ethnic classification.

Take "If A > C, and C > B, then A > B." That's OK as a categorical statement if categorical statements are impervious to schematization shifts. But because a schematization is such that any logical terms can be refocused due to a shift in perspective, then an existential analogue of the categorical statement only *seems* valid. Thus, "If seeing blue flowers is always accompanied by sneezing, and sneezing is always accompanied by a drippy nose, then seeing blue flowers is always accompanied by a drippy nose." Maybe. "Seeing" can be redefined as computer vision, and then sneezing can be provided without a drippy nose. Or for purists, the "drippy nose" as a computer production may not qualify as a "drippy nose" on account of chemical differences from the person's "real" drippy nose.

These incursions into a "pure" validity are a function of the semiotic principle that any representation is an icon—with "grounding" in the individual's perception and with dynamics affected by the individual's action and adaptation. The "grounding" does not mean we stop at perception as a rock-bottom predetermining set of experiences and parameters. Philosopher Mark Rollins, writing about Stephen Kosslyn's theory of images and perception, discusses perceptual frames and the detection of properties. Some properties depend on a frame of reference; some have to be inferred. Perception has "regulatory effects [that are] part and parcel of perceptual strategies, which should be factored into explanations of perceptual experience" (2001:280). Rollins views the process of perceiving as subject to "layers of meaning." The semiotic significance is that icons are of various types at different levels of sentience. This is logically important in that transaction from one layer of representation (sign system) and its relations to another requires rules of equivalence and identity. These rules are necessary for self-regulatory transitions that can utilize signals, patterns, perspective, categories in memory, attention, and "organization that has been assigned [as] essential for attributing perceptual content and for explaining task performance" (Rollins 2001:280).

In short, some judgments of perception may appear automatic; others are open to the individual's awareness of the inductive and deductive issues of regulation that Rollins describes. Moreover, the factors—if not the motivating force—of "task performance" and, as Piaget would have it, "adaptation" not only lead to the coordination of the many layers of meaning and sign transactions, but also call for variation of the interplay of consciousness, regulation, and perceptual experience and organization.

In general terms, icons, which include logical forms, are themselves bounded by the individual's thinking. Therefore, icons are themselves logical forms—natural logic, of course! And when they denote logical forms—for example, symbolic ones—we see a case of the icon as a set that includes itself. Icons can include many types of representation, and, therefore, transactions can be complex. When Ella Fitzgerald sings scat, a musician on the trumpet (Charlie Shavers) picks up her melodic lines. The resultant "dialogue" and counterpoint require rules of equivalence via notes, mood, theme, pitch, and other variables. Counterpoint is less the issue when the replication of voice is by translation to a less melodic reflection. (A great example is Buddy Rich on the drums with Ella Fitzgerald singing [Rich and Fitzgerald (1947) 2001].) In the mind of the listener as interpretant, what is sometimes a simple delight and other times a source of humor is the lack of identity in the equivalence transaction of one set of representations to another.

In Peirce's terms, the logical form of icons within the individual's thinking is a matter of how the icon relates to the individual's consciousness. René Jorna and Barend van Heusden (2000) discuss this sentience/ awareness aspect of a representation. Invoking the Peircean requirement that any sign be interpreted relative to an *interpretant,* they offer two points. First, an icon can be a different sort of logical form in accordance with how much the person's awareness is involved. And second, the interpretant's role affects logical validity. With these terms, representations—even those brokered by abstract logical symbols—cannot be expected to be exact fits to the slots and nodes provided in schematizations. *Schematizations,* after all, *are themselves icons.* Jorna and van Heusden discuss extremely rigorous requirements for a representation to achieve the status of a "notation," including semantic unambiguity, syntactic disjointness, and finite differentiation. Within the terms that set icons in an interplay with sentience and consciousness, such requirements would be a tall order for any one mind! *Who* can measure up—be held to account for the deployment of, the conveyance of, and the interpretation of representations characterized by unambiguity? Who can also be free of occasional syntactic crimes? And who can present meaning with pinpoint accuracy? H. L. Mencken, perhaps? What about William F. Buckley Jr.? Maybe William Safire! Comedian Joe E. Lewis probably had it right: "Nobody's perfect!"

The waiter calling out, "One turkey special," may be referring to you as a "turkey." Not very flattering! But the waiter's reference to you may be meant to go to the syntactic slot for an indirect object, not for a direct one! Therefore, even where we have very abstract representations, icons and signs cannot achieve a status of a "pure" notation if an interpretant is in the picture.

A major conclusion, however obvious, should be stated: *a schematization is an icon at least in the respect that it requires an interpretant.* How easy it is for a computer programmer to objectify a schematization as a script, yet how equally easy it is to forget that the script remains dependent on the programmer and on whoever (or whatever) uses it. But the problem is not isolated to computer programmers. Interpreters of laws, rules, rituals, or myths can easily be lured into "fundamentalist" doctrinal scripts. They have roles as interpretants, whether they admit their choice in adopting a doctrinal view or not. It may logically follow from a strict constructionist read of a law that certain courses of action are permissible and others not. The routines to implement these rules may well be highly consistent with the rules, yet there is nevertheless a cascade of interpretation and choices of roles and routines. Semiosis is part and parcel of the schematizations. In sum, we do like to think of "terms" and "premises" as logical structures, but to the extent that we agree to work with logical structures as natural forms, they are not completely separated either from vicissitudes of meaning that may make schematic placement awkward or from the perspective they cast on schematic placement—a way of seeing things that can create occasional or grievous absurdities.

Perspective is all but indispensable for a claim of validity. The prevailing perspective depends on the definition of terms, the semantics attributed to their denotations, and the relations of these denotations to external context. It would be compatible with a semiotic view of logic and its symbolizations to say, "Although it is possible to define things such that there is theoretically no external context, this definition of a universe of discourse and its terms is penultimate only." *At the least,* when schematization is involved, the relation between the denotation "internal" to a term or category is *not* securely isolated from external context. It is interdependent with it.

Although I have stated these points in modest terms, they have broad implications. Semiotician Uwe Wirth asks if we have entered a "postmodern hell" in which "objects are no longer objects and signs have lost their power to represent" (2003:35). If so, then no macroscopic view of a sign can capture the range of meanings and applications of the experiences we might want to code, classify, and index. Solomon Marcus makes a similar point when he points to a crisis for a semiosis of signs: "For the first time in human history, the main concern of science, art, and philosophy is situated beyond the macroscopic world, while human semiosis, as it was projected and developed during a long period of its history, had remained limited to the macroscopic world" (2000:abstract).

I take all this to mean that signs are under a "microscope." Only by reviewing and examining various aspects of signs—unseen *at the moment*—can there be further utility to a sign. Wirth (2003) points out that this approach might lead to despair or desperation because meaning would be so elusive that it would not exist per se. Nevertheless, he recalls the Peircean principle that a sign has replicability, even if its interpretants can color meanings in different perspectives. What remains stable is the process of refining meaning by a continual examination of the way to interpret signs. Your knowledge progresses such that you can do more with your representations. This progress keeps your meaning linked to action—and to your initiation of action. You do not simply embalm the signs (representations) you use, prop them up, and roll them out to cope with a newly presented problem or task. Accordingly, schematizations continue to have a live function in the redefinitions of signs and with regard to the recategorizations and vicissitudes in classificatory organization. For example, a decision to use an analogy to promote a newly schematized social or political agenda requires a choice of metaphor. Old metaphors and representations can be wheeled out, but they must undergo modification, at least in their interpretation and relation to context. When the George W. Bush administration recently chose to analogize the war on terror to World War II's struggle against the Nazis, it chose powerful signs that have been used previously to symbolize. However, it gave these signs new values by exchanging the time and place of the 1940s horrors in death camps that had gas chambers for the twenty-first-century political visits to the camps. Even though they had coiffed grass and apartment houses, the reschematized symbols may then have had pragmatic value in convincing other nations to join with Bush's more current endeavors.

So it is a matter of finding the right metaphor—where what is "right" requires an analysis of how it fits anew in a new analogy. But the decision to find the metaphor requires some abductive thinking, and the logic of the analogy requires protological juggling of categories via rules of and attributions of equivalence. Therefore, the thinker has to have a goal and a determinative form of action. Nevertheless, the thinker's representations in relation to past and future schematizations have all sorts of sign replicability, even though the resultant meanings and products are issuances of new and, it is to be hoped, improved outcomes and scenarios.

With all this in mind, let us return to the question of encountering a schematization and what logical patterns it opens for us to recognize.

Where the "I" is in the background and the organism is on automatic pilot, the validity of a logical form is syncretic; it is enmeshed within the particular schematization. If you regularly drive on a four-lane highway or turnpike from one city to another, your identity is a part of what you do in that drive. It is intertwined with the nature of the road, the car you are driving, and the negotiability of the curves. The logical nature of your view of "you" is not as a "you." It is more in the form of a "driver." So the logical form—to the extent that we can view it as separate from its place in the schematization—is a *species* categorization. Now ratchet up your awareness a bit. Imagine that you go on a vacation to England. You have never driven there or on the particular one-lane country road that leads to your lodging. With problems staring at you, and with no real way for you to go on "automatic pilot," your driving on that road can be accompanied by your awareness of yourself. This is an awareness of "you" as enmeshed. The "you" is not limited to the specific schematization, but here and there it infuses the category of "you" with schematization-inspired and related characteristics. Thus, you think of "you" as someone who is not sure of right and left, who is gripping the steering wheel with two hands, and so on. Your awareness of you is limited to a "you" syncretized with the schematization of the British car and winding road. Yet you have focused in on personal dimensions of your actions. You include the realization that you have feelings about the right/left discrimination. You become aware that the "white knuckle" gripping of the wheel is yours.

In either of these cases—the U.S. turnpike or the British one laner— the schematization of the driving and the driver is subject to change from one situation of particulars to the next. Even so, the form that would correspond to your "self" would be less specieslike in the British one-laner schematization. There, where your awareness is called into play, you would be a "you." That "you" would be tied to specific schema-related, albeit internalized, experiences—not being sure of right and left, gripping the wheel, and so on. I must conclude that this "you" is not too far away from the "driver" identity. Each form occupies a rung in a hierarchical progression, which as a function of the degree of consciousness moves up farther toward its subordination to a more abstract genus-type category of "you"—say, as a self.

For many reasons—development of efficient habits, quick judgments when necessary, communication with others—it is necessary to obtain some sort of stability in a category and in categorial relations. To do this, the schematization has to be articulated. That which is "the driver" fits a slot. We can hope that it does not spill over to something like "the unseen other driver on the road." The driver, the driving conditions, the vehicle and its characteristics, and the outcomes have to have proper slots like those for *agent, verb, object, outcome,* and so on. Terms such as "driver," "driving," and "unfamiliar roads" have nodes at which they pass through, transact, and interrelate. These nodes optimize the effectiveness of the schema, and they have to be available for coordination of term functions. The dependency on a stability of roles for terms and categories is to promote and optimize communication with others, but it is also so that we don't have to throw a schematization away. We can recognize it when we encounter it again—provided the categories within are stable. Does our recognition provide some logical pattern within which the categories are useable and can be related to the values and capacities of actions possible within the schematization? What logical patterns are produced by our recognition of our encounter with a schematization?

The predicate terms afforded by the slots and nodes of the schema provide a fit for inclusion in relations that mark off the category within the schematization. Likewise, the predicate terms define the inclusion relationships among the categories.

Thus, if we have the term "driver" in a schematization of "driving," there are possible propositions based on schematization slot relations or predicates. One such predicate is the syntactic relation Agent (S) → Verb (V) → Outcome (O). This causal sequence defines the relation of the terms "the driver," "drives," and "arrives in Philadelphia." However, if these three terms become categories, the sequence is one of successive inclusion: thus, where S is the driver, V is his driving, and O is the outcome of reaching a destination, S > V > O. The "fit" of terms with categories and the "fit" of scenarios to propositions always leave room for some ambiguity or "fuzziness" to accompany the aptness of location and function within a schematization.

We are talking about some set of self-regulations by which an organism maintains an "automatic formula." This formula is a logical structure. It may have inclusion rules that refer to terms, but it also may have a causal-sequence predicate that constrains the terms and categories. These rules and constraints would govern the application of the "automatic formula." So the effectiveness of that application is dependent on certain operations to maintain the sort of balance that supports the "fitting." For the "automatic formula" to hold, there have to be operations such that the fit of the relevant inclusion relations to a schematization is evaluated as the context changes. This implies a two-way street. The context—driving or entering a restaurant, and so on—may also call for juggling terms or categories and for altering the predicate.

Psychologist B. F. Skinner developed an automatic formula by which what was a "stimulus" had to be shifted into a slot at the back end of the causal sequence of "behavior" and "stimulus." This description is a stripped-down interpretation of Skinner's idea that the organism's responses should come first in a sequence. The stimulus should be seen as a reinforcer instead of as a cause initiating the behavior. However, the example is apt because the behavior is also affected as a "category." Instead of the behavior's being a class of "outcomes," it is replaced in the predicate as a class of "ongoing behaviors." The result in Skinner's time was an automatic formula that reigned over many schematizations that included animal training and human learning. Well, I won't get into Skinnerian theory here. The point is that because of the specific way events were to be operated upon within the training schematization, the formula that emerged as automatic was varied from a more usual predicate. The formula was then nonetheless automatic in that it could be applied to a variety of schematizations of the same class!

Keith J. Holyoak and Paul Thagard (1997) argue for "multiconstraints," and they name *structure* (such as syntactic role), *similarity* (correspondence of semantic elements), and *purpose* (the thinker's goal) as three major "constraints." The evidence for the search for coherence seems quite compatible with this triangle of constraining factors (Gentner and Holyoak 1997; Thagard and Shelley 1997). A person has to pull together (regulate) by way of the forces or rules within him, which push in the direction of viable predicates, logical organization of meaning, and motivated task resolution. The evidence can be reviewed to see whether empirical data justify the concepts offered. However, my interest is in how those concepts clarify Piaget's formulation that regulatory dynamics are set in motion by unresolved negations—a logical state of affairs.

The "I's" role is critical because the "automatic formula" has to be engaged in a conjoint fashion with varying orders of sentience. Unconscious dynamics and calculations toward self-regulated categorization as well as classificatory organization and its fit within schematizations are nice to count on as some sort of evolutionary "given," but persons are neither shoved around by archetypes of thought nor eternally bound to hallucinatory perceptions. Although predictable thought patterns and ways of seeing things can help persons navigate through the trials and tribulations of the "schematizations of everyday life," they remain subject to vicissitudes in conscious determination. You may go "on automatic pilot" to drive the turnpike, but you are responsible for the decision to continue with or alter automatic pilot when you see a variation you hadn't the last time you were on the road. Sometimes a "911" phone operator remains on automatic pilot when the caller is acting strangely—a decision that can be very unfortunate, and the phone operator can be held responsible for poor judgment.

The "I" is at some quasi-pinnacle to oversee the adequacy or aptness of the regulations, and hence to evaluate and determine them. In this oversight function, awareness of the schematizations and of the self can reach a limit, unless the "I" retains awareness of the *recursiveness* by which it can look at itself. With some order of a "sense of self," the individual's employment of the automatic formula and its operations can "fit" when the "I" is included within a schematization. But oversight and change require an order of the "I" such that it can view itself within that schematization!

At this *"I am aware that ..."* order, there can emerge a necessary perspective on the "I's" capabilities to assign a definition or to categorize at some level of generality. A formula might include a category of intolerable danger—with checkpoints to determine what a level of intolerability would be. An inviolate rule can be predicated on a series of checkpoints distributed in a causal sequence within the predicate. (If at point A, you find that B, then check C, or otherwise go directly to D). These checkpoints can certainly relate to an intolerability category and to other possible criterial categories as well. The defining and categorizing process permits new categories of events that have *not* occurred, that are *not* occurring, and that constitute *future* possibilities. In this way, the "aware I" can look for exceptions that have not yet occurred because its schematizations are now multileveled (spatially hierarchical and temporally sequential, including past, present, and future possibilities).

Conflict and Contradiction

Let's now try to differentiate unconscious from conscious structures. What happens when our thinking less than consciously leads to behavior that we would much rather consciously think about first? But there is also the view that what happens unconsciously *is* rational and calculating—at least in certain respects. To begin a discussion of the two apparently opposed concepts, look at two terms: *unconscious* and *unconscious reasoning* (see the glossary in Atkinson et al. 1996).

Although Freud was dependent on the great German physicist and mathematician Hermann von Helmholtz's ideas on the conservation of energy (see Thornton 2001), he struck out on his own in conceiving the role of unconscious thought. As a departure from Helmholtz's idea of unconscious reasoning as inference making, Freud proposed an unconscious logic, particularly in his theory of dreams. His version was that such logic suspended the basic rules of contradiction and identity. Helmholtz's idea was that "perceptions may contain many experiential data that are not immediately represented in the stimulus" (Boring 1950:308). Through a combination of habit, repetition, and association, the person perceiving an object analogizes to the learned generalization. The process is apparently inductive, but the form of thinking is compatible with the deductive conclusion-making of conscious thought (Boring 1950:310–11). If Helmholtz's idea implies that the logical laws are alive and well in unconscious processing of propositions, then it would appear to be difficult for Freud to be right also.

Many would agree that Helmholtz *was* probably right. I would even go a big step further to assign the notion of the automatic formula and its operations to his conception. Helmholtz himself, in his tribute to the chemist Michael Farraday, imputed such an unconscious intuitive grasp to account for conclusions that you might suppose required mathematical calculations. Thus, the unconscious inference-making applies to a form of reasoning where the product is not only perception, but also conceptualization. So I raise here the issue of the "automatic formula" involved. I am positing that schematization, as the perceptual basis for categorizing and calculating optimal adaptations, is a kind of reasoning that can be clearly automatic.

That which is automatic does not require conscious monitoring. I learn how much time to leave for driving a certain distance. After a certain period, I do not have to think consciously about measuring space and gauging time. If somebody asks me for directions, though, I can tell her about how many miles this destination is, how long it takes, and what driving speed applies. Yet this way of looking at how we represent perceptual-based and perceptual-bound icons in schematizations with automatic formulae leaves room for the Freudian dynamics of a logic that suspends the very principles that allow inference making! The point is that *both the inference making and the Freudian dynamics in the dream processing of icons are not conscious processing as we would ordinarily define it.* How can both Helmholtz and Freud be right about the logic at its "automatic" work when thinking is not conscious? What the dynamics reveal is a different point in time, one that, in comparison to the adjustment of thinking to referents, is at another end of a causal sequence— namely, the creation of icons.

Hence, in getting at the logical structure of unconscious thinking, we run into another basic assumption that I make. *The originating of an idea or a goal is something different from the schematization of achieving the goal or actualizing the idea.* The "something different" includes differences in logical form. Therefore, to describe unconscious logical structure, we get further into the idea of how symbols are assigned in the first place, as compared—for instance—to how definitions and the limits and contents of concepts and categories are consciously assigned.

In this approach via the comparison of logical form, we should keep in mind that Freud argued the coexistence of contradictory thoughts and ideas in dreams and unconscious ideas. In fact, one way he put it is that negation functioned either not at all or to cancel out the negations we ordinarily apply in "day logic" or conscious processing of ideas and propositions (see Werner and Kaplan 1963). I summarize these functions by saying that the negations applied to *categories* are canceled or somehow not present. I use the term *categories* to cover the instances of both ideas *and* propositions. This is in line with my general viewpoint that not only can categories include ideas as possible members of classes, but also, within categories, propositions can be cast as members of classes. Anyway, the cancellation of negation is a tricky thing to puzzle out if the categories that are contradictory remain. So to remain faithful to Freud's observation that contradictory thoughts coexist in dreams and unconscious thinking, I put it that *in nonconscious logical processing, there is a state of conjoint categories in contradiction that can be symbolized*

$$A \neq \bar{A} \cap A \cdot \bar{A}.$$

But in what category relation are the ideas (or symbolic elements) A and $\bar{A}$? Are they in opposition? In what categorial order are they? Are they subject to resolution as subcategories within a more general category?

The difficulty presented by Freud is that he does not address, let alone answer, this question. Contraries can logically coexist without contradiction when seen from the perspective of a superordinating category. Can this mean that the absence of contradiction in the dream implies that the hierarchical ordering is present and that the contraries are subcategories within a category? But how can hierarchical ordering be inherent in dream thinking? This structure to the dream is hardly likely if the model of development of thought holds or if repetition of a developmental sequence of thought holds. We would ordinarily regard thought as moving in a direction from primitive or immature to complex and mature. The dream, as unconscious thought and logic, should appear to have structures that are *prior* to hierarchical ordering. This point can be referred to Vigotsky's and Piaget's descriptions. But Werner's concepts open another possibility: the dream is *not* just an anachronistic moment at which some protoevent returns from a time warp. Instead, it is a retrogression, yielding the preceding level of order as the degradation goes the step back. In fact, as the degradation deepens, so should there be a match in the lessening degree of a mature state. This, you recall, is Werner's notion of how you can make the comparison of the stages of maturation with the level of pathology that "resembles"—*but is not the same as*—the immature state.

Daytime thinking provides the categorical hierarchizing. In the dream, which psychoanalysts find particularly dependent on the thinking of the previous day, there is a degradation of the daytime logic. This degradation entails retrogression from a *preexisting* hierarchy; the result is a descent into *a syncretism of superordinating and subordinate categories.* In short, when we review a dream's logic, we see the contraries as if they were still—after having already been—subject to superordination. We infer that the contradiction within dream logic is *not* conflictful. To clarify this point, use the notions of the "previous day," "the dream of the night," and the "next day." We review the dream logic the "next day"; the degradation takes place from the "previous day." What is imputed to the logic of the dream of the night is a "prior" status of superordination—as if a residue of the "previous day." Thus, we attribute to the "dream of the night" a contrary status, and we presume that negation of negation takes place because we attribute, probably rightly, a "prior" superordinating category and hierarchization within which contradiction *would* exist.

Take the case of the person who dreams that she is someone else. First of all, this *can* be seen as a metaphor, especially when the meaning of the dream is retrospectively unpacked. However, in the work of the cognitive linguist George Lakoff (2001), this attribution of metaphor is conflated as to its form in the dream vis-à-vis the "next day's" unpacking in the waking state. In waking logic, however, the dream's juxtaposition of the person's identity and someone else's identity is a contradiction—if it is not mitigated by a day-logic sense that the equivalence is by way of a figure of speech.

The contradiction *should* disallow certain juxtapositions. In "real life," a person can't simply go on, in A.D. real time, claiming that he *is* Jesus Christ, as in Milton Rokeach's (1964) examples. Yet if the contradiction were moved up to a higher logical order, can't the contraries coexist? Bateson (1972) suggests that a move to the next step up in the logical order is necessary to solve conflicts. A conflict can go on interminably unless there is a new way of resolving the oppositions. In the example of the person who would "be" Jesus Christ, it is the trope that can allow the claim to skirt the issue of identity. Figurative or "symbolic" levels of meaning and category processing can come to the rescue. Metaphoric or metonymic forms allow the kind of assertion that would say that "from a certain perspective" the categories can be shuffled, and an equivalence consideration can for the moment be "figure," whereas the identity issue can be "ground."

Now let's work the other way. We see what happens with form when daytime living has to be optimized. You can use perspective to rise above conflict. When you say, "A plague on both your houses," you rise above the conflicts therein, yet you might want to visit and make use of some of the good things in those houses. So you dream about the good old days. The dream can free you to reconsider and reexperience what you have left behind. But what has to be done if, in comparison to the strict distances your daytime logic dictates, the dream is to be more free, negations negated, and figurative constraints of perspective suspended?

In the dream itself, the status of categorical order is syncretic. This implies a primordial state in which objects of all sorts are mixed in the same goulash. To get at the logical forms involved, I ask that you "play along" with this idea of transition. Something has to be done to remove that daytime censor that would deny contradictory symbolizations and/or demand awareness of the juxtapositions of ideas *as tropes*. When and where does this awareness take place?

Freud ([1925] 1959) offers two solutions to the removal of the daytime censor: one, negation of the negation; and two, the assumption that negation itself doesn't exist at all in the syncretic state. Which of these solutions is the more comprehensive way of conceiving the transition and the state of affairs within the dream per se? Which idea can pragmatically explain more of the differences between the logical forms? Do you have to take the position of either or of both of these two either-or solutions? If you decide to play along with an elastic timeline, moving back and forth from dream to mature thinking and from mature to dream thinking, you do not have to take seriously the either-or questions. This is a considerable payoff; it allows open exploration of the apparent rules transacted and consideration of these rules in terms of vicissitudes of the "I" and its relation to unconscious and conscious thought. The "openness" has *its* cost: the timeline question is deferred. In the next chapter, first I refine the logical dynamic structures in relation to the dream and the trope. Then I introduce the star issue—the relation of the logical dynamic structures to the "I" and consciousness. Only at the end of that chapter can I return to the time issue.

6   What Are the Role and Function of the Self Vis-à-vis Consciousness?

I continue in this chapter to question how negation functions in the transformations of the dream vis-à-vis day logic. The perspective of time weaves in and out of these logical modes and the array of thinking governed by each. The problem of the negation of negation opens a wide array of different ways of thinking, and the fate of logical transformations in these ways of thought instructs us as to the requirements and the role of the "I" and the self. Ultimately, the self and its form, agency, powers, and logical function make the difference between figurative and nonfigurative thought in its great varieties.

Negotiations of Negation and Its Effects on Logical Forms of Contradiction

If you assume that dream thought entails a systematic degrading of waking thinking and logic, it follows that *both* of Freud's solutions—the negation of negation and the state of "no negation"—take place and in this order of descent. Assumed is that time moves forward not only in the case of the maturation of thought and its logical structures, but also in the case of degeneration. However, the future of the birth of the dream is the inverse of the maturation of thought. As you proceed to unfold the layers of logical organization, you encounter more and more syncresis. At such a point of decomposition, meanings and orders of classification easily cross boundaries. And all sorts of contrary and opposed entities, events, and phenomena might run together or be coequal in any comparison of categorical order. All of this evokes Werner's assumptions concerning primitivation. There is the trek toward a stage-by-stage dedifferentiation—from a differentiated state (Werner 1948; Werner and Kaplan 1963). The greater the descent, the more "you-as-you" disintegrate into forms that embed and absorb you. Lear becomes a shadow of his former self. The symbolic peak of assimilating all the thoughts and passions of the man in the role and power of the king is lost in a cloud of incoherence. His emotions and passions become so interwoven with his sense of the "I" and the "you" that his identity and his daughters' identities become less the issue than resolutions to schematic themes of his own hatred and fear and the blurring of just who it is that possesses or is possessed by them.

Just as in primitive thought, in a dream *you* can be *not-you* in some form. In primitive thought, you can hypostasize a power or a passion; in a dream, your emotion can be imbued within a series of events or other persons. The form of the not-you may be you as someone else; it may be you as something else; it may be you as a reduction to a name or some other metonymy, such as a thing, an emotion, a force, or a symbol. *The dream symbolism may well resemble the form of a figure of speech—although the interpretation of this form, when the dream is recollected, is from the point of view of consciousness.*

In light of this resemblance, what about Lakoff's (2001) argument that the linguistic form of the dream is metaphoric? His interpretations are, so to speak," daytime" symbolizations of the protologic of the dream. Freud, in his book on dreams, points out the difference between the logic of the dream *as dreamed* and the nature of the dream *as we become aware of it*. A basic distinction is his division of "dream-work" into *"dream-thoughts"* and *"dream-content"*: "The dream-thoughts and the dream-content are presented to us like two versions of the same subject matter in two different languages. Or, more properly, the dream-content seems like a transcript of the dream-thoughts into another mode of expression, whose characters and syntactic laws it is our business to discover by comparing the original and the translation" ([1900] 1958:277).

Lakoff's insight that the metaphoric forms are present in dreams neglects the difference in structure between the dream as dreamed and the dream as interpreted from some conscious point of view. As the great scholar Henri Ellenberger points out, many dream theorists recognize this difference (1970:303–11). From it, all sorts of intermediary levels of representation and transformation can be adduced, interpolated, and/or brought to bear as interchange between the more "mature" logical structuring of meanings and the unconscious syncretisms. Some would argue that this access to the multiple possibilities of organization, conceptualization, and representation is by way of regression to a "gold mine" for creative thinking (Kennard 1997; Luther 1997). I cast these transformations as functions of "raised consciousness." The individual becomes aware of many alternative ways of classifying, which can take her to new ways of synthesizing categories. These new ways can retrieve valuable elements lost if abstract categories are as if cemented into place. I agree with Bateson: all sorts of conflict can be avoided by figuring how to restructure stultifying contradictions to contain them as contrary elements of a *subset*. But in a regression to the goldmine of possible organizations that would retrieve lost valuable elements, the dream can, when interpreted, present new opportunities for organization. The method, though, involves combining considerations of the structure of consciousness with the structure of its classified and categorized objects. That is why I talk about "raising consciousness." If you become conscious of being conscious, you're at a new level for the inclusion of contraries of consciousness itself!

In the symbolization example next given, fA is my added symbol for the individual's action of elevating the term or icon A to that of a superset, which *includes itself and its complement;* thus,

$$fA \ (A > [A \cdot \bar{A}]).$$

Yet this formulation is primarily for the "dream interpretation," which transforms it for "daytime logic" and offers the understandings that make it accessible to reasoning and the articulated interpretation of figures of speech. I previously referred to gaining perspective on a representation of the self by considering the representation to be a metaphor for the self. It is also the case that another tropelike form for the self in dreams would be metonymy. Nonetheless, the question is, Where does the figurative perspective inhere? When does it appear?

More basic than metonymy as a figuration of dream particulars, *synecdoche* may be the categorical form intrinsic to dream thought. The dreamer's awareness of particulars requires a sensing of meanings. If

you dream of rotten apples, there can be all sorts of "sensing" by association with taste. But also, the awareness of the existence of particulars (i.e.,
of particularity) requires an imputation of meanings. Thus, *a metonymy
refers to exchanges of parts for wholes and therefore implies prototypic organization of particulars.* In contrast, *a synecdoche implies more skeletally the
exchanges of categories.* Both are matters of logical classification. In fact,
synecdoche may be less complicated as a logical form. Therefore, by itself
within a dream, it stands isolated from the meanings that only a later
interpretation can imbue. A metonymy, if based on contiguity or action
features, is still a way of classifying, and it therefore remains as something with a logical, albeit more primitive, form. We have to be careful of
that word *primitive* because, as Lévi-Strauss (1963) pointed out, it actually
denotes more complexity of meaning embedded within the structure.

We are a step closer to a "real" daytime trope, though we are not really
there yet. The difference to underscore is between the apprehension of a
trope as such and the part-for-whole logical structure of the more primitive thinking. Part-for-whole thinking may be like the primitive classification of animal powers by the wearing of animal skins. The idea is of direct
transfer of powers via a "part" of the animal. The primitive classification is
"species" making, based on a "motivated" category. This is Lévi-Strauss's
(1966) argument. From the particularization of an "animal-skin power"
can evolve a logical classification system involving binaries—oppositions
for that category. The net results of such a system come after a "history"
of newly derived subclassifications—all of which provide a great deal of
traffic and exchange between particulars as species and the development
of genera.

However, with a trope, the exchange of genus and species is limited to
the trope's order of the classification, and the particularization of a "species" is limited to the domain of meaning constrained by the categorical
ordering. Thus, the trope "Richard the Lionheart" is recognizable as a
figurative expression that does not require any hunt and surgery to remove a lion's heart.

Although probing the levels of meaning and how they are funneled
from one semantic form to another, we are also examining linguistic and
conceptual matters. Here, one can take a page either from the cognitive
linguists or from various depictions of the "mappings" of meanings from
one set of subdomains to another (see Gentner and Holyoak 1997). The
domain of meaning in focus is the "bravery of Richard." You can look at
this domain as a conceptual domain with a format like the comparison of

an epithet to a person—namely, Brave: Richard. If such a comparison is seen as a conceptual domain, "bravery of Richard," we have a conversion of the comparison to an "inclusion relation." "Bravery" includes "Richard" as an example or as some sort of subset. But if the concept is "Bravery of Richard," then the organization is less set and subset and more "species" dominated. More succinctly, the concept, if a logical form, is a prototype. All of this is to say that the rhetorical and conceptual analysis serves to examine the semantic vicissitudes from one level or domain of meaning to another. However, the question of the logic of a trope is still subject to logical form and the function of the "I" and consciousness.

Anyway, when the phrase "Richard the Lionheart," is used, it is to be interpreted. The concept of "Richard as brave" can become a template within which you can have a figurative perspective on the phrase "Richard the Lionheart." This way of seeing the metaphor is much in line with Lakoff's reasoning about the metaphor's general domain into which specific phrases or instances might fit. In relation to dreams, Lakoff theorizes, "The general metaphors are sets of correlations between source and target domains at the superordinate level" (2001:metaphor and dreams section, para. 17).

Bear in mind that an ordinary person—let alone an untutored robot, not raised with reinforcements of others' interpretations of English literature—can "see" the phrase "Richard the Lionheart" differently when the person is dreaming, daydreaming, or awake and ruminating about figures in British history. With different degrees of consciousness, the "I's" decisions select a perspective from which to "see" the phrase. For a trope, there has to be a figurative perspective and understanding.

Daydreaming, which Freud saw as something very close in structure to the dream, is an excellent example of this requirement: "[E]xamination of the character of these day-fantasies shows with what good reason the same name has been given to these formations as to the products of nocturnal thought—dreams. They have essential features in common with nocturnal dreams; indeed, the investigation of daydreams might really have afforded the shortest and best approach to the understanding of nocturnal dreams" (1900:chap. 6, sect. I, para. 6). Computer scientist Erik Mueller's (1990) analysis of daydreaming illuminates the literal/figurative difference that would differentiate between a "daytime" and a dream mode. This distinction is not made in the Lakoff analysis of "basic metaphors"—a concept that might be better applied to daydreams. Moreover, claims are made by Deirdre Barrett (1993) and others that dreams *can* be "put to work." But, again, the joker in the deck is the role of awareness—the "I's" access to the thought and the consequent accessibility of figurative meaning. *Figurative understanding is an interpretation.* It is made from a "viewing" of a fantasy. In the same way, figurative understanding is made from a "viewing" of a phrase such as "Richard the Lionheart."

The person interpreting has to view icons, their relations, or a phrase adapted to characterize a person like Richard. Accordingly, he has to select a pertinent and meaningful function from an array of meanings. This is the case over a range of phenomena. You select a function when you dig for Lakoff's "basic metaphors" to interpret a dream. You map meanings to selected variables, as Dedre Gentner and Keith J. Holyoak (1997) describe, for the coherent application of a metaphor.

"I," Self, and Negation

In general terms, what is selected is raised to the level of superordination. It becomes figurative when the "I" understands that the species organization is not merely literal. Hence, "part-for-whole" thinking becomes allowable and in some cases poetic or in others pragmatic. The "part-for-whole" thinking that might be summarized

$$> A \supset A$$

becomes recognizable as a metonymy-type trope, *iff* (if and only if) the logical structure becomes

$$fl\,[fA\,(> A \supset A)].$$

Read fl as the operation of the self in relation to the recognition of a superset, namely fA.[1] That superset fA marks the A as functioning relative to its subsets (> A and A), and the inclusion relations and implication operations ($\supset$) for the set A. Focus on the

$$(> A \supset A)$$

in the previous formula as something like the skeletal phrase "Richard the Lionheart" as it is put before the untutored robot, which may or may not engage various levels of awareness on the road to an interpretation in converting the phrase to the status of a trope. Now move backwards from that skeletal phrase, toward the left of the page. The first "awareness perspective" is

$$fA \ (> A \supset A),$$

and the fA appears as governing or "including" the skeletal phrase. Next, "order of inclusion" as awareness expands or consciousness is raised is a function of the "I":

$$fI \ [fA \ (> A \supset A)].$$

If I were Robin Hood, I would have to know, in relation to knowing that Richard is Richard, that bravery (or fierceness) would be something I would have to see as evocative of some behavior in Richard—that is, in comparison to me (and others) and as similar to a lion's qualities. In addition, I would have to accept not only that Richard, others, and I have hearts, but also that I can regard these hearts as symbolic of a lesser fierceness than that of lions. All of this requires that I have a sense of the "I" and of a comparison to others and to the Other. Moreover, the "I" would have to be compared to more than one symbolic level of icons in order for the classificatory transfers in metonymic forms to be apprehended.

This is not to say that you chart all these steps consciously. It is to say the comparisons and differentiations have to have been made and reapplied in an aware way if you have confronted any new metonymy or any need for a new application of an old one. If you get up in the morning having dreamed that you were "less than yourself," upon perceiving this equivalence as a contradiction, you would need room in the day and night's logical structures to see figurative aspects to what you can now interpret retrospectively as a figure of speech. There's also the issue of retrospectively attributing the contradiction to the dream and *only then finding a heuristic value in the symbolism.* If you accept that you are "you" and, in light of this categorization, that you also have the insight that you are "less than yourself" in the dream, you now have a more expanded view of yourself. You can look at yourself in at least the two different ways—and you are still "you." In this sense, you might be more than you previously conceived; the very commonalties that make you *both* "you" *and* "less than you" can thus be empowering. In relation to the dream, all this is

accomplished by the "I" in a retrospective view of the dream's organization of contrary elements. *The particular mark of the figure of speech is that
the heuristics are not only in the coexistence of the contraries in the subset elements, but also in the commonalties that permit equivalence status between the
contraries.* In sum, reading the fA formula in terms of a retrospectively
attributed equivalence relation, we get

$$fA \ (A \equiv \bar{A}).$$

The usual supposition is that the dream logic does *not* know that a contradiction is introduced. A Freudian would reason that by way of the dream,
either a contradiction is resolved, or the contradiction exists—without
the sense of conflict. I'll try another step—a distinction between these
structures of *nonconflicting contradictions* and those of conscious thought,
which I call *conflicting rules for contradiction*. The conflicting sense prevails
when you feel you can't say that a thing is the same as something not that
thing. At this point, the logical rule of identity *does not* accommodate a
negation of a symbol, yet the operation of equivalence *does* accommodate
the negation of a symbol. Cast these conflicting rules this way:

$$([A \neq \bar{A}] \cdot [A \equiv \bar{A}]).$$

This says that on the conscious level of thought, icons of the same order
can be subject to equivalence transformations (Matte Blanco's point is
similar), which would allow nonidentity of symbols and legitimize transactions between them *at the same time* that identity is held to its logical
criterion. A titillating example is in a television classic of the 1950s, *The
Prisoner* (McGoohan and Tomblin 1967). Each week the show would feature the operations of a superintelligence agency, but each week the name
of the official in charge could change, and the primary way of referring
to him was by the symbolic title "Number 2." The persons filling the bill
for the symbol Number 2 were exchangeable. The presence of individuals
with unique names would symbolically be a matter of their unique names
being subordinate to the role (Number 2). Where A stands for the role,
and B for the unique individual marked by a unique name, $(A > B)$.[2]

Thus, exchanges of identity (unique identification of a whole) make
sense when symbols do *not* include other symbols not within their scope
(denotatively or connotatively or both). So if $A > B$, the presence of B does
not automatically preclude identity for A. In *The Prisoner* example, Num-

ber 2 is Number 2. If, however, there is no B included in A, then (A ≠ A > B), and (A > B) can be relegated to Ā. In fact, this was an intriguing aspect of the series. The more Number 2 was replaced, and the more the fate of the previous Number 2 would be the result of unique plot changes, the more it became evident that the unique factors or characteristics of the person filling the role of Number 2 were not relevant. So if B symbolizes the presence of a strangely mustachioed Number 2, the B factor would not be relevant. But even if B signifies a whole unique "person," complete with "personality," it might be irrelevant. It's as if the prevailing of the "role + new 'unique' person" is as a category in opposition to the person's uniqueness. So, in a sense, "role + new 'unique' person" *is* category A, and "the unique person with his unique name" is still B—only it looks like "B" was Ā! Another way to make this statement of identity at the conscious level is to say the following: to attribute identity, conscious thought does not discriminate between symbols that do not include other symbols, and conscious thought does not discriminate between symbols that include the same symbol or symbols and itself.

Two processes are at work here, and, to an extent, they counteract each other. Growth and maturity in thought seems accompanied by our ability to differentiate. Joe Smith is not Joe Friday. However, there is a movement toward hierarchizing—both these guys are "Joes." We have to watch that the grouping function of the hierarchization does not cost us our ability to know that Joe Smith made amusing film reels, whereas Joe Friday was a fictional ace detective!

Last, note that Japanese psychiatrist Kimura Bin (as reported by Phillips 2001) theorizes that the subjective self (the "I" experiencing awareness of itself) converts the self to an I as an object. When this happens, there is a sense in which the I as object is not the "I." This conversion from experiential process to experiencing a product converts the process into a product. Hence, "I" becomes "Ī."[3] If the process of this conversion stops at this point, and the individual regards the object-I merely as "not-I," then, Kimura theorizes, we have a hallmark of schizophrenia. This is easy to see as pathology, when the objective I *as "not-I"* is considered either as another person or as the Other. The late comedian Flip Wilson's character "Geraldine" says (to avoid responsibility), "The devil made me do it!" The Other is the agent when the "I" has been converted to an object—and then is seen as not included in the individual's identity. The point is that the conversion of the object I to a "not-I" is logical, but it stops the logic process short. In this form, matters of identity are *unhierarchized*. Therefore, *the pathology is not seeing that the "I" has identity in that it is a superordinating category with subcategories that include the "not-I."*

In clear Aristotelian terms, the rules of identity and contradiction depend on certain categorical forms, which turn out to be a matter of what is included in what. Accordingly, in a way, we come full circle back to von Domarus's (1944) formulations wherein dysfunction in schizophrenia relates to distortion of the law of identity and disengagement of the law of contradiction. These things occur, he explains, when there is a failure to regard what is not common to entities as not relevant to identity. In short, the common intersection of entities—say, A and B—is what's identical. Von Domarus discusses these entities as propositions. But it is also the case that the entities could represent concepts and their context. Accordingly, I view them in terms of their category structure, and I make the following point in these terms. What is common to all the subcategories of the "I" is bumped up to the next highest class, where identity can be logically—and apparently psychologically—established.

A brief demonstration can serve as the foundation for the analysis of the logic of consciousness.

The conversion of the experiencing "I" to a product I—which is now not the experiencing "I" and thus is "Ī"—can be considered more closely as

"I" → I

I ≠ "I"

∴ I ≡ "Ī"

The "Ī" category is the complement of "I" and logical as such. But the essential way in which the conversion is set in motion is by the predicate ("I" → I). Von Domarus (giving credit to Vygotsky for this point) argues that schizophrenic logic is based on the equivalence relations of predicates and that these relations become the basis for identity. In contrast, mature logic bases identity on commonalities of the entities—or, as I would put it, that which is abstracted for inclusion in a next highest category. Thus, for A and B, there can be cases where (A $\not<$ B) and (B $\not<$ A). Such cases would not be included in AB, which is what I am calling the next highest class. (The AB is easy to picture as a Venn diagram showing the overlap of two circles, A and B.)

A mature logic would require an organization such that

"I" > "Ī."

In this ordering, the "I" is the class for which the complement can become a member. Even prior to the necessity for this "hierarchization" by way of making a superordinating categorization, there is another ordering step. When the predicate ("I" → I) is made, the "I" is also a superordinating category; hence,

"I" > I.

With this ordering, when the "not-I" does appear on the scene, the individual who is feeling or experiencing the self can retain or recover his identity. The full proviso for maintaining this stability of identity is present if the logical form

"Ī" < "I" > I

pertains—that is, if the complementary category is subordinate to the "I," which subsumes the object I.

These formulations are meant to be consonant with very general trends toward logical organization. Here, they are focused on the way the "I," the self, and agency function in relation to logical and dynamic forms, but they are also instances of more general trends in development that can help to specify the differences in pathological and "nonfigurative" forms as compared with figures of speech. These trends also show up in the differences between tropes and primitive forms, such as totems, myths, and images.

Werner describes a general developmental direction toward differentiation and hierarchization of structures: "The development of biological forms is expressed in an *increasing differentiation* of parts and an *increasing subordination* or *hierarchization*" (1948:41). Werner's view is applied to the whole organism. He is concerned with more than just cognitive events. I focus on his concepts in terms of logical structures of thought—even though the idea of logical structures, as not only Werner but also Piaget pointed out, may have general biological significance for the total organism. Further, the case of pathology is one of *primitivation*. For example, schizophrenic thought appears to lack differentiation between concrete and figurative levels of meaning.

With the description of what may be at work in the instance of the ordering of the "I" and its general developmental significance, I now turn more specifically to the issue of hierarchization, differentiation, and dedifferentiation as these processes are reflected in the logical-structural differences between the unconscious and conscious states, on the one hand, and in the differences between less mature and mature forms, on the other.

The Self in and out of Schematizations

The self (or the "I") can consist merely of its actions. Such a self would not be an entity-in-itself, nor would it be in a state of awareness of itself or its actions. Instead, it would be entirely enmeshed in a schematization or in a series of schematizations. Picture the person who acts like the proverbial "bull in a china shop." This person might be someone who just moves ahead with an agenda, without thought of self, of the meaning of her actions, or of its consequences to herself and others. Such a person—or creature—does not become "self-aware"; she works "on automatic pilot." A more articulated state of the "I" exists when the "I" can look at itself as an object (I) within a schematization, thus:

"'I' see that [*I* am breaking dishes]."

In Kamura's sense, the observed object I is reclassified as "not-I." The "not-I" can take the form of a specific member of that class. The self within the encompassing brackets is an object:

"I *feel* like [*I* (a bull in a china shop)]"!

Self-awareness of the object is outside those brackets. Absence of the self-awareness makes a difference in the governance of whatever action is included within the schematization. This difference, like the comparison between daytime logic and dream logic, is critical for comparing the logical structure in unconscious processing—whether in an immature state or a retrogression (dedifferentiation)—to more hierarchized structuring.

The Logical Forms of Consciousness

In a conscious state, I assume that various representations of the self, of others, and of the Other are subclasses of the "I." The subclasses are discrete foci of the "I's" reflective view of the self or its aspects. Each subclass has its own uniqueness or particularity. Hence, in relation to each other, they cannot be without some order of contrariety or opposition. The "I" as superordinating class and its subclasses look like this:

$$\text{"I"}$$

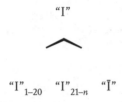

$$\text{"I"}_{1-20} \quad \text{"I"}_{21-n} \quad \text{"}\bar{\text{I}}\text{"}$$

Instead of listing or naming the clusters of self-related subcategories, such as self-confidence, self-absorption, self-satisfaction, and so on, I have arbitrarily assigned one subcategory of twenty members and another subcategory of all members except the first twenty. An example of "I"$_{1-20}$ may be the self as an actor who has had roles in twenty movies. I mean to indicate that "I"$_{1-20}$ and "I"$_{21-n}$ are different representations (Rs) related to the "I." These representations are likely crosses between perceptions of the self—both by the individual and by others—and apperceptions. Consider calling these representations *imagos,* which is Jung's term. Jungian analyst Paul Kugler points out the imago's value: "The imago performs a synthetic function, integrating both external sensory experience with internal psychic reactions.... [T]he imago is not simply a *reproduction* of the outer world (that is, a copy of an historical event), but rather, a psychic *production*" (1993:para. 2).

As representations, they form a class R, which is the class of all representations of the "I." Essentially,

$$R \equiv \text{"I"}_{1-n}.$$

Yet when you consider this relation as ("I" ∩ R), not all "I" is included in R, and not all R is included in "I." Instead of an equivalence symbol, I use the intersect symbol here because the superordinate status of the class "I" marks a greater inclusiveness than all the subcategories can comprise. This is no doubt a controversial statement, but it is made with the idea that the logical form of the "I" is itself so uniquely a part of its conscious and reflective nature that, as a class, its members cannot assemble without reversing a basic direction in categorical ordering. My notion is that an order of a natural category brings together a scalar and a vector consideration. Another way to look at my formulation is in terms of the features of an "I" as a subset of the set R, where the structure of the set and sub-sets is that of a lattice.

In reference to my argument, because the "I's" complementary class "~I" is a subclass of the "I," it also is an R (or related imago) of the self. With this hierarchized structuring, we can read the structure of con-sciousness to be a function of the self ("fI") as follows:

$$\text{"fI" } [(\text{"I"}_{1-20} \neq \text{"I"}_{21-n}) \neq \text{"}\bar{\text{I}}\text{"}].$$

As a function of "I" ("fI"), its discrete subsets ("I"$_{1-20}$ and "I"$_{21-n}$) are dis-junct from each other, but the conjoining of these subsets ("I"$_{1-n}$) is prop-erly subordinate to "I." We see the lattice features here if we go along with the idea that the set is different from the sum total of the subsets. Anyway, in the "fI" formula, the negation operations are marked such that the nonequivalence of discrete subsets—as a function of the superset "I"—are not equivalent to the "not-I."

This formulaic way of looking at consciousness, by marking the role of negation, can also be used to derive the representation of a transition from waking to dream logic—or, more broadly, from a mature self-aware-ness to a dedifferentiated or retrogressed state. We carry negation outside the brackets to the "fI," thereby degrading consciousness as follows. The formulation

$$\sim(\text{"fI"}) \equiv \sim[(\text{"I"}_{1-20} \neq \text{"I"}_{21-n}) \neq \text{"}\bar{\text{I}}\text{"}]$$

obtains when the self in its form and function of self-awareness is *disen-gaged*. In that case—namely, ~("fI")—the rich internal discreteness of the different possible distinctions between subsets of the "I," such as roles, self-confidence, sense of efficacy, and so on, is broken down. This break-down of discreteness is expressed in the part of the formula that says,

$$\sim(\text{"I"}_{1-20} \neq \text{"I"}_{21-n}).$$

Regard this state as syncretic. In it, a person confronted by criticism can't separate out the aspect of the self to which the criticism might apply. Hence, the whole of the person might be engaged in the scenario even though that might be inappropriate. When the antihero in the fairy tale "Rumplestiltskin" is called by his name, that name becomes more than a name. It becomes something that attacks his very being, and what follows is that his behavioral and emotional response literally tears him in two! In addition, a part of the formula indicates not only that the subsets of the self are not discrete, but also that whatever the stew of subsets of the self, they do not necessarily constitute the "I." Thus, if the "I's" identity depends on stable categorization of the "I's" instances and roles, then in

$$\sim[(\text{"I"}_{1-20} \neq \text{"I"}_{21-n}) \neq \text{"Ī"}]$$

it is *not* the case that the individual's identity is intact logically. What may show this phenomenon is adding up all the evidence that a person *is* aware of his actions, parts of his self, and his nature, yet finding out that it all still does not add up for that person. In this way, we understand that Oedipus did not know that he was he. Anyway, if a person argues she was blinded by rage, we can see that state as disengaged awareness of self. I show this in the formula as a negation of the "I," hence, ~("fI").

The result of the negation of self-awareness is a representation of the immature or undifferentiated state. In the form I present later, this representation may correspond to the idea of unconscious structure that has not progressed to (and hence not retrogressed from) any hierarchization of the "I" or consciousness. This undifferentiated logical state would be as if the "I" were at an object status—perhaps from the point of view of an other as observer.

$$I_{1-n} \geq \sim I.$$

Although there may be a basic self here, there is nothing I mark as an "I." Here, we continue with a transformation operation by negation. We do not even have to mark the "fI"; it just doesn't appear. Therefore, in the undifferentiated and immature state, we can talk about the absence of the "I." An example of this absence may occur when an infant's empathy is so great that *who it is* that is experiencing a given sensation is not clear. It might be the infant; it might be the mothering person; it might be a syncresis of both.

However, the pathology introduced by organic damage to a mature adult may proceed to the point of no awareness of the "I" as agent, as in the way Kimura (Philips 2001) pictures it. However, for the retrogression, you would want to retain the negations begun with the ~("fI") and the formula given earlier that reflects this retention.

One last point here.

Recall that Velásquez's masterpiece *Las Meninas* (1656) is a painting of the Spanish royal family, with the king and queen portrayed in a mirror. The framed painting is organized such that objective viewer would actually be the king and queen, or the figures standing at such an angle that they would see themselves portrayed in the mirror depicted. As "viewers," the king and queen are either absent or figurative. In *Las Meninas*, the artist removes the viewer's ability to enter the painting via an immediate porthole—namely, the position of the "actual" viewer relative to the view depicted within the frame (Searle 1980; LeGrady 2000). Accordingly, the viewer is presented with the painting as a *possible* trope in the sense that it stands in some relation to its internal content, but not in relation to the viewer. The painting attains the level of a trope, *iff* the viewer separates himself as experiencer ("I") from himself as an objective viewer (the "I" standing and looking at the painting). Only by doing this can the viewer undertake to separate his own identity from that of the "persons in the painting," who also occupy the position of viewer when the organization of the images within the frame is considered.

If the trope is to work, the person actually viewing has to create a porthole from himself to the intermediary objective position—filled by the "persons-in-the-picture." He then is viewing "them" viewing the painting. Or, more correctly, he is viewing them as his own representations—*as if* they are viewing the painting from a position that would be objective—*if* the trope were not a trope, but instead an objective position from which the perceived image of the painting would be consonant with his objective I and his "I."

The Selection and/or Intention of Conscious Forms and the Role and Nature of Self

If the "I" is functioning on several levels of awareness, all sorts of interaction can take place. I can review how I see myself in a given action schema and even how, from within that schema, I view my viewing of my self. If I argue with an agitated clerk that he is being too compulsive, I can look at myself doing this and evaluate how effective I am in that situation. But within that situation, the "I" arguing with the clerk may have contempt for too much self-inspection. That schematized "I" may "believe" that more intuitive action and less self-review is called for when a clerk unnecessarily delays things. However, the "I" can look at various "versions" (subsets) of itself. Therefore, whatever the status of the beliefs of the schematized "I"; after the "I" views *that* self, the "I" within a clerk-encounter schema may change. It is as if the schema that was a script of the person's beliefs has now been subject to a rewrite. So the "I" who is citizen of the schema can thus become more tolerant of delaying open criticisms of clerks. *The vicissitudes of the self involve two-way transactions from each order of awareness.*

Moreover, these vicissitudes affect and interact with states of consciousness. In line with this idea, I hope that the logical structure of the "I" shows the way to define the unconscious and conscious modes such that their relation to the logical forms of tropes, immature thinking, and pathological thinking and expression is clarified.

No, not everything is the "I"! So it is one thing to specify the logical structure of the "I" and its form variations, which appear as tropes, immature conceptions, and pathological senses, but it is another matter to think of the form that the "I" attributes to things other than itself.

Think of the child looking at herself as both a horse and a rider and slapping her side to begin a gallop. Is the "I" the child contemplating herself as "horsie," or is the "I" at one with "horsie"? This child's contemplation likely involves an "I" that wanders from one logical form to another. Sometimes it appears that the form of the child's "I" is immature and that the "I" is at one with the horsie. Sometimes the form is like a ":" relationship. It is a clear juxtaposition of the "I" at one category level with a second category at which the "I" is simply equivalent to a "horsie"—on a play level. That level would be "I_H." So the logical form would be

"I" : "I_H."

In this case, the two forms of "I" might be equivalent, except that the subscript of the second marks the "I_H" at a different categorial level than the "I." Then, the logical form is tropelike, and thus we can specify the logical form of the "I" itself. This form can be subject to the function of the "I" (f"I") and therefore, by way of such transformations as are made by negation, can be at various modes of consciousness. A child can be aware (or not aware) that she is exchanging roles and that she is looking at herself in different ways.

However, it is a different task to talk about the occurrence of a figure of speech as it relates to the Other. If I say a city is a gem, that's a metaphor that needs a specification. How is the city a gem? What subset characteristics and features are consonant with the comparison? This specification does have some relation to an "I's" consciousness, but not in the same way as specification of the "I" itself would be. The "I" is recursively related to some representation of itself, but not so related to a thing like a "city."

Think of another geographical example like the metaphoric representation of Italy as a boot. That's a trope form. If you want, there is some relation to the "I" of the person perceiving the trope. Therefore, the trope "Italy : boot," *is* to be understood in the context of an "I." If there were no "I" as agent to interpret and select characteristics of the two terms, the comparison might get bogged down. "It doesn't make sense because I think of boots as walking, and I can't see Italy walking." Unless the "I" steps outside its usual categorizations of objects, a new metaphor can be overly concrete or egocentric—and hence fail as a trope. And so we're back to the fact that this "I" can itself be variously organized logically.

Another way to consider the search for logical form when applied to the Other would be to look at the form of a belief that is some version of homeopathic magic. Frazer describes those who believe there is a "magical efficacy" in the specific qualities of inanimate objects such as *stones* ([1922] 1950:37–38). This sort of belief also involves a logical form based on an equivalence relation. Once a belief like this is held, the "I" too is likely to be swept into the schematizations of the magic. Thus, the "I" within the schematization is experienced as participating in the magic by some doing that facilitates the stone's powers or by some undoing of it. If the "I" is overly concrete or limited to semantic domains and categorizations that are egocentric, the "I" is trapped within a schematization. We all know such persons—they are scripted, not open to any rewrite!

How do we specify the logical structure that expresses the role and nature of self relative not only to its own status as a trope, but also to the trope that the "I" is forming or communicating? The trope that the "I" is forming may well concern an object, event, or relation having little in common with the "I." Yet the "I" is entailed in the production of the trope. Thus, it is one thing to try to specify the "I" of Shakespeare as his mode of experience when writing a figure of speech, such as "Juliet is the sun." It is another to try to express the form of "Juliet is the sun." And it is a third thing to spell out the logical structure of the relationship between the first two.

Let me outline my view of these forms and their relations by summarizing the concepts and logical forms I have advanced. The differences between the logical structures of consciousness and unconsciousness are a function of the "I" categorizations. (I have expressed this idea in the formulae, which obtain with "fI," on the one hand, and with ~["fI"], on the other.) The differences in self-awareness are shown as changes in the logical ordering of the "I" and in the effects on the organization of the self, its subparts, and its experiences. In short, the changes in the orders of the "I" are matters of categorizing the "I" in various degrees of articulated organization relative to hierarchization of order and transferability of categories. In addition, these changes affect the operations and rules of thought, such as the operations of equivalence exchanges and the status of the rules of contradiction and identity. An easy point of reference is Freud's intuitive grasp of the two separate sets of rules for cognitive systems. Psychologist Wilma Bucci writes that

> Freud's focus on unconscious processes is related directly to the nature of psychoanalysis as inherently a dual process theory. The duality of the primary and secondary processes of thought has been considered by many psychoanalytic scholars, as by Freud himself, as his most original and valuable contribution and as central to the psychoanalytic account of the mental apparatus ... Here we focus on Freud's identification of distinct forms of thought rather than [on] their differential access to awareness.... The psychoanalytic observations supporting a dual system model speak directly to current issues within the cognitive science field, providing evidence for dual or multiple processing systems rather than single-code or common-code propositional models. (2000:"Dual Process Theory" section)

I presented the forms formulaically. In a nutshell, the structures of consciousness and unconsciousness are a function of the "I" categorizations and *self-reflexivity relations*. So that we can take on the issues of the "I" and the trope, within an unconscious mode I symbolize the "I" categorizations as "I"$_U$ and within the conscious mode as " I"$_C$. If I assign the role of "functions" to each of these modes—as they are subject to the two "I" categorizations—then each appears respectively as f("I"$_U$) and f("I"$_C$). Fittingly, we can once again turn to dream logic. Bucci points out that "[t]he features of primary process thought are spelled out most elaborately in Freud's concepts of the dream work, the varied mechanisms by which the images of the dream are generated. His identification of the operations of the dream work constitute[s] viable hypotheses, well ahead of their time, concerning the forms and processes of nonverbal or unattended thought" (2000:"Dual Process Theory" section).

In my look at the logic of the dream, I made a two-way assumption. On one hand, the logical structure may be seen in an unconscious mode—as if from "within" the dream itself. On the other hand, the logical structure may be seen retrospectively—as it were, by the "I" looking in at the dream. Depending on which mode is at work, the symbolization of what happens in the dream should be specified. What should appear is the relevant "I" function.

Consider three possible dream forms. For each, I assume that the logical format is "from within" the dream and that the "I" is within the unconscious mode. Hence, the function of the self is marked $f("I"_U)$.

1. *The dream as a part-whole relation.* This form would include metonymies and synecdoches. (For example, the person in the dream is symbolized by one of his features, say, red hair.)

$f("I"_U) [(> A \equiv A) \cdot (> A \supset A)]$.

2. *The dream as metaphor.* Here the example would be a dream in which there is an interchange between one identity and another. The symbolization would be:

$f("I"_U) [(A \equiv B) \cdot (A \gtrless B)]$.

3. *A dream logic for equivalence of class and complement.* Here the example would be some representation in the dream of a thing or person in the terms of its "opposite." The logic would be

$f("I"_U) [(A \equiv \bar{A}) \cdot (A \supset \bar{A})]$.

The Structure of Tropes in Relation to Dreams, the "I," and Time

Is there a way of showing that "daytime thinking" influences the dream form?[4] The proof is in the relation between meanings and themes that you experience either the day before the dream or even much before that day. The problem with the proof is that you generally identify those meanings and themes after you have had the dream! If you dream you are "flying high" in the sky, you may recall after the dream that on the previous day you received a promotion. Lakoff (2001) might argue that a *basic metaphor* in such a dream would be based on the relation "achievement is ascension." To the extent that this "basic metaphor" might explain things, you would have to schematize it. The "ascension" would have to correspond to some image of upward movement or direction. This would be visual or spatial. Then, to fit the "template," the concept of "promotion" would be expressed syntagmatically as an object moving in a particular direction (up!).[5] The Lakoff approach is mired in schematization. His proposal is apparent: *schematization is the analogue of conceptualization*—a rule guiding any transformation of meaning that would transpire in the dream.

However, relative to the "flying high" dream, the "promotion" received the previous day is something that you focus on *after you have had the dream!* It may well be that the "metaphor," too, is constructed after the fact. This would be the case whether identifying a dream comparison, "promotion : flying high," or more abstractly, a guiding "basic" metaphor, "achievement : ascension," as the purported metaphoric structure. I wouldn't doubt that it was a fact that you *were* promoted the day before the dream. But whether that event was the impetus for the "flying high" should not merely be stated as if backwards causation of the after-dream interpretation or recollection.

A simple experiment is to attempt to predetermine the thematic content of a dream. When you are promoted, can you note it and say, "Wow, my feeling about this is so great, I will transform this tonight in a dream with some variety of equivalence between promotion and flying." Such ability to preselect, focus, and predict would be rare—and would be counterindicated by various Freudian principles of repression, unconscious irresolutions and motives, and the like. The main objection is the dual process theory itself. Freud cogently describes the process of *condensation* in the dream work. Many meanings and cross-categorizations appear tectonically within a symbol. The very syncretic and economic way dream categorizations are "condensed" would hardly be subject to the daytime mode of thought—let alone predictable or conceivable within a selected prematch of a specific prototype metaphor to a daytime fact. In fact, in his review, Ellenberger (1970) chronicles *as the rare case* the individual who seems to achieve "control" over and direction of his dreaming. More recently, Barrett (1993) reports some success in subjects asked to employ dreams to incubate solutions to problems. Some do report that "the answer came in a dream." However, we are still left with the confounding factors of recall and interpretation. When and how does the subject "know" that the dream, in its literal iconic mode, is a figurative answer to the problem or to the question the subject asks? How do you "read" a dream's icons as "answers to problems"? It is a little like the way characters in a *Twilight Zone* episode ask a "fortune-telling" device a question, read the device's literal words, and then attribute meaning to the words as answers to their questions.

Even with all these doubts, when there is a "daytime" combination of the dream icons, a series of relationships appears that does appealingly cohere. The problem is not to look at the relationships merely schematically or irrespective of the complex nature of changes in consciousness. These changes affect the structure of icons and the rules of logic that come into play relative to transactions from dream thinking to day thinking and/or from day thinking to dream thinking.

Shifting back and forth from dream thinking to daytime thinking is obviously fraught with influence arrows all over the place at different times, different conscious modes, and different modes of time. If causal arrows move in different directions and continue to move in different conscious modes, how can you structure an explanation of the causal functions of the dream itself and of the person as agent? We have a situation in which different time modes and temporal locations produce causal patterns as multidirectional. Also, if you conceive of an individual's conscious versus unconscious needs and imperatives as *values,* then you have competing vectors as determinants of causal patterns!

Freud (1900), in chapter 6 of his *Interpretation of Dreams,* is very aware of this shifting from dream thinking to daytime thinking. In *On Dreams* ([1901] 1971), he does distinguish between the logical rules of these thought transactions, particularly in terms of the dream logic of connections by contiguity. But Freud does not fully spell out the relation of these rules to the logical organization of awareness. Conundrums of causal relations are marked by multidirectionality and competing vectors. Conundrums of spatial location are marked by questions of *what* influences *what* and just *where* the rules are that appear to inhere *somewhere* relative to the dream. How do you salvage logical structures from these conundrums of causal relations and "spatial location"?

A key to resolving these problems is to realize the need for a synchronic mode within which all time is present, and hence predream, dream, and post-

dream thought is coordinated. For this coordination we need a picture of the various orders of consciousness that descend to an "unconscious sensibility," but also can ascend to the "I's" awareness of itself in relation to the various forms of unconscious and conscious thought and experience.

In order for the dream material to relate to the previous day's experience, rules of identity and contradiction—daytime rules for logical processing of thought—have to be suspended. So if we accept the idea that the "promotion" is an impetus for the "dream," inhibitions about how happy you are to surpass others would have to be unblocked for the dream to be a vehicle to express feelings about promotion. The icon exchanges for the symbols available as dream icons would have to be made on the basis of rules other than those of identity and contradiction. Therefore, although it is the case that you were not actually "flying" and that being promoted is not the same thing as flying, an *equivalence* (as in the metaphor suggested) can nevertheless be made in the dream.

Now, I would like to put things in terms of the key issues of synchronic time, the "I," and its capacity to straddle the day and night logical forms and rules.

What is the form expressing the transaction of icons between day and night rules? Say that the day rule is that of logical identity. The night rule of thumb would be that of equivalence of icons. So the transaction involves a relation between the two rules—namely, suspension (or negation) of the first rule. In a word, the identity rule that functions to govern *day icons* is suspended to permit the equivalence rule for *night icons*. The form of this relation between identity and equivalence appears to be dynamic—something like a metaphor! Consider this relation as a comparison between the day and the night icons. But in following the logic of metaphor, view the icons of day and those of night as members of categories—Day and Night:

Day : Night.

The genus status of each category is exchangeable. This metaphoric relation does not completely account for the way that day and night rules are transacted relative to a dream. The full form necessary to accommodate that transaction is dependent on the NEXT day's aware interpretation. Thus, after the flying dream, you awake and exclaim, "I dreamed that I was flying. I suppose that I didn't want to admit it, but now I feel I can accomplish more than those who were not promoted."

So we have the previous day, the dream, and the NEXT DAY. A prototype form for this relation would be

Previous Day : Night : Next Day.

But this formulation for the equivalence rules that may govern transactions of day icons and night (or dream) icons is very problematic. It can easily be sandbagged by *where* questions: Where in time? Where in phenomenal space? Where do the logical rules inhere? Where does their suspension takes place? And where and when does the motivation to produce the transactions and rule conversions occur? The villain responsible for these annoying "where" questions is easy to spot—overdependence on schematic units. (Freud [(1901) 1971] might describe this overdependence as a carryover of a logic of connections via contiguity. See Jakobson 1956].

Each of the three schematic units—Previous Day, Night, and Next Day—is discrete as a phenomenon, an episode, and as a diachronic unit of time. Among other things, a *diachronic temporal mode* leads to a focus on the causal arrows reasonable to interpolate within and between schematics. Where the arrows indicate cause or some sort of influence producing effects in accounting for—and interpreting—the presentations and representations of a dream, it seems promising to assume causal sequences such as

Previous Day → Night

and

Night → Next Day.

Nonetheless, the causal events and sequences are hard to locate securely because that which would tie them down—the person's awareness of the events, meanings, and connections—is something that is not always present. Consciousness is not always of the same order that would permit interpretation at each discrete moment within the scope of the set of the Previous Day, Night, and Next Day meanings and events. Therefore, the set's formulation (Previous Day : Night : Next Day) is inadequate and incomplete. Something more is needed to display the trope-like structure of the dream. Changing the temporal mode better illuminates the dream's trope formation.

Adequate formation is dependent on the *synchronic temporal mode* (Ts) of an "I" who assumes different orders of consciousness, which span over the time that includes the dream (*d*) as a focal point. I can put this,

$$d < (\text{``I''} \cdot T_s).$$

By this expression, the dream is included in a sweep of the time that the "I" can observe from a synchronic point. The "buck stops" at the "I," where the coordination of the causal arrows and the judgment about what's connected to what is a phenomenal act. However, this expression only fuzzily locates the dream—as categorically included within the ("I" $\cdot$ T_s). It remains for me to show how to express the dream as a particular focal point.

Let us define the terms needed for the picture of the structure I want to draw, and then we'll summarize the formulation.

A. The first term is of the "I" as a function. It functions in its various unconscious to conscious states. I express this function as an ascending order of awareness: "'I' am aware that 'I' am viewing my 'I,' who is interpreting the 'I' in the dream." This sentence expresses various orders of the "I." However, in relation to temporal mode, I express the function of the "I" in relation to a series of states—when there is awareness of a dream, awareness of the "I's" role in the dream, awareness that "I" am thinking about the "I" in the dream, and so on. The "I's" status may be one of relative unconsciousness in relation to the experiencing of a dream, and the degree of consciousness achievable is various. I express this function as

$$f\text{``I''}_{1\ldots n}.$$

The subscripts refer to the various degrees of consciousness.

B. There is a union of this function of the "I" with the function of time—as a synchronic mode. Label this function fTs. The union is expressed

$$f\text{``I''}_{1\ldots n} \cap fT_s.$$

C. The union of the "I" and T functions governs a dream in terms of the relation of the following five variables:

C1. Daytime logic, which utilizes the classical laws of identity and contradiction. This logic is symbolized as D_L.

C2. The array of daytime meanings or themes over a period of diachronic time. In the example, the promotion occurred the day before the dream, but incidents that may be represented in the dream can also refer to events such as competition with work peers, and the competition can have occurred at various—and discrete—times over a course of years. Such an array of predream meanings, I label

Pre $D_{M_{1...n}}$.

C3. The dream itself, although made a focal point by way of its relations with the functions and variables given earlier, is a set of dynamic relations of icons conjointly related to what I label NL, "night" logic. That label is for convenience, but it does not differ from Freud's concept of the logic of the dream as involving the absence of identity and contradiction constraints and the presence of such combinatorial possibilities as "condensation," which can be thought of as involving synecdoche (Jakobson 1956:81). Hence, the attribution of trope forms to the dream is not entirely inaccurate; it simply fails to differentiate these forms from nondream forms. Anyway, I label *the dream* d *in conjunction with night logic* as follows:

fNL · *d*.

C4. Now enter the variable I call the "Next Day," meaning that after the dream there is an aware interpretation. So I refer here to a "postdaytime logic," symbolized as

fPost D_L.

As I see this logic, it maintains the classical rules, but in a dynamic readiness to be applied to an interpretation of the dream's meanings. So it is different from the Pre D logic in that the rule of identity, for example, can coexist with the equivalence attributed in the dream comparison. With the postdaytime logic, the dream juxtaposition of "flying" and "promotion" is seen as figurative, hence the "particular-to-particular" equivalence rule of the dream is then suspended! Emerging fully formed is the need for the "I" and the Ts mode to be present in the formulation of the categorical structure of a trope—as distinct from the structure of the dream per se.

C5. The "daytime interpretation function of the dream" is the final term of my proposed formulation. The trope form results when the "I" can assume different orders of consciousness. From the vantage point of a synchronic temporal mode, the "I's" orders of consciousness extend over time, yet include the dream as a focal point. I call this

$$fPost\ D_I.$$

I add this term to the picture because the rules of interpretation require the "I" to ascend to a consciousness of the "I" and Ts functions and to oversee the presentation of tropes and their relationships. For example, such a $fPost\ D_I$ function can be one that asks, "Is this interpretation—I am so excited by my promotion that it's 'like' flying—consonant with all I see in the dream and related events as conceivable about me?"[6]

Thus, the postdaytime logic function ($fPost\ D_L$) in conjunction with the daytime interpretation of the dream function ($fPost\ D_I$) would be expressed,

$$fPost\ D_L \cap fPost\ D_I.$$

All right, we're ready for the whole formula, which is,

$$[(f"I"_{1\ldots n} \cap fTs)\ (D_L \cdot Pre\ D_{M1\ldots n})\ (fNL \cdot d)\ (fPost\ D_L \cap fPost\ D_I)].$$

A final word in this chapter is necessary to make the concept of synchronic time less mystical. This explanation should be by way of relating the issues of the "I," consciousness vicissitudes, and logical form to this book's major quest for a differentiation of and a comparison of logical forms for tropes, primitive, immature, and pathological thought. We have probed the dream to shed light on these differences. But the dream itself as a language and logic of signs appears both primitive and immature. On the issue of *language*, Freud ([1901] 1971) has made it clear the dream is quite different from the daytime way of expressing icons or representations. Elsewhere he writes:

> The dream-thoughts and the dream-content are presented to us like two versions of the same subject-matter in two different languages. Or, more properly, the dream-content seems like a transcript of the dream-thoughts into another mode of expression,

whose characters and syntactic laws it is our business to discover by comparing the original and the translation. The dream-thoughts are immediately comprehensible, as soon as we have learnt them. The dream-content, on the other hand, is expressed as it were in a pictographic script [Brill says hieroglyphics[7]], the characters of which have to be transposed individually into the language of the dream-thoughts. ([1900:277)

By observing differences in the nature of the signs, we might see that pictographs are more immediately loaded with exposable layers of meaning than are words as we use and view them in everyday communication. Yet all sorts of censoring prevent us from remembering a whole string of dream icons in their syntagmatic architecture. Differences in language may separate the dream from everyday thinking. But, then, "daytime" thinking includes fantasy, daydream, and art. The language route to differentiating dream from day thinking is not as articulated as the difference in psychodynamics, nor is it as basic to the difference in form as the logical structures I described. Look at the psychodynamic differences to see if they help to account for the various instances, which *appear* to show the same logic, but which can be distinguished by the "I" functions and temporal mode.

It is a canard that if you were to think during the "daytime" in the same thought mode as that of the dream, you would be psychotic. This thought mode would cancel negations that block equivalence of opposites and protological equivalence and identity. The logical form of the dream and of pathology can appear to be the same. Freud shows the difference between day and night forms, but those differences are psychodynamic. Dream formulations are "obscure" to us. They have been "distorted" because "at night there also arise in us wishes of which we are ashamed; these we must conceal from ourselves, and they have been consequently repressed, pushed into the unconscious" ([1908] 1959:149).

Tracing our thinking and finding our selves in the transition from the night logic of the dream to day logic is not simply an on-off switching of unconscious mode to conscious mode and of dream logic to day thinking. The "I" as observer can surface at some degrees of consciousness to view the dream. But, also, there are ways that dreams and the logic that appears very like night logic can occupy us during the daytime. And this phenomenon does not sink the "I," who can consciously entertain "phantasies." Look at Freud's statement about "phantasies" as "day-dreams":

"Our dreams at night are nothing else than phantasies...as we can demonstrate from the interpretation of dreams." There is the idea that interpretation is an extra step needed to view the form. Yet Freud capitalizes on the similarities of the forms: "If phantasies become over-luxuriant and over-powerful, the conditions are laid for an onset of neurosis or psychosis" (([1908] 1959:148).

Although recognizing dynamic changes, Freud's observation would tend to leave the forms looking very alike. The problem remains not being able to distinguish the logical differences between dreams, daydreams, and degradations into neurotic and psychotic forms. We can see the difference between the person who believes an equation of terms that are factually not identical and the person who realizes the significance of a comparison by seeing the figurative way the terms or images can be related. A person might believe dream logic without the perspective by which she can compare it to the factual. She might not have a sense of the "I" as arbiter of that judgment. She might not see herself as making a selection of some figurative way to look at the dream. With this person, you have something different from the individual who can make use of tropes. If you follow along with various psychodynamic ways of looking at these differences, you *can* recognize the differences of forms, but they would be a taxonomic distinction. My approach is to look squarely at the different forms in logical terms. Thus, in my discussion so far, *synchronic time, consciousness vicissitudes, and the ascending order of the "I" would appear to be constants in accounting for the differences in form across the different modes of thought.*

Semiosis of Icons in Art: Perspective on the "I," Consciousness Vicissitudes, and the Comparison of Different Logical Forms

I close this chapter with a semiotic analysis that takes us beyond the dream, perhaps to the primitive and the sublime. The perspective I focus on is how the artist, as a viewer and as someone creating icons to present to a viewer, may utilize visual images. Art historian Mark Roskill (1992) presents the concept of *sign formation*. He argues that twentieth-century artists such as Paul Klee and Wassily Kandinsky create "reference-carrying" sign formations. His examples include Klee's *Above the Mountain Peaks* (1917) and Kandinsky's *On Gray* (1919). In Roskill's terms, the visual icons of these works make the art object "independent of the fixed habits of perception" (187).

If you look at these artistic works, you have to scramble those fixed habits to reassert the order of things. In this chapter's terms, we can say the art object is one that has logical rules that depend on the cancellation or suspension of diachronic time. I say this because the very idea of *habit* depends on repetition, which denotes discrete units. These acts can be discrete ways of conceptualizing, ways of seeing, and so on. When icons become "sign formations," they take on the task of referring to all sorts of schematizations. One of Roskill's own associations with this mode of referring is primitive art.

Roskill shows that the "sign formations" of artists can incorporate reference to social and historical contexts and to what he calls "the 'world' outside the picture" (1992:186). But the manifest organization of the artist's icons is quite different from how we organize the facts and experiences of that outside world! I quoted Freud's observation that dreams and their interpretation involve two different language and syntax systems. Interpretation is an issue whether from the point of view of the primitive totemistic constructions of categorial perspective that Lévi-Strauss describes or from the point of view of the present-day artist who utilizes symbols to stimulate the viewers' interpretations. Perhaps what Lévi-Strauss would call "species organization" is one way of conceptualizing a set of rules by which transactions between one language/syntax system and another can take place. According to my hypotheses, the transactions, interpolations, and transformations involve not only different logical forms and their rules, but also the conjoint coordinated functioning of an aware "I" and a synchronic time mode.

Bear with me and relate these ideas back to the problem of coordinating the meanings of the dream by taking into account the Previous Day, the Night—which would be the dream itself—and the Next Day. The idea of the "I" viewing from within a mode of synchronic time is well illustrated by considering the symbolic reductions of the artist's icons. Roskill theorizes that the artist's reductions and simplifications serve a purpose something like condensation of schematizations—three-dimensional perceptions compressed into two dimensions. The syntactical expansions and denouements of living experiences are not presented in verbal and sentential forms. Thus, the schematizations of perceptual, social, cultural, and historical rules, practices, and situations are condensed—"compressed" is perhaps a better way to put this.

All this is a nice segue to "regression in service of the ego"! The artist views these condensations and compressions. So do other persons who examine the art. For these viewers, there is the absence—or suspension—of habit-ruled ways of perceiving. There is the presence of symbolic sign formation. All these things are major ingredients for an iconic frame of mind. The viewer piggy-backs on the artist's "penetration and expansion of consciousness beyond what can be gauged from the surface formations, [and] such practices keep the viewer experientially at a distance; but they imply in so doing a hieratic ordering of sensations and apprehension that correspondingly enhances, by distillation, the sense of the timeless and of the universal" (Roskill 1992:186).

Before going on to the next chapter, contemplate one or more of the paintings by Klee or Kandinsky I cited. Perhaps they'll arouse the sense of the "timeless," and the experience of your "I" in a synchronic time mode will give some life to the formulation

$$[(f``I"_{1\ldots n} \cap fT_s) (D_L \cdot Pre\ D_{M1\ldots n}) (fNL \cdot d) (fPost\ D_L \cap fPost\ D_I)].$$

7 Development in the Logic from Immature to Mature Modes

Primitive and Primitivized Logical Forms, Motivation, and Reasoning

The "excitement of equality" is a key motivation in the vectors and scalars of the dynamic nature of categories and their logical forms. This notion is based in conceptions of unconscious "resonance"—an internal process bringing together a pleasure or sense experience and the reduction of drive. (See Robert S. Woodworth on the ideas of Claparède, Selz, and Duncker [1938:776–79].) Cognitively, this process seeks out and results in the resolution of negations. But if the resolution is overdone, the outcome can be ennui, and the pleasure can simply be "too much." This key motivation and its potential boomerang in the psychology of logical forms are my cues to discuss "Cicero's law."

Sir Ernst Gombrich (2002) relates a wide scope of art to psychological topics of optics, organization, and motivation. He extracts from Cicero's law the idea that *too much fulfillment brings "disgust."* Gombrich interprets "too much fulfillment" as too rich a display of sensually gratifying images not only in the expression and products of art, but also in forms in architecture and in various kinds of expression in oratory. I approach this relation of fulfillment and disgust in different logical variations and contexts, but related mainly to the psychology of forms.

Is a psychological motive like a "drive"? "Drive" has its vicissitudes. Its psychological construct is built on biological observations and phenomena.

Nicolaas Tinbergen's (1951) consummatory response idea is of drive satiation along with some period of refraction. In psychological theorist Clark Hull's (1943) work, refraction, after drive satisfaction, entails an inhibition of further activity associated with the drive. The scientific observations that led to these versions of drive are in a lineage with Darwin's observation of a basic opposition of action and withdrawal of action inherent in the nature of the organism's nervous system. Although all of this is theoretically biologically based, human motivation vis-à-vis drive spills into psychology. I cite two pertinent major views.

First, Freud's psychodynamic view utilizes the tendency to an opposite reaction in various psychological mechanisms. Examples are repression, reaction formation, the opposing phases of conformity and acting out, and the back-and-forth exchanges of negation and its cancellation in "doing" and "undoing" acts. The extreme of sensory stimulation—beyond a satiation point—is basic. It occurs in traumatic exposures and exposures too complex for an immature psyche to organize and process. Gombrich's term *disgust* is almost overly commonplace. It does not sufficiently accommodate a range of reactions, such as contempt, moral opposition, and sublimation. The psychological character of the moralist implies more than his putting a clamp on his tendency to excesses of sense exposure. To say that the missionary in W. Somerset Maugham's story "Rain" (1921) has a moral code based on his revulsion for his own sensuality insufficiently explains the code's power and social value. Freud's model is more encompassing. It specifies conscious and unconscious modes in relation to negation, and it describes psychological mechanisms that turn interest, ideas, and feelings in opposite directions.

Second, when human "motives" were tied to comparative psychological investigation of animals—"rat psychology"—and to drive reduction as a desideratum, an appropriate reaction in the 1950s was to look at the alternate possibility: "human motives" are *not* aimed at some point of satiety; they exist "for their own sake" and are free of drive reduction. They include self-fulfillment (Abraham Maslow) and the hunt for novelty (Daniel Berlyne). Curiosity would be *a drive independent of the rules of drive reduction.* When a new novelty is presented, satiation does not occur; you continue being curious. Although this explanation is superficially sensible, novelty almost entails its own opposition. Once achieved, it's no longer novel. Perhaps this is placing the cart before the horse! The main—and prior—point is that *searching for novelty may be built on satiety with the non-novel.*

Focus on the artist and art should help reveal how factors of satiety and the novel affect tropes in light of their semiotic context and hence iconic nature. An artist's search for a new form can lead toward complexity and its organization. (Penrose 1981; Picasso [1957] 2005). Picasso repeated certain themes executed by other artists, but he made their perspective points more complex. Aware of Velásquez's theme in *Las Meninas,* he continued searching for complexity in the theme of the perspectives of the viewer, the artist, and the viewed. Thus, "On the 17th August 1957, Picasso shut himself up in his villa for almost four months to fight with Velasquez; this was known as the 'battle of the Meninas.' Essentially, Picasso completely stripped Velasquez's great baroque masterpiece, *Las Meninas* (1650) completely down and turned it into studies" (Packwood 2002).

Even characteristics of complexity can become non-novel. Then the new search in the next aestheticism in favor can be—as they say in Hollywood—for the "dumb and dumber"! When artists solve the problems of representation they decide to face, we have analogues of disgust and satiety. Disgust may be seen as overexposure to stimuli that have already solved the problems at hand. It acts like a drive; to reduce it is to shun the resolutions that have worked—resulting in a state of refraction. But refraction from resolution is not peace—it's an armistice characterized by simmering countervailing dynamics (see Rookmaaker 1994) .

Too many artists present installations with many discrete elements that are hard to rectify. The consumer then develops a framework within which she can view such installations as allegorical. The shock of many uncategorized and contradictory representations gives way to a tolerance of multiple stimulations and conceptual levels of stimulation. The consumer of the "overstimulating" art is numbed into an as-if state of comprehension of complex stimulation. Disgust emerges. The feeling of being exhausted sets in—along with a lurking suspicion of missing some unsolved issues. Inertia and restlessness coexist. Present accommodations and resolutions of perception and conceptualization are swept into a destructive vortex as there is a search for the unresolved, the lurking, and, to some extent, the forbidden problems. Rejecting mechanical resolution as alienation can be too great a cost.

I think Gombrich *does* have a point. Based on his interpretation of the changes in the historical development of art (2002:27), one can derive a psychological extension of Cicero's law: w*ith too much mastery over a way of expressing oneself or in experiencing what another person has to offer as a set of solutions, there is motivation to primitivize.*

Cicero's theory is a psychological observation relative to oratory, art, and architecture. However, by way of applying his psychological principle to the artist's and the art consumer's reactions to pictures and forms, we should be ready to go further into the trope and the logical form as icons.

For a trope, form can be primitivized, whereas sense references can be optimized. The figurative phrase "The idea hit me like a ton of bricks" is more innervating than "I was excited by the idea." The trope is intended to be figurative. If the conscious perspective is there for its interpretation, the meaning is viewed at a level not concrete. Although the form of the trope is protological, the specifics reference the senses "with a punch"! So when a trope is used, one primitivizes the logical form, yet oversatiates the need for sense excitement. This paradox appears self-defeating and circular. It would be nice to think the resolution is in the difference between a higher consciousness of meaning and a pathologically impoverished one. In fact, this paradox may be present only if the literal interpretation instead of the figurative applies semantically—as in the case of a pathological interpretation.

To say Juliet is a source of warmth is close to boring. To say that she is "the Sun," however, introduces a deep problem, perhaps partially because the logic of equivalence is more primitively presented in this trope than in ordinary speech. The order of categories is—at least temporarily—primitivized. A regression to the primitive can open the possible routes to the novel. Just as the poet's look at his own daydreaming can become a literary presentation (Kennard 1997; Luther 1997), the protological form in the eye of the viewer (the "I") can become a trope. Both are cases of regression in service of the ego. In each, the direction of the drive is to get away from the sense of satiety and the balance occurring when there's no apparent problem to solve.

If the poet decides to declare how special his hero's love is for Juliet, he *might* do so without the extravagant comparison of Juliet and the Sun. However, the overtones provided by breaking out of the constraints of "day logic" and prosaic containment in constrained and articulated categories provide the listener with a sense that the drama unfolding has asserted a love exceeding bounds and that a constraining comeuppance is at hand. Both anticipations move away from satiety. We are dealing with a drive not quashed by satiety, not calm in the face of resolution, and at a point of the "ultraparadoxical," when there is an excess of satiation or resolution. ("Ultraparadoxical" is Pavlov's [(1927) 2001] categorization for behavior likely to go in a direction opposite to the order of stimulation.)

Take the case of a pre-Renaissance artist faced with the contradictions involved in translating a three-dimensional percept into an icon in two dimensions. No satiation there! But the available resolution by way of a given form—say, location in the upper portion of a canvas to represent past locations, distant facts, and distant sights—may become an adequate schematic to transact meanings from one viewer (the artist) to another (the art consumer). In the 1600s, Velásquez might have enjoyed the accommodations of two-point perspective—at least for a time. Yet the adequacy of the schematic produces satiety, disgust sets in, and the viewer-artist begins to fashion other forms and perspectives.

One can make this very point relative to the Velásquez perspective in *Las Meninas*. The artist created a problem in a "viewer's form" of his own identity. The form (defined by angles of projection from the viewer to the objects in the picture) could have been much more easily achieved by not interfering with the known ways of relating the frame to the viewer's perspective. However, Velásquez, if sated with his knowledge that the viewer could identify where he was and where the king and queen in the picture (KQp) would be, had to change the "form" of the "I" of a viewer viewing from a specific point of perspective.

Calling that point of perspective "angle a," I refer to the "I" at that point as exercising a viewing function, $f"I"a$. So the change Velásquez made was from

$$f"I"_a (KQ_p)$$

to

$$(f"I" > fKQ)_a (KQ_p).$$

This last formulation reads, "The 'I' as viewer includes the 'I's' view of the king and queen's view at 'angle a.'"

So we conjecture about the role of satiety and the issues of imbalance: *when protological forms are introduced, more specific meanings can be accommodated in more complex forms. The classificatory inclusion incurs a state of satiety, but the logical order is in imbalance. Through the awareness of the "I," artistic goals can be served, but without this awareness the satiety/imbalance state signals a pathological state.*

The Place of Self in a Metalogic of the
Motivation to Resolve Negations

"All roads lead to Rome" is a fertile metonymy. Rome had a sociopoliti-cal structure that was constitutive. But the idea of "roads" as symbolic lends itself to a wide array of analogies. It is not just the architects of the Roman Empire who tried to build into their articulations and organiza-tion of parts (the Many) the resolution and reevocation of the One. In the Roman Empire trope wherein analogies extend the figurative meanings, "roads" become the "rules" of inertia. They hold a whole structure togeth-er and permit a variety of movements maintaining its form. Take apart any subdomain rule, and the basic premise reappears. The rules presume to accommodate the realization of constitutive hierarchical structure—in any case, of the finite vicissitudes, identified as moving in the direction of perfection of the form.

Aristotle based his conception of movement toward the potential for a final form on the biological study of life forms, and his basic metaphor for the understanding of form was "organic" (Gombrich 2002). However, the organic transforms to a "machine" metaphor when specified in terms of rules, premises, and a constitutive logical order. Lewis Mumford ([1964] 1970) describes the special circumstances within which this transforma-tion takes place. Persons and societies fall in love with their machines and the perfectibility delusion; everything in sight is swept into a "megama-chine," including the machines' acts of origination. Here's an analogy, spun off the "roads" symbol:

"Roads" *are to* cognitive rules and paths *as a golem* is to a person![1]

Whether superordinate or subordinate to this "roads" metaphor, the machine as individual human nature is within the metaphor's com-pass—a gigantic metaphor reminiscent of Mumford's "megamachine." The search for perfectibility of form can be a powerful force in reinforc-ing and expanding our beliefs about our nature and purposes—writ large as a state's and small as an individual's. How far can the metaphor go if its premises, rules, and various constructions—roads, routines, or bounds—shape our beliefs?

This machine metaphor can mean that social and cultural beliefs are at the center of a wheel moving societal purposes in their predicated di-rections. It not only can refer to the beliefs that bring about the history of the construction and destruction of social structure, but also can move "inward" and be an account of individual agency. (Although the "machine" is not always explicit, deterministic effects of social and cultural beliefs are present in postmodernist accounts [Harré 1984, 1991; Taylor 1989]. More phenomenology oriented, Paul Kugler [1993] accounts for the flow of cultural trends in determining the individual's ideas about her self and agency.)

The beliefs would have to be reinforced by all sorts of problem solving—not merely in engineering, both social and architectural, but also in basic conceptions of human nature and purpose. Roads are built and laws are made so well coordinated with rules of reasoning, evidence, and dialectical categorizations that internal dangers to the state and its individuals are solved, offset, or hedged. "Golden Age" phenomena come about at historical, political, and cultural junctures, when, as Alexander Pope intoned, "Whatever is, is right." Such junctures include the Ptolemaic cosmological beliefs, the Age of Rationalism, and, in hindsight, the recent Cold War period—especially with its political philosophy of mutual military deterrence. I give these examples to make the point that *within a system of premises and beliefs*, external dangers are seen to be held at bay. That's the upside!

The downside: if you believe that your accepted world model "works" and "is right," then an iron curtain shuts out visualization of unresolved negations. If the Ptolemaic conception of the universe cannot be wrong, then the search for negations proceeds inward—by inquisitorial methods of crushing any internal contradictions. I relate the inquisitorial efforts and effects to the beliefs of and by the state (a unit smaller than the universe) and of and by the individual (a unit smaller than the state). Within the inner drama of models and beliefs, the search for negation moves toward the inner circles and deeper into the individual mind. How far can an inquisitorial direction and its effects go in reaching to the inner core of beliefs and their logical forms?

Galileo veered from the perfect enclosed world of Ptolemy—as if he went off the cliff of the known universe. Although he publicly recanted, he apparently did retain his own beliefs. "Inquisitorial effects" seem to have stopped at a barrier to inner circles of thought—the individual's beliefs about the parameters of his own beliefs. Some individuals go further than Galileo. In recanting or adopting a "party line," they may accept officially held beliefs, their own beliefs about these official beliefs, and as well beliefs they hold about themselves. Either the barrier to "inquisitorial effects" is penetrated, or it did not exist in the first place. Where the barrier *does* work, as apparently was the case for Galileo, once the "beliefs about belief" allow inward belief and outward expression to bifurcate, we have fragmentation. The effects of this fragmentation on the inner needs for rectification, wholeness of a categorical structure, and the resolution of the One and the Many ultimately diminish the differentiation of logical structures.

In either case—the person who totally substitutes official beliefs for his own or the person who only publicly does so—the individual accepts the metaphor of machine-inspired regulation of thought. That metaphor has the form, "I" : Machine. The megamachine of state or official rules moves into inner belief and thought. To consider how this movement crunches thought and the self, I propose this dictum: *the machine metaphor is doomed to internal entropy.* Entropy, which implies natural forces, is a devastating concept. I should explain why I invoke it.

When looking at the machine regularities of thought, I make an assumption not too far from Aristotle's view. Regularities are not merely the course of cognition, but also the biological context of organismic dynamics and forms. Therefore, *any search for perfectibility has a disclaimer in the organism's countervailing internal pressures to realize the presence of imperfections, to grapple with the negations of order, structure, and being.*

Negations of order include the effects of contraries—immaturity and epigenesis of the logical capacities of mature thinking. Negations of structure include aging and negations of being. These negations creep into the texture of thought and memory. They affect the organism on many levels—namely, death of cells, functions, and ultimately the organism itself. To sketch the downside dictum *(the machine metaphor is doomed to an internal entropy),* keep in mind the distinction between the belief in the machine metaphor and the distance between it as a product and the natural and biological forces of its origination.

The Outer Reaches of Rules and the Empire of Resolution

For the state and the empire, we seek routes leading to the solutions of all its problems, yet the more these routes lead to their destinations, the more they correspondingly produce a search for potential roadblocks. An example of a solution is dependency on a constitution or fundamental body of laws. The constitution's rules seem designed to cover all sorts of contingencies. Yet the search for opposition and contradiction goes on, particularly when we think we have solutions! *The search is more frenetic when one believes deeply that the solutions are all at hand.* An odd case is the so-called strict constructionism with which some believe that the U.S. Constitution in its original form and with its original intent held intact can solve all legal problems. (I won't go into historicism arguments. Anthropologist Emiko Ohnuki-Tierney [1991:183] provides a nice footnote on this topic.) Strict orthodoxies too often leave all sorts of contradictions outside of accessible range. This applies to the strict-constructionist view of the Constitution. The idea behind "strict" is not to go beyond original intention. To maintain the original intent, the scope of application must be kept narrow. But the many additional meanings that accrue can easily lead to exceptions to the rule and thus subvert it. In regard to freedom of speech, issues of modes, venues, and range of outcomes expand implicatures over time, which then overflow the bounds of strict original definitions.

Strict-construction believers entrench themselves in more and more streamlined views, which turn out to be "philosophies of everything," an effect having its own desperate side. When biological research advances points of information and understanding of the function and structure of stem cells and the genome, definitions of life and human rights can either expand and change or seek to adhere to meanings and contexts at hand when the Constitution was written. Conservatives and strict constructionists work to toe the mark. Convictions to keep within bounds become more than resolute; *desperate* is not a bad choice of word, considering the emotions expressed by conservative judges. However, this form of reaction is contained within an increasing dependency on rules and codes. There are radical groups who in the name of moral codes enter into lunatic acts that include violence. I regard this fealty to the strict order as a desperate solution to all problems.

Some persons do not support the strict readings and do want to expand the rules to apply to new values and changes over time, cultural text, and social mores. Searching for ways to go beyond the socially organized but tight set of rules occurs in a *frenetic mode.* That's a mode of moving ahead to interpretations of meaning that take into account present-day circumstances and meanings in disregard of gaps in language and wording. The mode is frenetic because the gaps open wide-ranging implications. For example, no one really knows where or how far stem cell research and genome specifications will lead. They are so fundamental to basic concepts that they threaten changes that may upset whole zeitgeists. *The frenetic mode results from the impossible search for the resolution of negations that should not even exist within a strict model.* Freedom of speech should not involve an issue of excess, yet the contemporary access to other persons' expressions through mass media and the computer create unanticipated downsides to any absolute premises. If you want to be "strict" about following the letter of the original tenets, excesses are hard to recognize. Where freedom of speech infringes on other rights is difficult to say if you put "'other' rights" into a category in opposition to it. The original intent can be subverted. There are downside effects of such reclassifications, which can be made worse when subject to rationalizations. When "due process" rights are suspended by the PATRIOT Act, the changes not only subvert the original intent and expression of rights, but also the "strict" interpretation! Filtering exigencies from rationalizations suspends both the strict and the "liberalized" views, and the "machine" grinds with friction.

The machine metaphor's schematic ("All roads lead to Rome!") has extended too far; all alternate routes to external problems are blocked. The overblown conception does not recognize *an* other—*some* other—domain as a source of negations to resolve.

I propose to state these battles with machine organization in terms of the dynamics of logical form. *To block a process of creating new subclasses of opposition is to prevent information and meaning from entry into a complementary category!*

I am still steering the machine metaphor, wending my way to Cicero's law via Gombrich's interpretation. My point about the "empire" concept is not only meant as a description of a cultural or sociopolitical state of affairs, but also aimed at the individual's interpretation of these states of affairs and how constraints filter into the individual's thinking. As far as rules, laws, and other sociopolitical structures are concerned, the excessive success of the empire produces disgust, and internal paralysis and destruction become inevitable directions. All roads lead there!

A Psychodynamic Consideration

What analogous principle relates to the individual? When the machine metaphor expands to the megamachine, whatever self and agency the individual has is part and parcel dedicated within the machine's rules and routes. To the extent that the individual has no self separate from the machine domain, everything works! Any question within this domain is answered by following rules within the closed system of premises. What about the sense of self as the experience of one's unique identity? Here are my basic assumptions about that which characterizes this sense of the self's experience of identity and agency:

> The self and the "I" interact with a conscious and an unconscious mode.
> These modes impact the degree to which the individual is aware of and can exercise her own agency.
> The subcategories of the self and the "I" subsume all sorts of cognitive and affective experiences.

To identify an analogue to a machine constraint on the self, let us take the example of the way shame affects the self, agency, and the individual's presentation of self.

Shame is a major affective influence on the access to usually unconscious meanings, the consciousness of the self as an agent, and the self's capacity to allow and exercise its agency. Shame makes it difficult to recall dreams and makes for the elaborate aesthetic compensations with which the artist and writer imbue dreams, daydreams, and fantasies (Freud [1908] 1959). Shame dynamically acts as a drive to negate the negations of the self that the individual wants to present to others and to represent to the "I." So shame relates to its negation through mechanization of hierarchically proficient organization—and to the point of disgust that Gombrich describes.

You can hold your "self" in check if you believe you have a set of codes and rules that cover all contingencies. If shame is submerged simply because it is not necessary to encounter it, you can go on producing work and other effective means of adapting and contributing even when the self is out of the picture. However, with the feeling of "disgust" over the degree of non-novelty and the absence of any identification of problems inciting new resolutions and ideas, the situation calls for creativity and the invention of new forms to block entropy. If the shame feelings are still present along with new awareness of the self as an agent for the creativity, then two unattractive alternatives loom.

First, the shame can be too great to proceed. I wonder if something like that occurred when "Brahms did not finish his first symphony until he was forty-three. The delay is [*sic*] not due to a lesser degree of talent. Rather, Brahms put off the task because he was intimidated by the example of a previous composer" (Glesner [1996] 2000). Extremes are the result if the shame is so great that a frenetic need to expose negations is part of the "disgust drive." I cannot resist saying that I fear this happens when terrorists and their mentors decide on suicidal courses.

Second, the individual's projections invent new forms, which neither save the self nor require the self to emerge to consciousness. These new forms can be stereotyped categorizations, and they are directed at or attributed to others—for better or worse.

All this constraint and isolation of the self from its thinking involves forms quite similar to those of primitive thinking and the trope. However, to bring the relations of the different types of thinking more into focus, we should move from psychodynamic considerations back to the linguistic and logical issues that can more definitively differentiate their forms.

Basic Meaning as Grammatical Constraints and the Mechanization Effects

How does Cicero's law work when linguistic and logical forms become overefficient? An example of this outcome is the language acquisition device (LAD). Noam Chomsky's (1972) concept holds that inherent in a person's capabilities are sets of syntactic rules governing the learning of a language. Extend this concept to fit a full-fledged theory. Picture a general set of rules for the layout of any cognitive product or expression. The rules wed syntactic place, role, and any dimensionality of syntagmatic schematization to basic meanings. This sort of theory is not new (see Chafe [1970] 1973, for example).[2]

Consider the general set of rules this way: all meanings must be within the constraints and guidelines of the "basic meanings" as encoded in their rule-governed and determined grammatical formats. A basic meaning like that of "existence" is welded to copulative verbs such as *to be*. In the sentence, "I am that fellow," the grammatical format unfolds how the basic meaning plays out when the verb is articulated. The syntactic pattern dictates subject/object relations. Also, with this verb denoting "to be," what's on its one side has a form that looks like a logical identity with what's on its other! If the sentence were, "I am angry," you would have only a "partial" identity. Yet the implication is that "angry" defines the "I" in this limited case. And if the items belonging in a list on that side of the verb were added up, we would have a stable form for identity with that "I." "I" on one side and the list on the other make an identity something like:

$$\text{"I"} = \text{Angry}_{ss,\,sa} \cdot \text{Sad}_{ss,\,sa} \cdot \text{Communicative}_{ss,\,sa} \cdot \text{Empathetic}_{ss,\,sa} \cdot \text{Selfish}_{ss,\,sa},\, \text{etc.... } n$$

Each item in the list has subscripts for specific situations (*ss*) and degrees of strength of the particular attribute (*sa*). Each item is interactive, which I indicate by the "·" symbol. In order for the list to equate to an identity with the "I," it is necessary to add "etc.... *n*" to the items because of factors like the "I's" different levels of reflectively aware functions.

So with a rule of identity in our sights in this tight rule-guided universe of meaning and grammatical form, we go one more step—to logical trans-actions. They, too, are part of the constraint system. With a logical rule of identity, syntagmatic forms are not the only grammatical constructions affected by the combination of syntax and meaning. Paradigmatic aspects of meaning are determined by categorical logic.[3] Whatever belongs to the meaning of the subject is bound into its categorical form—which also demands a complementary category. For the "I," this complementary cat-egory is the "not-I."

The categorical rules and bounds—and their effects on the definitions of terms—are organismically programmed to conform to logical forms. So the subject of a verb can have a set of connotations that are held in check by its logical denotation, constrained by its syntactic place and role. All this machinery, setting the bounds for the definitions of objects and the rules for their transaction, applies to any meaningful category—in-cluding the logical and psychological rules of identity—for the category of self.

This modest extension of LAD is the basis for assuming purposes, beliefs, and domains of meaning (including the meaning of "self") to be part and par-cel of the governance and constraints of a closed mechanical system of premises. Any intention or perceived question (or problem) is solvable according to the rules of the closed system. The procedure affecting perception and subsequent schematizations of agent, action, and outcomes would solve problems as they are predicated. Feedback loops may be required to reset qualities of the identity or definition of terms and hence the nature of the schematization. However, these variations are in service of the basic meanings and their grammatical constraints. Within these constraints, purpose is achieved, as set forth in the major premises for being and function. All the interrelations of being and function—part-by-part *and* whole-as-machine—are reflected in the rules of governance, production, and expression.

If the machine is a "closed system," any query leading to negations should find none. I do not refer to the neat classification of terms and definitions that allows the negations in the form of complementary categories. Nor do I refer to the person's capacity for trivial negations that allow feedback within the defined terms and purposes, and that call for modifications of the schematized action patterns. For the individual's "unresolved negations," as in the proposition "I am angry and loveable," there are available the tolerable resolutions by "specificity of situation" and "degrees of strength," which I indicated by the subscripts ss and sa. There are also "mechanistic resolutions." A great example of these resolutions is Freud's concept of "defense mechanisms" as a way of resolving the coexistence of apparent opposites in a statement of an individual's psychological and logical identity: "I am angry and loveable. I handle the antisocial aspect of my personality by the development of endearing characteristics and habits. People then accept me."

With all such built-in routines to knock out instability and utilize negation to achieve resolutions, should our "closed system" be happy to find nothing left to negate? Even with defense mechanisms or with other sorts of logical rectification and attendant wondrous systems of constraints, however, the individual still reaches a point of "disgust." There is a need for primitivizing as the only way of relieving the ennui brought about by the absence of negations to resolve. Maybe ennui masks the irresolvable anxiety of unuttered feelings or suspicions that something may go wrong. This masking may be why a person's anger or sadness may not dissipate even though Bateson's idea of jumping out of the conflict box to "the next higher class" of logical formulation of a problem leaves the person feeling "OK."

I realize that logical forms are *not* machines! They are natural forms, and they take into account transmutations and the vagaries of awareness, motivation and negation resolution. Consider these transmutations and vagaries the terms of my proposal of logical distinctions for the forms of tropes and pathology (primitivations).

Some (e.g., Kugler 1993) easily agree that the person is a subject (who experiences) and also an object (a definable action, structure, or function). I wrote a book about this (Fisher 2001) arguing that William James mistakenly set the subjective self aside from "natural psychology" (also see Fisher 2003). With some support in Husserl's descriptions, the view of a subject-object binary is Kantian in nature. I propose that, logically, *what the person does and produces is in a category of the person as object. The self (as the experiencing "I") cannot be within that category. The "I" is not an object; it is an experiencing of the object.* The "experiencing of an object" neither means that the experience is not fuzzy nor implies that the experience is codifiable. In Peirce's terms, the person can be aware of a representation of an object, yet not be aware of the sign signifying that representation. I may know I like the taste of a spiced dish, but have no idea what the spices are—or that they are ingredients. The categorical relations of objects and our knowledge of the objects and their signs can be semiotically specified in terms of the complex triad relation of *object-sign-interpretant.* For semiotician Lucia Santaella-Braga, the interpretant is an effect. She therefore (2003) appeals to keep the triad as object-sign and sign-sign relations separate from the functions and issues of mind. But I nevertheless argue that the unconscious and conscious modes (the aware "I" and the nonaware sentience of the self) interact with the way the individual views and can utilize logical forms. Within a logical format, *the aware "I" of the self relates to the nonaware sensing, perception, and thought of the nonself as category and complementary category.*

Resonating Ideas in Freudian Descents into Dreams and in Mechanisms of Defense

Rupprecht (1999) describes the awkward reaching from the aware person to her dream as a brave attempt at an intersemiotic translation of two different iconic systems. This dialectical antagonism of the "aware "I" and the "nonself" rings a Freudian bell. The two states of awareness allow you to enter and return from an exploration of a "dream world." If the aware "I" is the more superordinating category, it includes itself and its complement (the nonself). In terms of classification, these two become subcategories, and the aware "I" is more the genus. As a genus, it is then more abstract, but you can expand its denotation with the species specifics of the subcategories. The logical form looks like the experience when we "visit" a dream, or rather, as I phrased it, "return from an exploration." The return is to the superordinating "aware 'I.'"

Freud, in his groundbreaking 1905 book on dreams, set the stage for unearthing the logical forms and structures of this "netherworld." The problem of access to the inaccessible—the access to what one was not conscious of—can be solved by the idea of the "subconscious" (Ellenberger 1970). The recall and analysis of the dream, rather than the dream itself, answers the question of access. What may become accessible are the aware "I's" understandings not only of the past and its paths to reforming ideas, categories, and relationships, but also of the logical forms needed for these changes.

Interaction with the Peircean triad comes into play here. To accommodate "the aware 'I's'" understanding, a logical order for representations requires sign-for-sign structuring—with a referential descending ladder that can reach the individual's "sense" of the objects of thought.

Braga (2003) interprets Peirce's semiotic view as "sufficient unto itself." As a biological foundation, the processes of representation—forming the sign, using it, referring to it, and so on—are embodied. Nevertheless, consciousness, mind, and agency are subsumed within the functions and structures of semiotic structures, their functions, and their relations. Although this phenomenological interpretation organizes the biological individual in meaningful terms, the absence of the self as an outside factor makes it difficult to see agency as other than predetermined—even in its purported role of "determining." Ergo: a mechanized version of the individual. At the sound of a Freudian bell announcing overexpansion of machinelike controls, we enter the internal tendency to primitivize—a netherworld of dream and mythlike thought.

According to Gombrich (2002), the ancients saw the perfection of art and architecture as a highpoint from which it would be a relief to "return" to a simpler set of forms. In our own day, we apprehend overcontrol by the computer. Perhaps M. C. Escher foresaw the degree to which this overcontrol might occur when he pictured evolution of life by the tessellation of forms (*Liberation*, 1955). But his pictures of impossible staircases and his ridicule of the robotic figures negotiating them (*Ascending and Descending*, 1960) are a turn to the absurd and contradictory as a doorway to the primitive. Although artists still attempt to use computer art and to imitate computational effects in art (Bridget Murphy, for example), there remains a continuous need to invent more and more outrageous forms of art. This claim is easy to document by simply visiting installations at museums such as the Massachusetts Museum of Contemporary Art or the Whitney. An example at the Whitney is Jonathan Borofsky's *All Is One* (1984). It is not difficult to find examples of these outrageous forms and of the dreamlike organization of representations, but we should be careful. The relation of regularization and disgust can be turned around; interpretations can easily be circular.

Circularity is easily shown by the flap about an artist's use of elephant dung to make his "point." As reported by ABC News, "An elephant dung–embellished painting of the Virgin Mary by British artist Chris Ofili is one of the works of art in the 'Sensation' art show that aroused the ire of then New York Mayor Rudolph Giuliani." Mayor Giuliani's outrage engaged a large cast of characters. His response turned the circle of reaction a notch in the opposite direction: "A federal judge ordered the city today to restore millions of dollars in funding to the Brooklyn Museum of Art in a dispute over the controversial 'Sensation' exhibit" ("NYC Ordered" 1999).

More enraging forms of art can lead to moralism; it is a hop, skip, and jump to the mechanical controls of selection and outcries for moral censorship in place of freedom of expression and its excesses. In a democratic setting, other voices join in, and the results can be offset or moderated. What happens when there is a totalitarian context? We cannot spend time here examining the fertile ground onto which the Nazi claims of "degenerate art" fell.

With circularity, there is yet another resonance with Freudian themes. Here is another instance of overregularization, the subsequent tendency to primitivize, and the resultant arrant pleasure seeking and indulgence of the senses. The 1950s were too formulaic in the United States. The popular mind turned away from "Dr. Marcus Wellby" to the likes of Abby Hoffman and the excesses of the 1960s. The movie *Pleasantville* (G. Ross 1998) neatly presents a theme of this sort. A young wife seeks to break out of the "colorless," predictable resolutions of social roles and return to the "primitive"—the sensual and sensational. However, the breakthrough can be too much. In the movie, you see the "town's" predictable reactionary stance: too much color! We are at a second point of oversatiety.

Return to Gombrich's formulation. With too much fulfillment of pleasure seeking and too much excitation by its outright expression, "disgust" appears. *In sum, the relation of primitivity and control is two sided—each side signals disgust as an impetus to transformation.* I have described overcontrol as a machinelike regularization. For Freud, this kind of control and its sequelae are functions of the defensive mechanisms that an individual employs. But the highly defended individual is simply holding the "lid" down and not really dealing with the dynamic needs of which she is unaware. This defense culminates in a need to express the unexpressed. We get a "return of the repressed." In a sense, the famous outcry in the Paddy Chayefsky movie *Network* (1976), "I'm as mad as hell, and I'm not going to take it anymore!" combines defensive control and disgust. Together, they trigger primitivations, which in *Network* result in the form of uncensored, unabashed appeals to the senses. However, the circularity again comes into play. The defended individual's "control" produces more and more need for the eventual "return of the repressed." But preceding that return, which you can picture like a color explosion painted by the Fauve artist André Derain, is a sense of unresolved needs. This sense is a premonition, as in Edvard Munch's painting *The Scream* ([1893] 1998). In the line of the circle of return, the sense of negations unexpressed leads to a "disgust" with the existing regulations. Call it "disgust 1."

Once we move to the next phase, the "return of the repressed" is upon us. Picture Picasso's *Guernica* (1937), wherein excesses of primitivation occur symbolically. But the drama of primitivation is real. Hannah Arrendt (1951) points out the absurd Nazi sociopolitical categorization of the "innocent" as "guilty." Whether the primitivation is by icons in art or by inchoate social happenings, the ensuing "disgust" evokes a phase characterized by regularization. When there is the full flush of primitivation—or the "return of the repressed"—disgust appears once more. Call it "disgust 2." Its upshot is the reintroduction of defense mechanisms like "reaction formation." Post–World War II Germany is devoted to democracy, but with ironic restrictions of freedom so that every one can be "equal." With such examples, we can show that "disgust 2" comes with an *overcontrol*—defense mechanisms that are tighter than ever. The cycle can proceed just so long; primitivations get more primitive and controls tighter, and then there's a reaction to overcontrol. Call that "disgust 3." The fate of this cycle is not very pretty.

The wheel continues to turn. The overcontrol itself is met by disgust ("disgust 3"). That is a severe danger point. What is left after both pleasure and its control are repudiated is a regression to disorganization.

An art installation fits the danger point and illustrates the descent to disorganization. The possible regularizations afforded by computer art may give rise to disgust and then to primitivation. Jon Thomson and Alison Craighead's (1998) installation *Trigger Happy* appears inspired by the phenomenon of computer games. The artists provide rules, but often these rules become mindless ways of shaping cognitive activity, such as strategizing, calculating wins and losses, and making decisions. However, the *Trigger Happy* artists react to the regularization by a deep assault on the effects of a "game." The game's predictable routines for automatizing the channels for decision lead to a deconstruction, which culminates in disorganization—and even disintegration. The Massachusetts Museum of Contemporary Art description of the installation is striking: "instead of shooting UFOs, the player must destroy a descending paragraph excerpted from Michel Foucault's essay, 'What Is an Author?' In destroying the passage word by word the player metaphorically deconstructs Foucault's text which itself deconstructs the idea of the author. After shooting a few words, a 'Yahoo!' search page appears on the screen with results defined by the eradicated words. This forces the player back into the Internet which is a platform for the death of the author on a daily basis" (Tribe 2001).

Circularity and the Search for Unresolved Negations in Psyche and Art

Inviting parallels exist between defense mechanisms and regularization, on one hand, and "the return of the repressed" and primitivation, on the other. Even so, the analogy between Freud's ideas of control and retrogression, on one hand, and Cicero's law of disgust and primitivation, on the other, is not as easy as it might first appear. Cause-and-effect considerations appear to be not only a product of two points of perspective on the Freudian psychology of dynamics, but also a product of two points of perspective from which to see art and reactions to art, when considering Gombrich's historical analysis. I summarize the cause-and-effect patterns and their seemingly opposite directions in the following lists. The oppositions in each set are parallel. The parallels are sequence 1 to 3 and 2 to 4.

THE TWO GOMBRICH/CICERO SEQUENCES

1. Too rich sensual forms → Disgust
2. Mastery and complexity → Disgust

THE TWO FREUDIAN SEQUENCES

3. Excess pleasure and impulse behavior → "Shame" and moral outrage (analogues of "disgust")
4. Excessive defense and control → "Contempt for control" (analogue of "disgust")

In these lists, the first set is of the two Gombrich/Cicero causal patterns, which appear opposite. The second set, which represents the two Freudian sequences, also appears to show opposite patterns. With this two-by-two combination and its permutations, you expect oppositions and suspect they can strain, if not ruin, the analogy. Causal directions often appear opposite if you do not take into account their circularity. Even though dynamic movements occur in opposite directions in both sets of descriptions (Gombrich's and Freud's), *the analogy holds when—and because—the search for negations continues beyond any points of resolution or presumed state of perfection.* But, in fact, this state of affairs rests on a powerful *"principle of circularity."* We see circularity in the psychodynamic revolutions of excesses and controls. We also see it in the "growth of signs," as Peirce describes it. There is semiotic movement toward the symbolic from the referential, and there is movement from the iconic experience of an object to its representation in a sign. If a sign refers to itself, the self-reflexivity is obviously circular (Nöth 2003). In sum, *the principle of circularity refers to psychodynamic, semiotic, and logical states of resolved and unresolved negations.* The principle is pervasive. I define *circularity* accordingly:

Circularity is a dynamic pattern reflecting a closed system's turns inward for its search for new states of unresolved negation.

The closest contemporary "system" concept for this circularity might be Piaget's *equilibration.* Although Piaget's concept is dynamic and involves self-regulation, it usually is weighted on the side of the reduction of unresolved negations. Let's go further into the issues of the closed system to characterize the dynamic outcomes of circularity. How do we understand a self-regulation that moves away from regulation in order to find new negations? In Aristotelian terms, a biological basis for perfectibility leads to maturity, but then also to the organic individual's death.

In artistic terms, after reaching for a "golden age," there follows the "death of art," its subsequent rebirth, and the repetition of the process with the vicissitudes of yesterday's maturity as tomorrow's primitivity. Consider Gombrich's statement that the Greek cradle of art constantly subjected artistic works to *contests* and that a point of perfectibility to be reached was mimesis. Trophies for perfection of replication! Together, the contest and the search for better mimesis led perfectibility in the direction of rich sensual forms, colors, and images. Hence, the "disgust" reaction—satiation and search for moral clarity led to the primitive in the form of the simpler and starker.

This direction appears the opposite of Freud's idea of *revulsion and defense*. In *Totem and Taboo* ([1913] 1971), Freud classically describes a primitive state that, after renunciation, is superceded by defenses—repression, various forms of reaction formation, and sublimation. The primitive *prehistoric* state is of unbridled pleasure seeking and/or impulse satisfying. I liken to revulsion the feelings leading to renunciation of the impulse to satisfy. Renunciation is accomplished through defensive control and, if you want, moralism. The prehistoric aspect is important to keep in perspective; it is primitivity as the unwritten and unconscious. Therefore, it is a fiction—what appears to be the case when one cannot be aware that it is the case! The view is possible by "looking backwards" to what one can construct to have been the case. With this "backwards direction" to a reconstruction, the fictional—or plausible mythical—cause-and-effect sequence is possible. The case can be made by faith in some sort of revelation or by scientific argument—but that's too diverting a topic for the present discussion. Although conceivable, the "story" of prehistory or of a past unconscious state has the form of a myth, and its temporal frame is set in motion by the narrator, who is looking backwards. Within this temporal frame, the Freudian myth pivots from a point of defended thought and psychological dynamics from which to look backwards.

By way of *Totem and Taboo* and Freud's other classical accounts, I cast the more "civilized" social stage and the more mature cognitive states as built by defense against impulses. Thereby, we can consider the reconstruction of primitive forces and themes in a myth as a case of *revulsion and defense*. Although the defensive maneuvers themselves are the product of revulsion in the face of the terrible effects of uncontrolled impulses, looking backwards from the high ground of defended and "civilized" behavior turns matters upside-down. Then we come across an analogue of disgust when contemplating the problems of overcontrol. The result is an impetus to achieve a breakthrough of the defenses so as to produce a sense of more freedom to experience the senses and the sensual, and to remove the moralism that has too tightly gripped pleasure seeking. Examples of such a breakthrough are the sexual and feminist revolutions of the 1960s. Instead of disgust, though, the operative sentiment was contempt for the defenses as false and oppressive beliefs. The return to the primitive was to pleasure and its satiation. The idea seems opposite to Gombrich's invocation of the primitive as less developed sensually. In the Freudian sequences (p. 193), although the impetus to change appears at parallel points of oversaturation, its resolution is not always parallel to Gombrich's.

The differences between Freud's and Gombrich's explanations have all sorts of permutations, so keep in mind that Gombrich's picture is *not* opposite to the Freudian sequence of *trauma and defense*. Thus, in the case of too early and too much exposure to sense excitement, the repression that takes place is presumably a force for the good of the individual's capabilities of control. Of course, when we look at trauma, we tend to be "looking backwards." Yet the look is from a point in time after which the defense of thought and psychological experiences is forever changed. Even if defenses can be knocked down, their traces include the ghosts of the controls. We stop at what we cannot help blocking from knowability. *The past as a myth is not to be reconstituted; instead, we move forward within the defense-protected awareness of self—even employing it to create beauty of form.*

So, at a point in the experience of defended thought, myths emerge. In them, we see the transformation of the functions of defense mechanisms. *Myth-producing defenses* is my term for the form that has come to be used to create myths. These defenses move the circularity in a forward temporal direction; they may even contribute to "higher-order defenses" such as sublimation and serve to produce rebirth of ideas that go beyond the primitive. In this account, the Freudian analogue of disgust goes through phases of controlling the sensual and pleasure orientation, hence this *trauma-and-defense* schematic and its denouement in different temporal modes does *not* appear in an opposite sequence to that in Gombrich's "disgust" scenarios.

The Freudian pictures and the Gombrich essays—with the various turns of their two-by-two permutations—all benefit from a principle of circularity. Perfectibility is a tendency within an individual organism because it is a form that is held together dynamically as a closed system. Accordingly, consider the following definition of *circularity*, although in a larger sense I may be describing entropy. *Vicissitudes in direction and value lead the individual's dynamics to points of perfectibility that seem to resolve all problems. They work to coordinate with the purpose, functions, and being of the form, but they also eventually counterreact, directing force toward the restoration of unresolved negations.*

The Self's Place and Logical Role in a Machine Model

All sorts of forces in reasoned thought lead to regularization through logical rules. All sorts of forces in scientific observation also lead to regularization through evidence and predictability. With these forces—not to say the regularization and predictability brought about through technological models and through the achievements of the "posthuman" era of information, computers, and robots—there is a reinforcement of congruent biological and mechanical definitions of a "closed system." How close do we come to the design of a machine when defining the system elements and processes of the human organism? Some may regard this development as a "crisis" in representation and a loss to the meaning of what an individual, an agent, and a self may be and have to offer. Semioticians Winfried Nöth and Christina Ljungberg (2003) provide a nice summary of what leads meaning and representation to the "posthuman" versions of self-regulation. Where would circularity in theorizing about the self take us? Cognizant of the dangers of underestimating or overvaluing machine theory, I take up how the self, placed in a machine model, affects its logical role vis-à-vis self-regulation and entropy.

Two Factors Optimizing Regression to Disorganization

Two factors seem to optimize regression to disorganization. The first factor is the *absence of an external or externalizable complementary category. Call this "the ~CC" factor.*

We have seen this factor in the form of the "megamachine" that sweeps the self into the mechanical process and its model. We have also pictured a primitive state of affairs in which the self and the nonself are not distinguishable. In prehistory, the individual, the gods, and the fates are and are not parts of the self. In the infant's early life, what is the mother and what is the self are not cognitively differentiated.

The second factor is the *exhaustion of internal routes to self-regulation.* Self-regulation as a dynamic process can be conceived as a function of values and their directions. One value is of a positive meaning and its tendency toward expression. An alternate value may be one negative in meaning. Its directional tendency might be refraction or retraction. The directions, as I conceive them, are within the individual as a *closed system.* Therefore, the directions are internal routes to self-regulation. (My discussion of monads should suffice for debate about how closed a system can be if it is interdependent. Also see my view of interdependency as an irreducible aspect of systems in Fisher 2000a, 2000b.)

A positive meaning might be that of the individual's sense of beauty—with due implications for its form being or becoming an artistic one. This is the form for a positive value; its inner route of expression might be in the semiotic movement from an image to a more complex representation. A negative meaning might be one of "badness." The inner route for its entailed value might consist of rules for avoidance or for the eschewing of its various forms, which might be conceptualized by an array of representations of the mythic, the symbolic, or the stereotyped.

Exhaustion occurs when there is complete mechanization of the alternate directions of the values. What I mean by "complete mechanization" is that each person may have her own built-in version of a "Turing test." "Do whatever set or rules I have adopted explain everything I believe, such that if explained in any other way, the explanation's power to regularize and predict wouldn't be improved?" If the answer is "yes," the mechanization is complete! For the value of "beauty," there is internal permission to develop and create meanings, icons, and all sorts of intellectual production. But as in the art constructed to achieve and extol beauty, there is also a defense against that which would evoke disgust. Thus, this "complete" mechanization provides a control by categorical opposition. The category of explanation you choose "opposes" any other, and in this way, dialectically, it is a coherent—yet dynamic—construction by way of a set of values.

The defense model is based on categorical opposition—with the downside of restrictiveness. This sort of rigid antagonism yields a control that only leads to overcontrol! *Paralysis* might be a better term. This way of depending on defenses is like abstaining is to an alcohol addiction. While you are defended, all sorts of things can be accomplished, but if you take a drink, you're sunk. You can't stop. Don't take a drink, and you're in great shape. But your whole belief system has to be mobilized and directed to not doing so. The model—whether Gombrich's, based on the rigid antagonism of regularization versus disgust, or Freud's, based on satiety versus defense—utilizes categorical opposition. The result is the periodic bind of overcontrol versus no control. A person in a bind like this can be aware of the bind, but the control at its highpoints is in such overdrive that there's no escape.

Consciousness is thus bounded and "overcontrolled." I use the abbreviation CO for "consciousness overcontrolled." It is the second factor in optimizing regression to disorganization. It works by exhaustion of the internal routes to self-regulation. I picture the CO factor this way: *the pleasure-obtaining mechanisms lead either to irreversible reinforcements or to reversal through overcontrol.* The irreversible reinforcements are protective principles, like defense mechanisms. Their function is to reverse a trend—such as the tendency toward oversatiety. How is a defense irreversible? After all, doesn't the alcoholic "fall off the wagon," and isn't there the phenomenon of primitivation? Once the control mechanism itself has been achieved, its function persists, irrespective of continuing pressures to revert to former states. There are vicissitudes; defenses fall away when the pressures are great. You might insult your boss even though your better judgment says, "This is a primitive thing to do!" Yet the control function remains there, and the "regression" is dealt with in terms of another vicissitude—namely, a more complex defense, such as rationalization. "It was for the best!" But the mechanism is irreversible not only because of a "programmed" move toward completion and maturity but also because of the affective and "value" urgency to be intolerant of return from reversal.

The very commitment to a machine model sets in motion the resolution of problems that would stand in the way of self-regulation. Even if we refer to a model of biological adaptation as it applies to overall mental and psychic functioning as the ordinary course of cognitive functioning, the tendency to regularize moves toward a mechanical form. (A takeoff point for this idea would be the Piagetian extensions of the model of regulations from Charcot, Janet, Claparède, and Ribot.) To adapt, to resolve problems, to develop ways of coping, and to develop guides for those ways, there would need to be a search for negations that might be stumbling blocks. Moreover, *the two factors (the ~CC and the CO) exacerbate the revved-up search for negations and their resolution, which follows the adaptation of a machine model to solve problems of self-regulation and ecological adaptation.* Let's look at how these two aggravating factors negotiate the relationship between the mechanization of belief and the experience of self. The mechanization of belief aims toward the state of having a model to solve all problems—that is, forming one's beliefs in a self-contained machine metaphor. The experience of self includes the individual's concept of the self and the beliefs about it.

The ~CC and CO factors aggravate the state of affairs in which an individual's machine model includes the self. An example is the person whose model of handling all problems is to line up possibilities and decide in terms of least cost. Oliver Sacks describes an Asperger syndrome individual who is a high achiever in chemistry. She reaches a judgment that intimate relationships are impenetrable in terms of the rules of social interaction. Her conclusion is not to try to engage in or develop such relations. The belief in decision based on "least cost" is possible if that idea is subsumed within some notion of consensual validation of "cost." So the chemist may well believe others will agree with her conclusion not to relate intimately, possibly because it will be less costly to them not to relate to her intimately and possibly because it will be less costly to her own sense of self. In this way, the same person may believe actuarial thinking applies not only to judgments about things and others, but also to the welfare of the self. If such judgments are matters of social consensus, then, to begin with, the self may be a myth created by others.

One way there can be a head-to-head opposition in the use of this method of reaching consensus is if the self were subject to nonnormative statistical thinking. Some people would want to do a different sort of statistical reasoning—or nonstatistical reasoning about taking chances. Even the Asperger-afflicted chemist can place the self outside statistical reasoning by claiming it is not a self in the usual sense because of the absence of capability to construct rules. Therefore, the self can be out of the loop of factors to be considered in a "least cost/least effort" analysis. So a key question for discriminating logical forms on the basis of self and its role in the dynamics of categorizations is, Does the person's model of a perfect set of solutions to problems include or not include the self?

The individual sees the self either outside his machine metaphor or else included within it. A person who feels isolated from others and from the world around him is clearly feeling *alienation*. But someone who is a dualist may also reach this position. She may regard her "self" as some sort of phenomenological entity—outside the domain of the objective things she can grasp. In a sense, the person who leaves the self outside the machine metaphor does not implode if the machine model "goes too far." Such a person can look to improve her "spiritual" life or somehow change her sense of self. However, to do this, the existence of the self would have to be admitted. In the case of the chemist, that might mean that the self "outside the cost model" *does exist*—albeit in a flawed way. Its experiences, if impenetrable and uncodifiable as rules, still have some effects, which swell within the constraints of the ~CC and CO factors. Picture the person who comes to grips with the idea that "I can be no other way, so I 'renounce' my needs." Unless this person is a saint, you might have visions of Maugham's missionary or perhaps Dumas's (1996) Edmund Dantes. What swells within the constraints seeks alternate expression or transformation. The missionary doesn't metamorphize so well, but Dantes turns into a complex character trying to make things right.

The outsider's search is a force toward rectification of needs or finding ways to open the complementary categories that would accommodate negations. Not to take this force into account would leave matters to some sort of internal implosion or degradation. Yet the difference between the missionary and Dantes is that the missionary apparently believes in his airtight system of rules, whereas Dantes lives in a world that believes in such a system, and he is forced to behave in accordance with the governing rules although he himself does not believe these rules of justice should prevail. The degradation can take place if the self *is* incorporated into the machine model and is itself considered a set of rules and regulations subject to the same constraints—namely, the ~CC and CO factors. The takeover of the self such that it is not outside the categorization of the rules and regulations is the ground for entropy.

Once again, the issue of the self as "outside" and the implications for an aware "I" make the difference necessary for agency and perspective. Edmund Dantes is in control of his planning and the ultimate execution of his plans. This control contrasts with the "driven" quality of the missionary's "acting out." The perspective afforded by a conscious mode is a way of creating facets of meanings. Dantes's choices can go in the direction of a "figurative" version of himself—although he continues to know he is Edmund Dantes. The missionary's choices, if not acted out, are closed off from consciousness; the danger is implosion. In sum, *the most dangerous slope to entropy is when the world of the "self" and the world of "problems solved" are coterminous.*

To express all this symbolically, where S stands for "self" and PS for "problems solved,"

$$PS \cap S$$

characterizes the first alternative—namely, when the person's machine model of ways and rules to solve problems is separate from the arena of self. The second alternative, formulated

$$PS \cup S,$$

shows the two (PS and S) in union.

Now return to the two "aggravating" factors. The ~CC factor blocks the individual from any concept of the self that would exist outside the model. Self is locked in. Some views hold the self to be nothing other than cognitive activity—even if of a special kind. A current-day version of this view is to model the mind by fitting it to the algorithmic governance of information processes and decision making. There is the "metacognitive" regulation of cognitive activity—which is no more than cognitive activity to the second power! So, yes, there might be a "real world" aside from the self, but there's no self out there in any complementary category!

And if this were not enough hemming in, the second aggravating factor (CO) pulverizes the self wherever it is located. The easiest way to see this is from within the consideration of the self as inseparable from the machine model. There is only one set of rules, and these rules are tied to the organism's pleasure system. Within this assumption, you cannot leave out the pleasure system. But you could assume that this set of rules and the pleasure system make up a discrete mechanical domain. Such an assumption is made by the suicidal terrorist whose beliefs are tightly bound to the after-life rewards of earthly pleasures!

Even so, without your falling into any logical traps here, suppose, instead of picturing a "complete mechanization," you somehow succeeded in proposing alternate machine models. Then, in some sense, the self might be considered external. For example, following the work of Michael Gazzaniga (1998; Turk et al. 2002), different attempts are currently being made to cite a part of the brain at which the self is located. An enclosed system is just that. The "realizations of self"—whatever they may be—would be subordinate to the physical nature of that portion of the brain and its determination of the "manifestations" of self (also see Pinker 2002; J. Ross 2003). Within these constraints, the CO factor—at whatever level of its functioning—would produce just as much neutralization, and so the inaccessible negations would pile up. Then you would have an explosion within the self and an implosion in the main machine model.

With the CO factor, there are psychodynamic antagonisms of defense and urge, as well as antagonisms of *form* and *function* at the morphological level. The physical nature and the structures of the neurological determinants of self would be an example of a form in an antagonistic relation to the functions aroused and unresolved within the total organism, if not as a communicative set of events somehow affecting the self. Therefore, *antagonism as alternation of control and its antithesis is the essence of the CO factor.*

Let's take the interrelation of information and in general the communication between self and nonself. One form, the alternate machines of self and nonself as metacognitive rules and cognitive routines, winds up with ultimately limited pathways to the resolution of problems of the pleasure system. The prototypical CO situation is one of adversary relations between the needs or functional requirements and any combination of rules and routines. The example we have examined involves the control/ expression dichotomy and its subjection to a competition for fulfilling not only the pleasure purposes, but also the purposes of a combination of self-regulation and adaptation. All roads of competition engendered by CO lead to ruin! The entropy occurs no matter which way you conceive the workings of the CO factor.

Foray into Dreamland

I have traveled along with Freud's descent into dreams so that I could explore clues to unconscious logic. However, to know about the logical forms involved is to acknowledge the "I." The "I" must be present to become aware of a logic that might exist without it or that seems to the "I" to exist when the "I"—at least, in its aware state—is absent. Nonetheless, one can dream about the self. In an 1850 song, J. Paton Clarke writes,

> I dreamt last night that I was wandering lonely
> Through fields the fairest mortal eye could see
> 'Twas summer evening and the land rail only
> Discours'd to me.

In the process of dream-image formation, just *who* is making these statements and inventing quasi-tropes about the self? Well, it is the self— but *not* merely in the form of the Cartesian *cogito*. How do we get to an expression of that self "who" invents the dream and dreamlike images?

When we take another look at "I think; therefore I am," we see the recursive and reflective nature of the "I"—especially if we take the Kantian point that for any statement there is a parenthetical "I" ("I" understood) to mark the place of the thinker or utterer. I merely extend this rule to a statement about the "I"—by the "I"—as follows:

(I [say/think]) I think; therefore I am.

But we're not there yet. Even this expansion of the "I" is insufficient to mark the aware state as different from a form that might govern the reflectiveness about the self in a dream. To show what is needed, I add an expression to the expansion of the "I." Note this expression to the left of the formulation below:

[I am aware that I *am* I] $\geq$ {I am aware that (I [say/think]) I think; therefore I am.}

The expression "I am aware that I *am* I" marks the "I's" awareness of itself as the necessary step. I call this step *the extreme of the "I's" reflexiveness*. I would define that extreme as the awareness of the *cogito*—in the form of a lattice operation that is reflexive. My point is that to express the "I" in union with a conscious state, *the extreme of its reflexiveness* has to be present. That would be the "I's" awareness of itself. More exactly (although a cumbersome construction), it is the "I's" awareness of the *"aware 'I.'"* In this sense, the "aware 'I'" can become an object of the "I's" awareness. Circular?

Well, for one thing, we have seen the need for a principle of circularity to account for opposition in causal sequences. In a sense, that circularity is penultimate; it accounts for the functions of and regulations by the "I"—given a unique individual. But that individual is unique and a closed system. Therefore, circularity is involved ultimately in the sense that the search for perfectibility in a closed system may be conceived in terms of the self's search for identity. Nevertheless, the principles of finding unresolved negations—even after identity is established from a number of directions—would continue to apply to the functions of the aware "I." Even with all this, in regard to *the extreme of the "I's" reflexiveness*, circularity isn't the immediate dynamic of operations. I should say that the macrodynamic of the closed system *is* identity. With this formulation, though, it is important not to lose sight of the interplay of the state of identity, its structure, its internal contradictions of force, scalars, and vectors, and its maintenance. However, our understanding of this interplay appears at a microlevel of the closed system. Identity at a microlevel involves a flux of logical regulations and transformations, and, therein, what we can specify is reflexivity as the "immediate dynamic of invention" of images and creativity.

The Immediate Dynamic of Invention and the Point of Perspective

I put this immediate dynamic of invention as:

$$I_a \gtreqless I_{1-n}.$$

By this, I mean that the aware "I" (I_a) has a latticelike relationship to all versions of the "I"-as-object (I_{1-n}).[4] The "I"-as-object is the self as seen by the self. As such, this I_{1-n} can be present in the dream, but also can be thought of as $f(I_{1-n})$—that is, as instrumental in the formation of the dream images of the "I."

Accordingly, the theoretical nature of the dream logic is

$$fI_{1-n}(I_{1-n}).$$

But this formulation is only comprehensible if I_{1-n} is a category that includes (or relates to in a lattice structuring) all images, icons, and representations of all objects appearing in the dream. The dream is the world according to the dreamer. I would put this relationship as

$$fI_{1-n}(I_{1-n} \gtreqqless \bar{I}_{1-n}).$$

We now have a state in which there is the presence of negation—along with the possibilities for absorption of the I_{1-n} by $\bar{I}_{1-n}$. We have seen this before. The lattice structure prototypically functions to create a suspension of alternate ordering. So, in this case, *the lattice structure is used to neutralize the categorical complement of the "I" such that equivalence relationships are permissible.*

As we take the logical form in a backward journey to its points of origin and origin of its being, we can take another step. At some juncture,

$$I = \bar{I},$$

which would be a state of *identity* of the "I" and the "Ī." It would be an odd state of the self; nonetheless, it would account for an individual's self at some point of its unawareness of itself. This identity is possible *as a function of the "I"*—namely, the presence of fI_{1-n} and the absence of the aware "I" (I_a). Of course, to know any of this, we have to be looking backwards from the perspective of an awareness of the dream. Clarke's ballad looks toward its end, which is to interpret the dreamer's own dream as a declaration of true love. But the dreamer acknowledges looking back to see all this in the dream:

My dream is told
And now thou may'st discover
Wherefore in twilights shadows I repine.

Thus, the "aware I" is the point of perspective:

$$fl_a[fl_{1-n}(I_{1-n} \gtreqless \bar{I}_{1-n})].$$

It's also necessary that I say a word on *this* logical structure. Otherwise, the term *point of perspective* falls flat. Each symbol in this logical structure, when seen as a separate unit or category, is an *icon*. That is, each symbol within the expression $[fl_{1-n}(I_{1-n} \gtreqless \bar{I}_{1-n})]$ is semiotically a representation perceived or conceived by the individual person. The boundedness of each unit or category as an icon simply *defines the scope of its particulars as an object*. I mean that the bounds of the object are set by something like "focus." You look at the sky, and even though its borders are penumbral, as a unit, the sky is bounded. There is a complementary category that is not-sky—and that includes a great deal. So the sky is one kind of object. You "narrow the scope" when you look at the Milky Way. It is another kind of object, perhaps its boundaries determined by something like a "pattern." You can narrow further and look at an individual star.

In addition to the icon's boundaries are its semiotic dimensions. These dimensions locate, measure, and delimit the results of the icon's dynamic "growth." Icons develop symbols, which change the icons, and so on. Therefore, we have a growth in semiotic articulation of what I label an *expression* space. In relation to the logical structure of the expression of the object, there are the individual's icons, symbols, and system of internal and external references.

Another kind of object requires a wider scope. It has more obvious—or manifest—elements to it. Perhaps a better way of describing this kind of object is that it is more articulated in its action, movements, and cause-and-effect relations with other objects. If we were talking about categories and units that were language based, I might describe the difference of these kinds of object as being like the structural difference between words and sentences—with the latter being more articulated. In logical terms, I'm referring to terms versus the propositional structure, which brackets would mark off.

The boundedness of the proposition (or whatever the forms bracketed together in some relationship) makes for *another sort of object*. In semiotic terms, *the bracketed form, denoting a relationship or relationships, is also an icon*. Hence, the symbol fl*a*, as it appears in the formulation, has a special position *outside the brackets*. In short, it serves as a point of perspective.

This outside position is subject to a regress. The brackets might be extended to engulf the fl*a*, something like saying that the "'I' experiencing" can be, in turn, an object of another moment of the "'I' experiencing." Then, the first "'I' experiencing" becomes an object of the second! Another way of saying this is that the *point of perspective* can be expanded by a regress of the brackets. Now, when we write this down, the regress appears to go to the left, constantly moving to a new point from which to begin a recording of the newly expanded perspective on the experience. But this picture of the direction of movement is a product of language and the way we unfold it in speech and writing. The way to conceive the regress is therefore not to the left or the right, or "backwards" in time or space. Nor is the regress either to pure experience of the "I" or to the point of "full" experience of all possible modes of the "I." So, when we say that the regress "continues," a linear scalar can be used to put the notation in some linguistic form, but the impact on logical structures available from the changing points of perspective is a matter of the expansion of the possibilities of *outsidedness*.

A regress can continue (or advance, if you wish) such that the individual's awareness "moves" farther "outside" by way of a process of first bracketing everything in sight and then "looking at" it all! This is a special point of perspective—one at which a synchronic view of self and time is always at the ready or always a possibility. Although I said I would not get into issues of a time mode or dimension, it seems obvious that the "aware I" must continue to occupy this special point of perspective. There's an *if-and-only-if (iff)* to this requirement. The special point of perspective must be *outside* any division (or diachronicity) of time because this perspective is necessary to the definition and articulation of relations of the objects in view—including the "I" if it is an object. This requirement of a special point of perspective and its *iff* contingency is one way of feeling the fl*a* level of awareness that I have placed outside brackets. The point of perspective of the "aware I" is thus outside any propositional structure and in a position of undifferentiated "nowness."

The Different Reflective Structures of Dreams, Primitive States of Thought, and the Trope

This inclusion of an ever-evocable felt point of perspective, located outside the brackets of any icon, distinguishes the dream as a temporary pathology—more appropriately, as regression in service of the ego—from the dream as a primitive form. The argument may be that the dream is not primitive, but rather that it is *primitivized* when in the form

$$fl_{1-n}(I_{1-n} \gtreqless \bar{I}_{1-n}).$$

To deal with this objection, one might say the primitive state is simply

$$(I_{1-n} \gtreqless \bar{I}_{1-n}).$$

In the form $(I_{1\text{-}n} \gtreqless \bar{I}_{1-n})$, I have eliminated what would be outside the brackets. Therefore, this definition would be arrived at by subtraction of one degree of reflectiveness of the "I." Nonetheless, though we can't prove it, we infer primitive states of dreaming. In this case, how do you account for the origination or formation of dream images? Expand this $(I_{1-n} \gtreqless \bar{I}_{1-n})$ structure to include a state of self that is and is not self—as we can capture and experience it as an object. We get

$$f(I_{1-n} \equiv \bar{I}_{1-n}) \cdot [(I_{1-n} \gtreqless \bar{I}_{1-n})].$$

In *this* statement, the "I" (with its categorical complementarity simultaneously intact *and* cancelled!) is at a penultimate point of syncretism with its functions and objects. If a trope, such as metonymy, were *not* subject to the reflexive operations of the aware "I," then it would have a similar form. Call that the form of an "unreflexive metonymy," and consider one that characterizes the "I." Think of the commonplace exclamation, "I am all thumbs!" In the case of the unreflective metonymy, the logical status of the "I" in the $f(I_{1-n} \equiv \bar{I}_{1-n}) \cdot [(I_{1-n} \gtreqless \bar{I}_{1-n})]$ state would be indistinguishable from the "I" in the metonymy. Similar metonymies include equivalence of self with something like "pain in the neck." Cartoons such as *Bugs Bunny* often show equivalence of the whole "animal/person" creature with a painful headache or a stubbed toe. The unreflexive form I am describing is reminiscent of primitive identities. Thus, in the manner James Frazer ([1922] 1950]) describes the principles of contiguity and similarity for the conceptions of primitive magic and myths, the logical form of the "I" in "I am all thumbs" would be identical to a self that is adorned with a bird's head or animal fur.

8 Pathological and Defensive Logical Forms

The Individual Without a Self

What would the organism be if not individual according to its own intrinsic judgment, but instead a momentary focus from the outside? It would be an individual not only who is defined by others, but also who defines himself and his perceptions by what others perceive. Like Polonius's clouds to Polonius—or like Polonius to himself when he's feeling fundamentally driven by what others want him to see and be.

What if logical forms were *not* reflective of the developments of the self and its various compounded levels of awareness? The ordering of the categorial system and the capacity for bracketing out the "I" would be subsumed within its functions (fIs). "Species" organization would prevail. All this puts a halt to coordinated decisions and internal locus of cause relative to the governance and selection of categorial orders, hierarchical models, and the transformation of values from pragmatic to aesthetic and enduring.

If a person either cannot look at herself because she has defended herself against certain feelings or she has accepted certain stereotypes, the categorical system is distorted—if in no other way than to obscure or bloat some complementary categories. These descriptions are on the level of logical analysis of phenomenological events—felt and experienced both consciously and unconsciously.

Many researchers regard all this as old-fashioned psychodynamic levels of description (Kandel 1999; Kihlstrom [2000] 2006). In light of the au courant turn toward the brain as the site from which to explain psychodynamic events, however, it's necessary to say a word about the idea of the self and its functions *as neurologically located.* Many studies and views promulgate the neurological inspiration—if not the neurological constitution—of self (see, e.g., Ross 2003). Science professor Joseph LeDoux (2002) argues that the self is "in the synapses"! According to J. Andrew Ross, LeDoux's model of the self is that of an integrated representational structure "distributed over the brain system as a pattern of synaptic connections" (2003:68).

If the self *is* located in specific neural configurations and neurological structures, then what is the status of my statement that "the 'species' organization will prevail"? How do you take into account subjectively reported phenomena? What organization do you place up front when you consider that the self organizes its concepts differently when one side of the brain is dominant and/or the other side is disengaged? Gazzaniga and his associates have identified specific neurological structures that must be present for a range of relevant psychological functions: recognition of self, theory of mind that recognizes the self, and theory of mind that recognizes others (Pinker 2002; Ross 2003). Not unreasonable theories follow the evidence of such correlates: *in the same individual,* more than one organization of a "self," more than one sense of experience of the "self," and more than one type of self—in terms of personality, moods, attitudes, and sensibilities—can exist. Where does all this leave the "logical organization of self"? If logical organization of the self is not coordinated, it is a more modular construction brought about by objectifiable "function packs." One part of the brain does this, and another does that. Subjective experience is also not coordinated.

There is a difference between experiencing the self-as-self *as a function* and experiencing the self-as-self *relative to the whole self* and its search for identity within its logical, metaphysical, and ontological bounds as a being. With the likes of "split-brain" selves and with the "function-pack" self, the self does not fill the shoes of a whole person. And if an individual person does not make that judgment, who does?

Internal and External Sources of Self-Definition

We come back to the bare-bones choice. If not internal, the focal point for a definition of the self and its identity becomes external. Steven Pinker

(2002) compares a social and/or cultural construction and definition of the self to the biological view of determinative neural structures and patterns. When and if the self is reduced to function packs dependent on structures unaffected by various sources of brain trauma or injury, the self can be considered "internal." However, others complete the story of the individual's identity not only in relation to her own organization of self functions, but also in relation to her unique individual body, life, and being. The observer of the split-brain self is the "other"—the scientist as objective observer. Accordingly, the observer/self experience dimensions are split to internal and external sources.

Nevertheless, we are back to puzzling about phenomenological events of experiences in different psychological spaces. There, another puzzle piece is marked by Freud's notion of the "it" (translated the "id")—an entity one can refer to as a self function, a form of self missing the subjective awareness of the "I" in relation to "it." As a self function, the "it" is not within a helical orbit that would spin outward to the newly bracketed position from which it becomes the object of the "I's" awareness. When this form of helix does not get realized, but instead closes off—and becomes an encircled function—we have a choked-off internal locus from which the self fails to seek out its identity as a superordinating logical category.

Expansions of the "Governing Belief" and the Belief Giver's Domain of Self-Defining

These points about the closed-off selves have far-reaching consequences relative to stereotypes and constitutional rules. On one hand, stereotypes and constitutional rules can be external focal points for the definitions of self and the valuation of beliefs. On the other, they are instances of icons and relationships of icons subject to tropes and their underlying dynamics.

Change in icons and in representations of self and others' identity can be directed from an external focal point. A flag can be waved, and the individual will feel that she's a member of a group called to patriotic acts. Alternatively, the grasp of icons and tropes can also be something the individual is aware of, originates, and transforms, and for which she takes responsibility. Under the call to arms in World War I, James Montgomery Flagg's "Uncle Sam Wants You" posters (1917) could stimulate individuals to join the service. George M. Cohan (1917) could create the lyric, "We'll be over, we're coming over, / And we won't come back 'til it's over Over

There!" The first line relates to identity. The two *overs* in the second line are related to patriotic *resolve*. There ensues a series of equivalences in the analogy "Over : over :: Patriotism : Identity." The songwriter originates a complex trope for which he consciously takes responsibility.

Change in the icons and tropes that govern and instantiate stereotypes and constitutional rules takes place over epochal historical time. Philosopher Stephen Pepper (1949) describes "root" metaphors that are thus generative; they govern scientific hypotheses. Before these "master ideas" are displaced, a "tide" of counterevidence or values is required. Once a governing idea in the form of a root metaphor takes hold, it tends to spread. In the case of political tropes, such as "the master race" or "the Fascist State is itself conscious and has itself a will and a personality" (Mussolini 1932), the spread is to myriad social and political functions that are not easily displaced. The twentieth century saw the rise and fall of *isms* such as Nazism, fascism, and communism. Change did not occur overnight. It required war and dramatically affected human life, well-being, and the concrete facts of property, persons, towns, geographic borders, political structures, and moral codes. The fall of the Soviet Union, although a peaceful dismantling, has still not displaced that empire's legacies—unruly dangerous infrastructures, defunct nuclear facilities, authoritarian dicta. One of the great problems of the twentieth century still remains in the twenty-first: the Communist legacy in the totalitarian rule in China and North Korea. Changes in *governing ideas* are "sea changes" that can have profound influences on the welfare, health, fate, and perfectibility of the individual and the human species.

The terms and meanings of these governing ideas are stereotypes. Stereotypes attenuate meaning. A partial account is made equivalent to the whole. They often refer to the Us/other dichotomy. In the case of constitutions, there appear distinctions about citizens and noncitizens. Garth Kemerling, reviewing Aristotle's concept of the "citizen," points out, "Although a good citizen is a good person, on Aristotle's view, the good person can be good even independently of the society. A good citizen, however, can exist only as a part of the social structure itself, so the state is in some sense prior to the citizen" (2001).

Constitutional rules may provide rights, but, like all rules, they also provide boundaries. The "citizen" can exist within a sociopolitical space and therein replace the more inclusive concept of the "human being." This is easy to see when stereotypes and rules interact. Stereotypes can diminish the meanings of a self or a group, and rules can restrict either the self's or the group's decision powers. Where stereotypes are involved, a reduction of the meanings and rights of a group affects their selves. This effect appears to be isolated to an *outgroup,* but the in-group is also in a crunch. Within the in-group, the self is in some way "special," but, insidiously, the question is, Do its members really believe this about themselves? They have a terrific double-bind. If they believe the "special" nature attributed to them, they accept the belief giver's designation, while suspending a wider concept of a person's rights. If they do not believe the attribution to themselves, they are acting internally contrary to the beliefs given by the belief giver!

However, dynasties, eras, "ages," and the like exist. There is stability in sociopolitical patterns such as those of "states" and their "citizens" because the support for certain governing ideas comes from all directions. This support can enhance social rights and suppress individual rights. For example, the political nation can codify the idea of citizens' rights, but the general characteristics of regulations given by rules can neutralize the self—pulverize it from different points of leverage—so that its powers to define its own beliefs are severely limited.

I do not want to overdo the point. I am specifically concerned with the diminished meanings and restrictive rules as regards the self. From the vicissitudes of historical time, many diminutions of the self have been swept right into the current scene. There is the tendency to see self as role only. There is the split up of the self into particularized schematic indexicals. There's the neurologically inspired split self—the brainchild of neural events and structures. There is the shunning of self in favor of religious obeisance. Governing ideas such as computer metaphors, robot competition, DNA possibilities, and genome specification continue to expand the domain of the external loci for self-definition. We live in a time when the progress toward the inner control of one's destiny is turned backwards, rushing toward the rule of the Fates and those who are—or who hold themselves to be—outside the individual's self-determining capacities for origination and regulation of meaning and motivation.

Historical Transformation of the Nonself

Ohnuki-Tierney (1991) views historical transformation of meanings and tropes as change by contiguity. Contiguity is a formal factor, and it is accessible from many points of observation. Insofar as changes in icons, meaning, and tropes are tied to the issue of contiguity of events, behavior, association, and so on, temporal sequence is an inherent determinant not only of the very observation of the changes, but also of the resultant forms and their variations. Whether it is thoughts or behaviors or external events that change, their perceived and conceived contiguity depends on the observation of temporal units, modes, and sequences or patterns. This focus on schematics is less problematic than assuming historical change is simply that of and by awareness. After all, "awareness" is hard to define and include within a scientific account. Yet it makes logical sense to think of history as a process somewhere connected to the idea of time, and it gives us some observational pegs onto which we can hook what we are aware of and to mark where and when that was.

Certainly, though, if we tie the causal forces of time to the individual's perception, we would make just as much an error of reduction of causal power as we would if we deny the individual's self and its causal work. Do I see and experience the thrill of a home run in baseball only because time moves from the point of the batter's swing to the end of the arc of the ball when it is beyond the fence? Was it Ted Williams's intense intent that caused his bat to "connect," or was it simply the coincidence of his observation of timing with the temporal facts of the ball's trajectory and speed?

In avoiding reduction, we should not confuse the observation of the milieu in which transformations take place with the individual's perception of the changes, what she thinks about those changes, and how she organizes those changes in her logic and beliefs. Merely casting time as advancing on the basis of contiguities should not be confused with perceiving historical transformations and subjecting them to the "forms of thought." This reduction subjects the forms of thought to the external forces of change in conceptual forms.

When a cosmological trope is in style—like a universe with the earth as the center—the contiguity machinery rolls on, and the shape of the individual's thoughts about himself and his experience is subcategorized accordingly. The view of the earth's importance in the universe provides a category relation of earth to universe. The practical upshot would be that short of some divine intervention or accidental genetic mutation, no Copernican idea would occur because the category relation of the person to the universe would be too bloated to permit the idea. The form of the person's thoughts about self, universe, and thought itself as well as its logical categories would follow suit—or in less causal terms, it would be subsumed within the contiguity of time and the effects of its products. The forms of thought would be conjoined with the forms of historical change through contiguity, and to some extent these two types of forms would be syncretically intertwined. An example is magic. If a cloud appears and then there is rain, not only can there arise the belief that clouds cause rain, but also, by analogy, homeopathic magic can become a "form of thought." A magician can try to act like a cloud and expect that rain will occur. Through a person's mature stages, the syncretism would last—in the form of "totems of thought." By this phrase, I mean that icons would be made stable points of departure for the accrual of new classificatory divisions and that logical ordering would remain, therefore, protological. Mind and soul would become devotionals to the cosmological "church rules." Thus, whole social and political belief structures can be predicated upon a totem—one such as "the royal family" and its proclamations and beliefs. The subject of a kingdom mirrors the dictum that the king's word is the law of the land, and the corresponding belief is that whatever the word, when uttered, it becomes truth and beauty. Shades of Panglossian thinking!

Rhetorical Degradations: Stereotyping and Propagandistic Forms

A syncretism of mind is virulent; it is difficult to shake off when it reflects a congruence of the individual's forms of thought and the forms with which he is presented. Certain tropes can grab a hold on your beliefs—even a hold on the way you form those beliefs. If the tropes you're presented with are formed by part-whole exchanges or by contiguity-inspired conceptual domains, you might (a) believe them and (b) go on to think by part-whole exchanges and be influenced by contiguity as a phenomenon and as a form for categorization. With rules of this sort, stereotypes are easy to transmit and difficult to get rid of. Propaganda and its half-truths, based on synecdoche, become easily successful when the recipient's thinking is syncretic—both in its reception and in any production of propositions falling within the sway of the forms of thought engendered by and intrinsic to the propaganda.

Ohnuki-Tierney (1991) describes the part-whole form of synecdoche as the basis for political stereotypes. "We" as a part-for-whole transformation expands its referential space from some anointed privileged group, person, or way of speaking to "all" persons. The "royal we" is one of these expanding stereotypes. The king becomes a "We," and thus the "we" is a stereotypic category for anyone in an asymmetrical relation with another person. Graham Scambler and Nikky Britten (2001) acknowledge sociologist Talcott Parsons's seminal description of the asymmetrical role relations of doctor and patient. Their review points out the importance of ways of achieving a helpful and nonpathological dialogue. When the doctor or nurse asks, "Are *we* ready for our medicine?" stereotypes and propaganda fly all over the lot. The *we* classifies the patient and the doctor as both members of the same set—those who would follow and believe the doctor's orders. But an irony is thrown into the recipe. In communicative and grammatical terms, the relation is "subject" and "patient." So the "royal we" becomes an "all"—with the irony that "some are more equal than others." Osama bin Laden and his followers have the same beliefs and mantras, but some give the signal for others to be the sacrificial lambs.

It's one more step to make that "all" into a class with either an opposite or a complementary class. Then you can have a really vicious stereotype. The members of the "royal we" are the good guys, but think of the Japanese as the emperor's men from the point of view of Hollywood movies of the 1940s. Stereotypes can exist as secular politics, with the master tropes tied to such organizing metaphors as "race." The "master race" of the Third Reich was the brutal protostereotype of the twentieth century. Religiosity can be brought to the process of stereotypy and political categorization as an emotional—not to say symbolic—motivator to accept the synecdoche structuring. One can go into current events and history for the many examples of religious wars and find the religious symbols combined with the nationalist symbols on flags carried to identify each "side" of the combat. Jihads and crusades are examples of religious/political movements that involve stereotypes of both the righteous and their opposites. When religious imperatives, papal bulls, and even the antireligious sentiment of certain Communist movements of the twentieth century governed political stereotypes, it was not merely the person's cognitive process whereby primitive beliefs were bifurcated into self and other. Instead, the religion-soaked ideals as stereotypes and their opposites—or targets—stoked that process of primitivation of perception and forms of thought.

The primitive form of thought may be something like

$>A \equiv A.$

This looks like the tension between categories that Ohnuki-Tierney describes. The categories >A and A are not identical, yet they are equivalent—landing us somewhere between their fission and their synthesis. The "stoking" of primitive-like beliefs would be like the invoking of a belief giver. In desperation and darkness, we issue the call for a muse, a prophet, a leader. It can be a teacher, a preacher, a mullah, or some version of a grand inquisitor. He might be, as in pre–World War II Japan, a divine emperor. Call the "belief giver" X. With X, we would have a form of thought like

In the name of X, >A ≡ A is the case.

The stereotype >A is equivalent to the class of persons, A. In the wartime movie *Five Graves to Cairo* (Brackett and Wilder 1943), Erich von Stroheim portrays the German general Erwin Rommel as a time-obsessed person and as someone with no regard for the individual feelings of those he subjugates. He is drawn as imperious—with a sense of the incontestable and insuperable "rights" of those in power and of the lack of rights for those who do not have power. Breaking the stereotype of the German soldier required some distance from the war, so more than a decade later in the American movie *The Young Lions* (Anhalt 1958), a German officer, Christian Diestl, portrayed by Marlon Brando, is a person with both a sense of duty and an awareness of caring for individuals that makes him veer from duty.

In the playing out of stereotypes and propaganda, the portraits and promulgations *de-self* the individual mind. The stereotype deprives the person pictured—and the person picturing—of a self that would otherwise perceive, attribute, and make judgments. Sadly, one or another of these deprivations does a rather complete job of negating the self.

Compounding the surrender of self to the promulgations of others takes the function of the "I" out of the structure of the form of thought, and therefore primitivation of form is a cause and outcome too. I have set forth a context that includes historical sociocultural and political factors. Philosopher Hannah Arendt (1951) carefully and thoughtfully accounts for how this was done in Nazi Germany. The state and its laws became the instrument of promulgation, ordering the surrender of the rights of others and of the way of having nonconforming perspective on thoughts, rules, and pronouncements. In the wake of these events, the role of tropes, totems, and superordinating metaphors in the transmissions of state ideals, imperatives, and values became greatly exaggerated. But independent thought, even in science, can thus fall prey.

Root metaphors are analogy based, and they govern the categories and accrual of evidence in science (see Pepper 1942). Although not easily susceptible to change, if stereotyping, especially state-promulgated stereotyping, were to assign a root metaphor governing a science of persons, this metaphor might be subsumed within a totemic race concept. Thus, during the Nazi era, some important work in philosophy and psychology was allowed to go forward pursuant to the "racial" exclusion of Jews (Ash 1998). Nazi legislation took academic jobs from such notables as William Stern and Heinz Werner. Many others were driven from Germany (Sheppard 2001). Historian Jacques Barzun analyzes the general situation, pointing out the "principle that Hitler exploited in his Third Reich. A nation is forged into unity by successive wars and the passage of time. When this result has not been achieved, some other means must be found. Pseudo-science and determinism suggested faith in race as a substitute; it is inborn, a 'natural' unifier, and it is present in each citizen; it can be made conscious, it bridges over religious, political, and class divisions" (2000:695).

Thus, sociopolitical causes can be shown to stimulate political and psychological primitivation. Propaganda can be a purposeful instrument to achieve an exchange of unconsciously inspired modes of thought for more conscious and mature ones. However, in describing something like the "return of the repressed," I do *not* want to overemphasize the role of social and cultural factors as delimiting the forward-moving evolutionary machinery of species change! Nor do I want to underestimate the vicissitudes that bring about primitivation—or, in Barzun's (2000) sociohistorical terms, "degradations" of art, philosophy, and styles of life. My purpose is to come to an acceptable picture of what a great "return of the repressed" and its devastations of the meanings and values of humanity might indicate. This picture should take into account not only the preference for the primitive in art, but the effect on the thinking and the self of persons. Here, then, is a broad assumption—something Jaynes-like: *the mind and self advance in evolutionary terms.*

One problem in making this assumption is the dubious notion of finding aboriginal persons so isolated biologically and culturally that they would be "primitive" (see Stanley D. Porteus's [1931] hypotheses). However, I simply claim that persons should have structural potentials congruent with the biology and evolutionary effects of their given era. If these effects are present at a given point in historical and social time, what ensues later on can be assessed to be incongruent with that era's values and with the capabilities of persons who lived in that time. As this assumption stands, it is one of social and biological interaction, and it would require analysis of the relation of the genetic given, the epigenetic unfolding in conjunction with supportive environments, and the malleability of human nature in respect to social and cultural factors. That analysis is beyond this book's scope. I can only briefly explain what is necessary to evzaluate this notion of the advance of mind in an evolutionary mode.

A "given era" would have to be specific to historical time and cultural context. So, in terms of purely counting centuries, we would have to exclude the society or civilization that "never was" in an advanced "era," and say that a "given era" would have to include evidence of accomplishments and products commonly agreed upon as requiring a certain level of cognitive maturity. The kind of thinking that advances concepts of individuality and presumptions of the individual's capabilities and that reinforces these concepts in societal and political documents may be evidenced in some societies and times, but not in others. *If there is a retrogression, the status of "once had been" would be enough evidence to assume an advance in "eras."*

The obvious example is Nazi Germany. Art, music, and philosophy can be seen to have progressed to a certain point. Then, due to a variety of factors, the civilized mode seems to turn backwards in time. We have a "once have been." Therefore, we would not suspend the assumption of an evolved mind and self reflecting an advanced "era" and therefore including characterizations of the products prior to the retrogression. In Nazi Germany, the sense of the self disappeared into power vortices. Stereotypes and their reinforcement by the propaganda machines, tropes such as "Arbeit macht frei," totemic symbols such as armbands and "patches" sewn onto clothing, and legitimations through "laws" portrayed persons as less than human and with no sense of the sanctity of their selves. And people behaved as if they believed the negations of self.

Yet the Nuremberg trials were predicated on the existence of the self as uninterrupted; responsibility could not be denied by deflection or displacement onto state propaganda or stereotypes or "orders." Barzun states that "[n]o period style affects the entire population. A majority remains untouched by what is most visible in the age, yet without changing the style that it declines to follow" (2000:786).

Structures of stereotypy as beliefs during this century may give the "era" a debauched status. Yet these structures, however deeply they penetrate style of life, transform but do not destroy the signal capacity for self-awareness and perspectivization of the forms of thought. *The function operators of the "I" work in conjunction with the various transformations of an "I" faced with sociopolitical negations. They may even be transformed within the individual and be subordinated to the individual's acceptance of rules prohibiting the "I's" functions. Yet these "function" operators would still be present and at work.* They would be dynamically engaging the primitivation process and thereby engendering the construction of stereotypes and other similar forms of thought.

Accordingly,

$$f\bar{I}[(I = \bar{I}) \equiv fX(W = A \cap \bar{A})]$$

can be read as follows: as a function of the negation of self, $f\bar{I}$, the self cancels into a nonself or other, $I = \bar{I}$. This state of affairs is the equivalent of the function of a belief giver, fX, in relation to the process of asserting a privileged class, W, to be identical to a broad class of "all of some group," A. Moreover, this broad class is separate and either opposite to or complementary to the class, $\bar{A}$.

In a nutshell, the "I" is there in a transformed logical state. Within this state, the $f\bar{I}$ is *not* the same as the other—the other person or persons. It is also not the same as the Other (any force or entity outside of the self). It *is* a role of the self, standing in *as* the self. To the extent that the individual accepts a definition of herself in terms of the roles she plays, the $f\bar{I}$ is an "understudy" for the self. (The belief would be that the self is still $f\bar{I}$, but an $f\bar{I}$ set aside for the performances of the "era.")

The Function of $\bar{I}$, the Negation of the "I," and the Protoself

A person having the belief that the self is a matter of the roles it plays has engaged in a part-for-whole synecdoche about the self. The only way around this would be an infinite regress by which all structures and experiences reduce to the actions associated with a variety of roles. The specification of actions would lead to a description of the role of recognizing the role the self is playing. The structure, which would be explained this way, would be something like *self-concept*—or the self's capacity to look at and characterize itself. However, even if we go along with this regress, if it comes to a stop at a concept of self available as the fI (and/or its under-study), we wind up with an illusionary absence of the "I."

Sociopolitical determinants and temporal successions result in "eras" with the manifestations of self-initiated thought and the recognition of self. Advances in perspective come with social adaptation, cultural accretion, long-term biological adaptation, structural change affecting forms of thought, and, in short, species change. So as a product of biological evolution and whatever long-term selection is involved as a partial product of social factors, contexts, and changes, there is the emergence of the auto-regulative system, its dynamic intentions and movements, and its sign systems and exchange regulations and rules. This system and its features and capabilities constitute a *protoself*. Thus, it *is* possible to have a self that does not reach the status of an "I."

When you can discuss that protoself, you are faced with the hard-to-explain assumption that your "I" is there. But look at the bright side. The "I," aware of the protoself, can observe its primitive state of affairs. Therefore, once the observation is made that the self is a collection of roles, we can stop the regress at the fact that it is the "I" who makes that observation.

The protoself can be assumed to exist as a "primitive" state of affairs. The sentience of the autoregulative system is syncretic to the entire organismic system, yet the protoself is *not* aware of the sentience as something that can be decided about. So that sentience cannot be a "next-highest" level of awareness! And it cannot become one. There is no "I" to be bracketed out! However, just as in the recall of a dream, as best we can know it, this type of sentience is only retrospectively apprehended as an experience without the aware "I" and its bracketing capability. The moment of the retrospection involves a *retrospecting "I."* Therefore, the primitivation assumed is also a retrospective product of the "I." Once the retrospection takes place, we can say we assume the knowledge of the primitivation is achieved by negating the "I."

Phantom "I" and Understudy Self

There are probably many ways to negate the "I." One way might be through the sociopolitical belief that the "I" exists only as a series of socially learned actions and operations, which come from a sociopolitical-cultural context that dictates values and rules governing the social transmissions and practices that make up a "style of life" during a given era.

This "I" would then be explained by a thesis like Barzun's (2000) "age of individualism." Within these assumptions, you can certainly "locate" individuals. But as sociologist Erving Goffman (1963) points out, although there may be linguistic acts of locating the individual subject as the actor playing out specific roles in given schemata or scenarios, the "I" may not be an inner experience or causal agent. Strictly speaking, within this "sociopolitical-cultural context" thesis, the "I" is not fundamentally an inner experience or a causal agent. Instead, it is a social command—with other commands about due allegiance to the particular social group or entity making the command.

In this way, at a nonprimitive stage of societal and sociopolitical history, the function of the "I" gets picked up by the fĪ in its role as an "understudy" for the self. An easy example is the highly "individualized" variegation of hairstyles that became popular with "punk culture" following the 1960s. Without this dubious, shocking, and yet socially "in" type of expression, the person's individuality would be in question. The very outrage of the hairstyle in its deviation from norms would permit social definition and attribution of individuality! So the "punk-and-particular hair" understudies the individual aspect of the self! This use of the "understudy of self"—something like an invasion of the body snatchers—is a reach for the destruction of the "I's" consciousness (at least, for its negation). Is the protoself (PS) the only remnant of self that survives the invasion? No; instead it's an odd nonconscious state of affairs depicted by

$$\text{fĪ} \cap \text{PS}.$$

In this expression is the dissonant chord of the fĪ. Without it, we would simply have the protoself, PS, as the nonconscious state of affairs, which would be more easily understandable. But in fact, the fĪ *is* perceived and placed in the position and role of a function that replaces the "I think" bracketed out of the Ī as a complementary category that includes all sorts of possible instances of a nonself. Hence, in a strange sense, there is awareness as an *out-of-bracket phantom*, and

$$\text{fĪ}([\bar{\text{I}} > \text{PS}] \equiv [\bar{\text{I}} \supset \sim\text{fĪ}]).$$

Of course, the fĪ in a bracketed-out position is a contradiction in terms. This is a major flaw in the attempt to define the self in terms of the other.

Anyway, the outcome is that the "I" *is* out of reach; the self is out of any schematic; and the search for negations to resolve does not come to rest in any way that it can be grasped. It cannot be grasped by the subjective "I" aware of itself, nor can it be grasped by the understudy because the understudy is only an *out-of-bracket phantom*. The fĪ is the organism's icon while it indexes some order of awareness; its state of negation cancels out its agency.

I can say that a social state of affairs teaches me a set of rules and promotes behaviors based on them. That's an attribution, but it cannot be confused with my act of attributing it. With the fĪ functioning, I can be aware of the attribution, but not of my having a causal role in doing the attributing. Some might say this is a case of a "me" without an "I," but that formulation doesn't do it because there is still the matter of my being aware of the "me"—which takes things at least one step beyond a sentient protoself.

So with all this relegation of the subjective "I" to a phantom status, the protoself machinery rolls on. There are now many reasons that the individual not only continues searching the existing negations of self, but also searches for new negations of the self that threaten the understudy from the shadows.

A serious example is the character Willy Loman's search for the negations of self, which threaten his image of "the well-liked ≡ effective person" he believes he once was (Miller 1949). The loss of this self by way of others who consistently do not find value in his continued presentation of the "well-liked self" decouples effectiveness from his self-definition. (The relentless consistency and the repetition of the instances in all spheres of his objectives bring the decoupling home to a denuded fĪ! He is now aware of who he is not.) Willy faces these negations of the "understudy self" and their relation to any self that he might have had alternately. In doing so, however, he encounters such a flurry of self-negations that self-worth expands only by the destruction of self.

Frenetic Search and the Binaries That Replace Accessible Negations

The phantom "I" and the understudy self become *frenetic* in their search for further negations of the self. Binaries are likely to occur as relief from the alienation from self and the inaccessible negations to be resolved. The binary for Willy Loman is the idea of appealing to others to be liked versus flirting with disappearance as an individual if he is not liked. If he experiences the negations of the self that others offer, he faces the "I" who has no apparent external source of definition. Thus, we come to the binaries "search for liked-self versus no individuality" and, more noxious, "social self versus alienation from the self who would be defined by the 'I.'" These binaries are exponentially raised to oppositional status by the inaccessibility of external sources of identification, by the absence of internal markers of identity, and by the neutralization or paralysis of causal power to find negations. Such a state of affairs is frenetic, and the proto-self can take over as soon as a route to destruction—as a way of defusing its urgings—presents itself. Loman's last passion.

Prototypes of Conflict of the "I" with the "Ī" and X

The agony of the fĪ and its endgame are perhaps a prolegomena for the twenty-first century. The age of individualism descends into decadence. Barzun (2000) isn't the only one to see the descent. The excesses of individuality can be ubiquitously documented in the disregard for classical social rules, roles, and customs in art, cultural "styles" of life, and mores. A conservative outrage, if not "disgust," appears and calls out for classical order and—yes, Gombrich's point—for good old "simplicity." The George W. Bush administration in the United States seems to fulfill the need for what a British journalist called "the politics of simplicity." In addition, Kugler's (1993) point about the return to theological concepts as governing symbols in the wake of diminishing the role of individual expression can be related to many trends in the conservative U.S. political and religious agenda. For a variety of reasons, in the twenty-first century, we have entered into a postindividualism era, where armies are important, ideology trumps the value of life, and religious zeal is in fierce competition with the expression of individuality. The conflicts between decadence and disgust seem unabated in their tendency toward dangerous entropies—such as the disregard for individual lives and the attacks on the innocent, to name just two upheavals in the impossible counterbalances being attempted.

The conflict between the "I" and the "belief giver" produces patterns on a very broad scale, and these patterns reach across many eras of mature and so-called civilized status of thought forms. They affect the nature of self, the awareness of the "I," and the power of the externalized figures of speech—tropes, stereotypes, propaganda, politically correct rationalizations and reasoning patterns, and, in general, the relation of the individual to knowledge.

The "I", the Belief Giver, and the "Ī"

Here, I describe classic examples—Socrates and Galileo. They are mythically packed into our archetypal conflicts between the internal "I" and the power of belief givers to define and bound the "I." But they also ontogenetically repeat themselves in the lives of those who have survived and moved from the prehistoric syncretisms of mind to the present era of differentiation of self from Other, of mind from external context, of species from genus in logical ordering, and of literal from figurative in perspective—with or without societal paroxysms of primitivation from time to time.

SOCRATES   As Plato shows, Socrates defines his self in terms of what he does not know ([1892] 1996:para. 7). He uses the negation of knowledge as a tool to extend to his own knowledge. By his own powers, he cannot "know" about himself, about any wisdom he might have, or about what his "I" would be. This leaves him with some sort of *"pure awareness" minus an "I."* Hence, from the word go, *Socrates' premise is that he is an Ī.*

With the support for the self as Ī in the idea that the God of Delphi recognizes nonknowledge as wisdom, the buck stops at Socrates' right to share this form of "wisdom"—to "give" it to others. This right is what he tries to get others to discover for themselves, just as he is convinced of it himself. He *is* convinced of it because it is not *his* knowledge; instead, the God of Delphi—the belief giver, or X—"gave" this belief. Plato in the *Apology* writes these words as Socrates': "I will refer you to a witness who is worthy of credit; that witness shall be the God of Delphi, he will tell you about my wisdom, if I have any, and of what sort it is" ([1892] 1996: para. 6).

What happens when the state assumes the role of the believer giver X? The state declares to Socrates that he has no right to "corrupt the minds" of the young with this idea. Unless he recants, he must drink hemlock and thereby die. He has to give up his belief in favor of the belief of another X—the state. He might convince himself to live, but that would violate his right to his ideas and introduce the gigantic conflict of having an "I" in addition to the Ī. He might recant the value of his teaching, but that would come to the same violation of his right. (If he recants his gift, it will result in not giving others a path to knowledge as examination of knowledge!) Alternatively, if he gives up his life, he can accept X's judgment and yet not violate his own belief and his right to have disseminated it. Here the Ī extends its base, clearly including the God of Delphi, the court, and Socrates' living self—as well as the "I." However, pouring into the Ī is also the dramatic consistency of Socrates' belief in combing out inconsistent beliefs. This drama projects the dialectical mechanics into a martyrdom machine that remains the legacy of anyone who undertakes to use the method.

Socrates accepts X as the belief giver of the proposition that he should die *if X so decides.* The state makes the decision to implement the fĪ as a replacement of the living Socrates. The "I" of Socrates will not exist anymore. Socrates' "I," as he casts it, is in the form of "Ī." As the living Socrates, however, this Ī is an "embodied" form. So, somewhere in a lattice-arrangement of protologic, the "I" *can* surface! But, even though it would be subordinated to the Ī, as such it would be *only one instance* of the latter. (A complementary class contains everything under the sun!)

With Socrates' death, the Ī can exist—certainly in the form of a function (fĪ), which can in principle be utilized by someone else's "I." This outcome may entail a significant replacement of many aspects of Socrates' "I." In sum, the state's judgment that Socrates should eliminate his life— and perforce his self and his "I"—leaves his teaching intact, albeit with some order of contrariety in his not being around to do his thing. Thus, Socrates gives over to the state the right to judge his life and will not argue that they should not have that power. Yet he does argue that the accusers should not be so sure of their position, and in his typical style he torpedoes their arguments, dialectically showing contradictions.

For Socrates, the search for rectification of self implodes to the extent that the death of the self becomes his teaching example—which apparently cuts short the Socratic Method of dialogic exchange. He argues that as "God's gift," his style of showing that any authoritative opinions have holes in their reasoning is not something they can replace. There is unlikely to be another Socrates—a point consonant with identity as Ī. There may be "Socrates machines," but no one to invent a new or superseding one.

Look at the fĪ as method. Where advocacy, counteradvocacy, and dialectical rhetoric can be codified, and where Socrates himself is expendable by the same rules, we have social, societal, and sociopolitical roadblocks. An individual such as Socrates accepts the rules, the game, and the codifications. But that individual, if expendable to the state—and/or to X the belief giver—is blocked from further articulation within the sociopolitical context. The roadblocks also neutralize Socrates' followers because they are bound to his subservience to the rules, to the rule giver, and to the societal game board. However, there is a crossroad. The dialogic method moves inward—not only for Socrates, but also for his pupils. It moves inward for Socrates in that the external dialogue is not necessary for him to reach a decision about what and who is correct and what his rights are. It moves inward for others in that they can learn his method as it is without interaction with the live Socrates. Although this way of analyzing this matter is inviting, Socrates depends on the state, as the belief giver, to engineer the transfer of his method to others and to make it possible by making his own life and self no longer possible.

GALILEO: SOCIOPOLITICAL POWER, THE BELIEF GIVER, AND PERMUTATIONS OF THE "I"   The voice of the Inquisition—presumably talking for the Catholic Church—tells Galileo that his heliocentric view (U1) is incorrect and that the earth is the center of the universe (U2) (see Nirenberg 1996; Machamer 2005). The church—via the voice of the Inquisition—is X, an external belief giver. Galileo is also told that if he doesn't recant; he will lose his life. He does not believe U2, nor does he believe that his view (U1) is a mistake. He decides, however, to *say* that U1 is a mistake, and by doing this he saves his own life. *What part of the self is violated by saying that what you believe is what you don't believe and by saying that you accept X's view even though you do not?*

Galileo accepts the authority of X as belief giver—or at least he says he does. So X's authority is as the giver of what you can *say* you believe. If you decide to recant within the scope of this sort of authority, other people may not believe your recanting—or restatement—of what your internal belief is and what fealty you have toward the believe giver. When a political leader wields power such that you feel you will lose everything if you do not say what he wants to hear, it has profound effects on how you and others may see you and view the truth of what you say. You are thrown into paroxysms of metaphoric exchange. You want to be consistent in your beliefs—as in a logical/psychological view of the self. You also want to save your life and your inner purposes—consonant with a fragmented identity and with what I describe as the "constituent" view of identity (see

p. 233). While you shuttle back and forth in thought, a person outside that sociopolitical culture or domain cannot determine whether you are doing so also in your actions or statements. Thus, in a culture where a leader or an institution wields such power, a person outside that culture may neither believe your position on an issue nor accept any protestation of your loyalty to the leader, institution, or culture.

There are many examples of this situation. Ash (1998) chronicles the difficulty of the great German gestalt psychologist Wolfgang Kohler in his attempt to navigate between a loyalty to values in science and loyalty to and/or fear of the state. The state wanted Kohler's stewardship over a prestigious journal to comply with racial policy against Jews, but Kohler tried to walk a line between not opposing Nazi rule and not making the journal into a political instrument. This one individual's travail is a mini-drama when compared with the epic expansions of such phenomena. The Nixon era and the "president's men" offer a recent example of individuals caught in such conflict. A study of the ambivalence and submergence of individual judgment at the hub of political power in a great state reveals a phenomenon that ballooned to worldwide significance. Thus, within the United States more currently, the significant waffling over facts in reporting what was and was not known prior to the U.S. invasion of Iraq produces tension between rule givers' loyalties and their belief or disbelief. The dramatic conflicts between Colin Powell's public proclamations, on the one hand, and his private views and sense of what his compromises brought about on the world stage have yet to unfold fully. When there are such dramatic conflicts, what is at stake is not only the individual's belief in the beliefs given by an X, but also the individual's belief in her own ability to form and express an independent belief. Both modes of deriving and accepting beliefs are assailed within the society, across societies, within the individual, and between the individual and the state.

On the world stage, we can see degradations of sophistry committed by clerics, like those of the Taliban. There are all sorts of explanations for terrorist murder of innocents. Some are put forth in religious terms, some in political terms, some in sociocultural terms. There are unexplained suicides by middle-level bureaucrats; there are depressing levels of comedic relief. At the end of Saddam Hussein's rule, the Iraqi information minister continued to make almost psychotic statements to support a state despite TV images showing it in obvious collapse. *At present in the geopolitical arena, such an awful situation exists that the fabric of truth—wo-*

ven together by individual logic, its subjective elements, and the interplay with consensual validation—is in tatters.

Yet to the extent that your "I" remains essentially defined in terms of what you say to yourself and what you believe internally, your "I" appears to emerge intact. Call this "the formula." In Eugene O'Neill's play *Long Day's Journey into Night* (1956), the author bases the characters on his own family. In the play, the father, James Tyrone, appears to the audience as a person having illusions about himself. He seems to accept the illusions, so his "I" appears to stay intact. In contrast, his older son, Eugene Tyrone, finds his illusions unacceptable. He continuously digs to deeper levels of self-revelation. In his doing so, however, "the formula" works again. He winds up believing his insights into himself, so *his* "I" also remains intact. In contrast, the mother in the play is so ambivalent about her extremely varied ways of feeling and acting toward her family and toward herself that she does not face the problem of believing her view of herself. Instead, she relies on a drug addiction to keep the conflict from tearing her apart. So "the formula" still works—but in reverse! Because she does not believe what she says to herself or to others, and she cannot entertain the idea that others would accept what she knows about herself, her "I" does *not* emerge intact.

Two issues emerge as paramount: How coherent are the individual's subjective beliefs, and how much are they at odds with observers' external or objective beliefs? Still another query concerns the dependency of the "I" and the self on a stable identity.

If the "I" is the index of the self and of its coherence of beliefs or propositions concerning the self, what a person says to others can be disjoined from his experience of the "I." Truth inherent in the person's inner harmonies *can* be separated from public statements. The bad side of this deal is that the individual can say what he does not mean. Not every dissimulation is acceptable as "for the greater good." The good side of this deal is that the individual's work can be judged objectively and aside from personal peccadilloes. A person's work can have its own meaning and can become an essential function of the self. We have the image of a scientist such as Albert Einstein, whose work is more important than his social presentation. Among the O'Neill characters, Edmund Tyrone, the younger son, can make a decision to cast the lot of his self-evaluation in the form of his writing. So in the play that character's social impact on others is "played down," and the character is funneled into his attitudes toward his work. Moreover, the character appears to do this within himself and to be seen—and accepted by others—as doing so.

Logical/Psychological Identity, the Integrity of the Self, and Two Forms of Entropy

For biological reasons evidenced in the dynamics of autopoiesis, for psychological reasons of integrity, and for logical reasons involving the coherence and bounds of structure, we try to keep on top of ourselves! No one likes to feel he's "not himself," "beside himself," "not in control of himself," "divided within his self," and so on. Therefore, the game objective of "self relative to the belief giver (X)" is to work out a way for the "I" to be in such a position that either the self merely stays afloat or some point of integrity is shared among the self, the "I," the functions of the "I," its beliefs, and the beliefs given by X. If all this isn't enough to balance, consider that the "I" has beliefs about itself, beliefs about its beliefs, and beliefs about the belief giver. I try to put these in some order before taking on the case of Galileo again because, as you know, he appeared to have decided to hang around, stay alive, and face the divisions between his beliefs and those of a belief giver powerful enough to condemn him to death.

Rules of Reflexivity of Reference and the Governance of the "I's" Loci of Action and Experience

The "I" is related to the self by various rules of "reflexivity of reference." These rules, in turn, govern another set of rules—namely, those of locus of origination of experience and action. "Reflexivity of reference" is marked by the status of a term I earlier referred to as a "shifter." The "I" is such a term because it can be in different positions or modes from which it makes perfect sense to shuttle meanings to the *same* term, "I," in yet another position. Thus, the sentence "I tell you I told her I was beside myself" can be read to mean there are different versions of "I" or that the same "I" can refer to itself at different times and in different modes, roles, and so on. Rules for such reference can depend on a definition of the "I" as agent. Thus, the "subjective" "I"—think of it as the f"I"—can decide to utter the sentence given. In it are at least three instances of the "I" as agent—relative to specific acts located at the moment of the action and at the place of agency. The set of rules of "reflexivity of reference" include a subset called "locus rules"; these rules govern attribution of specific experience as agency and action. They govern where and when the "I" is attributed its causal role for its experiences and actions. *In this set of rules, with their specifications of self-relations and the subset rules governing the indexing of loci of time and space, we find the logical structure of identity.* Willy Loman's rule moves backward in time to instantiate that others' acceptance of him adds up to his idea of the quality of his "I." The references to himself and his "I" have a particular pattern and route by which he pictures himself, shuttles meanings about changing values of his same "I," and denotes what begins and ends causal sequences.

Self-doubt and a series of negations may be psychologically haunting Willy, so there is the issue of psychological identity as well as the matter of its logical structure. How can we picture the fit and application of logical identity to the psychological phenomena of the "I" and its beliefs?

Conceive of the "I" as a superset, and the subsets of the "I" as propositions and beliefs across many domains. An individual has beliefs about the nature of the universe, about her country, about other people, and so on. These beliefs are different "domains of belief." On one hand, it's easy to see that the function(s) of the "I" (f"I") can affect the various domains of belief in different ways. Some even argue that in relation to the different domains, there are if not different "selves" of the individual, then different versions of the self or the "I." However, there's a more coherent way to view this. One thing to include within the domains of belief would be the subdomain of the beliefs about the self and the "I."

I now call the beliefs about the self a subdomain instead of a domain because of these beliefs' status as a subset of the "I." If we keep that subdomain in mind—along with the full range of the individual's belief domains—then the f"I" and the "I" can be brought together to produce logical identity. This would be the case when there's an equivalence of the f"I" as an operator as it crosses from one to another kind of belief among the various domains. If the f"I" *is* viewed as an operator, it would follow the rules of reflexive reference. It would have the same properties and effects when distributing the "I" and its agency to the different domains. Think of how Willy Loman applies his likeability/achievement formula to all the people and situations he encounters. With these rules, there is *coherence of propositions and beliefs about the "I, its functions, and its causal status across domains. The coherence provides a basis for the f"I" to be equivalent across the domains and therefore signals the "I's" equivalence in relation to the f"I."*

One of the effects of these equivalences of the "I" and the f"I" is a congruence of the "I's" beliefs and her sense that these beliefs are "truly" hers. On the one hand, an advocate can argue a belief, but not believe that she believes it. On the other hand, she can argue another belief and regard it as something she truly, deeply believes. Any cognitive module the individual adopts as her true commitment (say, a group of beliefs) has something so essential in common with any other group of beliefs that the "I" can *be* (or be coterminous with) a governing category (f"I") for the whole coherent lot. Let me symbolize this relationship as

$$f"I"[(B_{1\ldots n}) \cup (BI)].$$

Within the originating and causal functions of the "I" (viz., f"I"), the various arenas of belief $(B_{1\ldots n})$ are in union with the specific acts of the "I" that include its deeply held beliefs. I note these acts as BI, which I define here as the *"I's" perception of truly believing that "I believe..." is the case for "I."*

BI may include one or more or a series of propositions that the individual holds as beliefs $(B_{1\ldots n})$. For example, let us say that BI includes just two propositions. We can then say that the target series is $B_{1,2}$. The first of the two propositions is the belief that the state has the right to make rules that have an impact on the individual's self-expression. The second proposition is a more specific belief—that the earth revolves around the sun. The BI would be the person's belief that he truly believes both of these propositions. In terms of beliefs and attitudes, we can say that this sort of "true commitment" has the feel of "resonance" and is of the order of a "sentiment." Another perspective on the BI category is to say that it involves beliefs about the self because the sentiment that "I believe that..." is perforce a belief about the self and conjointly a sentiment that would govern any belief about the self. The net result is that where

$$f"I"[(B_{1\ldots n}) \cup (BI)]$$

is the case, the f"I" has achieved an equivalence with the "I" as the individual's logical identity:

$$f"I" \equiv "I."$$

This analysis is in the terms of "I" as the superset and the propositions and beliefs described as the subsets. The question of logical identity can be described in terms of the equivalence that obtains when the set elements are the same for each of the sets involved. We ordinarily think of sets as independent of perceptual vicissitudes. A logical concept of a set and identical elements appears to subtract the discrete and concrete features from a grouping such as a semantic concept. Abstract considerations are left, such as the numerical values or the common operational functions of identical algebraic terms. However, even though I use the terms of logical forms, my description is of a logical structure with its governance and its dependency on psychological phenomena. In sum, we are working with icons, and I call the identity that I describe *logical/psychological*.

The Logical/Psychological Account of Identity and Its Relation to the Constituent Approach

It should be helpful to account more specifically for the psychological aspect. For this account, we need terms closer to a more concretizable semantics of self than to a pure logic emphasizing abstractions. An approach to describing psychological identity is a *constituent* idea.

Identity is made up of a number of properties, and if all these properties are on hand, they constitute in their totality the individual's identity.

This account of identity is inductive, and it becomes a matter of observing and of the observable qualities of the full complement of properties (cf. Hanna's [2006] analysis). How does such a view fit into an account of the psychological character of identity? And how does it help in the search for the "I's" logical functioning and status?

Even with all good intentions relative to the constituent approach to identity, we know that the description of individual identity should still have a logical character and be subject to deductions of denotative bounds. *This understanding is essential to the description of Galileo's balance of identity, belief, and relation to X (the belief giver).* Thus, Galileo's psychological identity would have features such as beliefs and needs. His possession of features can be expressed not only in propositional terms of subject (Galileo) and predicate (believes that...), but also in terms of *properties* (beliefs about the self, BI) and *individual constants* (Galileo's, symbolized as G). In a sense, the properties add up to Galileo's identity, and the constant is the categorical requirement of the full complement of these properties for the individual identity of the person (G). So, for one thing, a constant for Galileo is his name or the observable qualities he provides for the attribution of that individual constant. There can be a "properties/individual constant relation," which can be symbolized as BI_G. The total property list has to be assigned to G, so G's identity is no more than the property list, but the list belongs to the person with the constant, the name "Galileo," and whatever aspect of the constant that makes the name attributable.

As far as the psychological identity of Galileo is concerned, it can be established but *not* merely by asserting G to be equivalent to his experience of "I." We also work with a buildup of the "I" by way of its equivalence with the sum total of its properties. To make sure of the equivalence, the properties would have to be those of the very specific features and subclasses of meaning consonant with the consideration of G as an individual constant. So we have on hand semantics and extensions by which to account for the individual person. What's necessary is a judgment that these properties are those specifically reflective of the person as an individual constant. Now for the most difficult point.

Whether it is Galileo or it is others who are making the judgment of that consonance, the properties (and their eventuating in the self) become an *object* of that judgment.

If Galileo is making the judgment, the proposition is, "I judge BI_G to reflect my 'I.'"

You can see that the "I" in the proposition is an object of the judgment. As an object, it is the "me." However, it is the object of an "I," rather than a social product, as Mead (1934) conceives. *Whether Galileo himself makes the judgment or others do, the judgment would correspond to an object as the index of the person and therefore define the "me."* A syntactic formula makes the point of equivalence clearer here: "The sum total of the 'me' *is* the 'I.'"

This sum total idea is to list all the aspects of self, including the beliefs as features, meanings, and ultimately properties that can be extracted as objects, yet in their particularity (and by way of their completing the category defined by the totality of these properties) characterize the individual constant. All persons may have beliefs about their destiny as contributors of knowledge. In the case of an individual, call that kind of belief D. Galileo's belief D might be particular to a qualitative view of what he has to say about motion. Put a subscript to D like this: D_{ML}. This might mean that Galileo believes his destiny is to contribute to the concept of motion ($_M$) on a level that would provide knowledge of a "law" ($_L$). The D symbol represents an important *property category*, and the D_{ML} represents a subclass.

There are many other property categories and their subclasses. With this sort of logical conception, if we can list all the particulars—properties and their subclasses—and take into account all of *their* particulars, then the self can be constructed (or reconstructed) by adding them together. In addition, quantifiers can represent some particulars. So where a quantifier is ∃ (there exists at least one...), we might symbolize D_{ML} with a quantifier as follows:

$$\exists \, (D_{ML}).$$

This quantifier would mean that there is at least one belief D_{ML}. In this way, different kinds of markers in different positions can particularize the features and functions of a property. I *do not* enter into a treatment of what types of symbolization to use or how to keep the symbolization directly in line with that of a usual symbolic logic analysis. What I *do* indicate is that the different symbolic possibilities can be fit to the richness of the particulars of a property such that the property "belongs" to the list that can "construct" the individual identity.

From the *logical/psychological view*, the logical structure of the "I" is that of a set. But along with the consideration of semantic extension, the class and subclass ordering and the propositional specifications make for an index of the self that would have to be reflexive. If we combine these viewpoints, we would have a nice way of establishing the self's identity. It would rest on a logical identity in which the sign or icon "I" would be equivalent in meaning to the referent "self," and all sorts of causal powers can be exchanged between the "I" and the self. (The assumption of the indexical functioning of the "I" does not mean it is not a natural form subject to the individual's wider powers, if not agency.)

The *logical/psychological view* is *not* a simple constructivist position. It would be in tune with scientific objectives to tie the workings of the self and the "I" to neuronal phenomena and to construction from a material base. Douglas Hofstadter's ([1979] 1999) earlier view appears constructionist, and his later book on the self also builds on the idea that the self compiles information and therein incorporates other selves. Yet his earlier descriptions of "strange loops" sets the groundwork for his concept that the self's cognitive processes are those of a logic of reflexivity (see Boden 2007).

From my point of view, the rules of reflexivity of reference accommodate dynamic movement and transformation from the "I" to the self. In addition to the dynamic nature of transactions and transformations, there are the bounds, constraints, and originating functions of logical form. The logical identity of a sign ("I") and of its referent (self) makes for a coordination of indexical and causal functions. Agency in the hands of an icon, which is a "shifter," can locate its powers in its attributes (the properties of the self) and relocate power over those powers in its agent (the "I"). *The dialectical swing of powers between the "I" and the referents of the "I" assumes if not the primacy of form, then a nature to the "I" that superposes its properties and features.* I refer to such a swing as "metaphorical," but the best way to conceive it is to say that it has the features and capacities of a lattice. Thus, logical exchanges of order can take place such that at times particularization is superordinating. A property—or a synecdoche—such as "my better nature" can superordinate the whole of the self plus the "I" in determining an action or a value. But then the superset can be returned to the whole—the "I" as encompassing itself and the protoself.

How should we look at the *constituent* view in the light of this "primacy of form," its features, and its dynamic and logical capacities? The constituent view would be based on the contrary idea that identity is logically established *iff* all the discrete features of the person add up to the person's identity. Form would not be a primary determinant. But the absence or marring of any feature prevents a statement of identity. If we had all the beliefs and cognitive matter of a person that would enable us to build a (copy) clone, we still might be missing the model's individual consciousness. Until and unless that individual consciousness can be defined and "added" into the copy, it is not part of the copy. Accordingly, for this constituent way of defining or establishing identity, *any change in a self characteristic or in a subclass of the self threatens the nature of the self and the "I."*

This constituent view does not assume any a priori nature to the "I." That's good news because if we avoid assuming something a priori, it pleases the ego and injects optimism into social theory. The a priori always raises uncomfortable feelings that things are out of our hands— that *there they are,* "fate" and "predetermination," calling the shots. But it's also good news because the constituent view can stand more simply on a pragmatic position. With the focus on process and on the specific steps and phases of construction, causal acts and events can be observed at different and specific points. It's enticing to think that if the "I" is in disrepair because of brain injury or neural cell death, a "self chip" can be inserted, and it would rebuild the "I." The big question is how far the deterioration of an "I" and a self can go before it is too late for a Frankenstein's "monster" to be electrified and replicate the "I" of the original person! However, so enticing are the specifics and the possibilities of replication by construction that the paradoxes of infinite regress just fade into the background.

Repair, Replacement, and the "Partology" of the Self

Within this constituent/constructionist view of identity, if one of Galileo's beliefs had been that he should be the kind of person who shares his true beliefs with others, the subclass of moral convictions to which this belief belonged would be damaged. All sorts of rationalizations would be needed to repair it, but they might produce conflicts and unresolved negations with other subconcepts of the self. One of these conflicts might involve Galileo's belief in himself *and* in his scientific credo vis-à-vis any prevailing authoritative view—even that of the church. However, as Ricardo Nirenberg (1996) reasons, Galileo's self did not depend on integration, but instead on the fulfillment of a professional role. Nirenberg calls this a detachment, but I see it more as a fragmentation such that it's unnecessary for all subclasses of the self to "add up" to the self's identity or for the self to be integrated. Consequently,

$$f\text{"I"}[(B_{1 \ldots n}) \cap (BI)].$$

Can there be good news for the "I" in the face of a divided self? Well, there's good and bad news. The good news is that the life of the self does not have to be sacrificed; work can proceed as the individual's high ethical value. The bad news is that identity is fractured.

So, in contrast to Socrates, who decides to salvage all the subclasses of self and keep the "I" intact *both* as a superordinating indexical and as an identity established by adding up the features or property subclasses, Galileo gives up on the second method. The "I" as an indexical remains afloat. The price is a continual bailing of water.

Yet come back again to Socrates. He gives up his life. The indexical "I" *is* stopped in time, and any originating function is also. The negation is severe, although, from a subjective perspective, only for as long as Socrates lives can the ordering of the "I's" subsets (and/or subclasses) remain the same. Moreover, from an external perspective, others can have a view of Socrates as a person with coherence within his belief system.

Now comes the issue of the protoself, or that portion of the individual necessary to *have* a self and that *is necessarily* a self. In the case of Socrates, he gives up this portion. Therefore, as a "pure" "I" he completes the negation of self because without the protoself, there is no "I"—from an internal point of view. An as-if "I" of Socrates appears in the expression and observation of others—for example, as he "speaks" in Plato's dialogues. Galileo chooses to keep the originating function of the "I" from negation, whereas Socrates does not value that function as much as he values the dialectic splitting of any belief. Remember that for Socrates, knowledge is not going to go beyond method. Galileo, paradoxically, can state that he rescinds a new conception—and new knowledge—yet believes mightily both in the reconstruction of what was devastated and in new constructions of knowledge. Socrates, in contrast, is careful to argue that he basically has nothing to teach. He does not have a professional identity. As Plato has him asserting, "the simple truth is, O Athenians, that I have nothing to do with physical speculations" ([1882] 1996:*Apology*, para. 4). Socrates has no investment in producing anything more that would be an originating contribution—except, as he puts it, to be a "gadfly." Yet one might conceive of a "Socratic questioning machine." Accordingly, *as a fold-out of Socrates' "method," it's possible to see the connection between dialectical methodology as categorization and recategorization, on one hand, and a view of cognition as a machine, on the other! If cognition is a machine, the rules of regularization do not depend on the individual organism's autoregulatory laws and dynamics and on the relation of these laws to the needs, unconscious logic, and the conscious orchestration of all of this.*

Galileo may have illustrated an awareness of the "I" as part and parcel of the individual's life and that, ethically, creativity should not be destroyed for the purpose of closing the subsystems to unresolved negations. A nagging issue is that Galileo acts to preserve the life of his own—unique—"I," yet his view of knowledge is that it can advance by construction. So why can't the "I" be replaced too? It strikes me that a commitment to one's own individual life is a choice of the entropy that would be the fate of a closed system. In this regard, the process of entropy is a degeneration of qualities. It proceeds in steps and is all part of the aging of a living system. And, as might be expected as an unsurprising—yet odd—feature of the simultaneous surging for adaptation in the face of such degeneration, the course of the degeneration has unexpected turns. The aging losses of memory, for example, seem to force syntheses and "wise insights." The story would be different for degeneration without a living process and form, which would be the case if the "monad" were fundamentally a cognition machine—no matter how sophisticated. There is a lack of a remedy for degeneration because the monad's cognitive machine leaves no room for variability in identity. The monad's identity is what it is via constituent parts, and, therefore, that it can be replicated makes it expendable via extreme negation reduction! Socrates is like the cognitive machine. His self remains isolated since his identity is in his methodology. He cannot work on the reorganization of self necessary where unresolved negations would be introduced into the subclasses of self. Thus, isolated, the self would move rapidly toward the end of its entropy and have to be coterminous with the Ī. The Socrates of Plato's dialogues and the Socrates in the memory of his students become Socrates as Ī. Then the negation of "I" can take place, and the life of the "I" and the self are disassociated from Socrates' identity. So one of the questions arising is of the value of the "I" as transformed into machine. *Others* can use the dialectical method to assess that value: if Socrates *does* recant, he has to come face to face with modifying that dialectic, if not reasoning. He has to take on the issue of fractured truth and the lessened ability of others to know what he "really" believes. Is he saying that he's recanting because he believes X (the state), or is he saying it, but not believing it?

Belief-Giver Power, Synecdoche, and Runaway Beliefs When the "I" Is a Sacrificial Lamb

We presently are involved in the awful crisis of a geopolitical situation in which the usual forms of advocacy are apparently inapplicable to ferret out the truth of others' beliefs and statements. Who can argue with the super-argument that any argument has a premise that is an assumption outside the machinery of the argument? A major terrorist argument is that crossing categorial boundaries by equating *murder of innocents* with *firepower to wage conventional war* can counterbalance power and human values. The two categories ("murder of innocents" and "firepower to wage conventional war") are equated in terms of the same premise—that a greater good can be predicated on lesser sacrifices. Picture this premise symbolically as

fGG(MI ≡ FP).

That is, murder of innocents (MI), as a function of working for the greater good (fGG), is equivalent to firepower of war (FP).

Oddly, the exchange of "murder of innocents" for "war tactics" has synecdoche written all over it. So the form of thinking becomes a symbolic exchange based on equivalence, and both logical and psychological identities are the victims. The symbols are categories drained of meaning and particulars. In that way, they are part-for-whole exchanges in which equivalence can be based on confabulated concepts. These concepts may be statistical abstractions, such as "the greater number." Such an exchange is part for whole in that the number in "the greater number" replaces the varied nature of the individuals included. In that case, the stereotype might include something like "the people," as in the phrase "the people's movement" or the "people's republic." But the symbols and concepts may also be the old-fashioned "Us" and "Them" types; "enemies" can be described as subhuman, and the positive qualities drained from the conceptual domain.

It's scary and depressing to turn back to the study of the American Agitator called *Prophets of Deceit* (Lowenthal and Guterman 1949). The current frame of reference is that of terrorist logic, its political contexts, and its effects on beliefs. Yet look carefully at this 1949 analysis and at the pre–World War II and wartime use of "Us" and the "other." Anti-Semitism and its war against the innocents had the same characteristics of stereotypy and category logic with synecdoche. Then and now these characteristics appear as techniques of drumming out any considerations of human qualities and individual identity and worth. Then and now the function of working for the greater good (fGG), once put in place as a governing idea, can be to advance the cause of "believers" versus that of "nonbelievers." To refer to this dichotomy, I again use symbols to show the logical structure. For the next formulations, B is the symbol for believers, ~B for nonbelievers, and R for religion:

$$R > B \cap R > \sim B.$$

With the above dichotomy in place, religion (R)—defining its total set of teachings and beliefs as applying to the nature of the "believers" (B)—can then be assimilated to the function and even to insidious equivalences such as (MI ≡ FP). This equivalence can now functions as a "next-higher-class synecdoche," governing the logical level of religion and its reduction to $R > B \cap R > \sim B$. We may have a lattice structure here. Let us see how it works out. The synecdoche washes out the particulars that would add up to individual identities and to expedite a juxtaposition of the stereotypes promoting the "Us" and "other" dichotomy. In this washout, concepts of the "good" and the "just" are sifted so that they apply to R > B and not to R > ~B. The invocation of a religious umbrella (R) for the "Us" and the "other" allows the "greater good" (fGG) to be that of "Us" prevailing over "the other."

The addition of religion *appears* to add another level to the logical order of thought. However, what results actually reduces the next higher class (R) to another synecdoche! This reduction at first can be read,

$$R > (R > B \cap R > {\sim}B) > fGG(MI \equiv FP).$$

But, the religion is also drained of its full and often ambiguous particulars of meanings and sentiments, and its logical order becomes exchangeable with the fGG—its subclass. Thus,

$$[R > fGG > fGG(MI \equiv FP)] \equiv [R < fGG < fGG(MI \equiv FP)].$$

Wow! It is like a lattice. But, more appropriately, I should point out that the form

$$[R < fGG < fGG(MI \equiv FP)]$$

is like the species organization that Lévi-Strauss attributed to primitive logic. The form is rich in its capacity to sweep up logical transformations within the governing protocategory that becomes such a powerful idea. Should we say that the transformations are *trope-type thinking gone bad,* that the thinking never quite reached the figurative level, that it is pathological or primitivized in terms of the place, role, and function of the "I"?

The salvage job for the dialectical method is in great doubt—even in the method's realpolitik and negotiation models. Fragmentation of truth apparently affects the protoself as well as individual identity and its logical structure. All seem expendable in a wild need to handle the multidimensionality of negation irresolution. Rhetoric outpaces dialectical examination of causal sequences—perhaps because there is no originating identity to superpose the degenerating spirals of negation that are felt emotionally.

Assuming Socrates had lived to consider the failings of and potential changes necessary to a logic of dialectic as it related to individual identity and had retained his powers of inquiry, the advocacy system might not be what we know as such today. The fractures in his beliefs would have been a great blow. Enter Winnicott's ([1971] 1988) idea that the individual, assaulted by his own sense that he did wrong, digs deep for powers of restitution. We can only speculate that for Socrates, the reparation would go so far as to call for new principles. The prepotency of knowledge as not-knowing would give way to the moral imperative to provide new knowledge.

When the loss of individual life is placed in the hands of caprice, unreason, power seeking, or the like, the tragedy can be enormous. The human tragedy multiplies with travesty, irrespective of the power of justification in reasoning, however seemingly dialectical—as I illustrated can occur via "terrorist" rationale. Socrates in the *Apology* warns about the damage done if unjust men take the life of a just individual! I relay this idea as we presently continue to struggle to find the psychological and sociological elements not only in political dialectics, but also in the law and in the various forms in which the advocacy system continues as an essential part of our individual and cultural lives.

The brute fact is that we *do* live in an age of terrorism. Rhetorical machines seem to grind on toward sociopolitical forms of entropy—almost in the light of the individual's intolerance of an internal movement toward entropy. Reason and the value of the individual life—even in its ambiguous completion of the life cycle—do not seem to enter as rhetorical values, functioning to stop the flow of extreme ways of resolving the negations of political thought and sentiment. So I end this chapter with a complaint. Our own systems of thought may well have constrained our abilities to find the rhetoric necessary to appeal to those who do not revere the value of the individual—a group that includes political leaders, people who accept those leaders as belief givers, people who can reason their way out of such madness, and those who cannot.

9 The "I," Identity, and the Part-Whole Resolutions

Originating: The Trope's Logic of Synecdoche as Dynamic, as Form, and as the "I's" Product

The power of creativity, especially its locus, is a major issue in the balance of the two forms of identity, *logical/psychological* and *constituent*. The creation of novel and new possibilities to supersede prevailing category systems logically favors a view of an "I" that can rise out of its superset slot by reflexively originating a new level of awareness. Look at several uses of the "I's" possible positions as superset—namely, "I"$_0$, "I"$_1$, and "I"$_2$.

If I ("I"$_2$) think about myself ("I"$_1$) thinking about a problem, I ("I"$_2$) can see what "sort" of thinking I'm ("I"$_1$) doing and perhaps change that thinking. I ("I"$_2$) can change the way the "I" ("I"$_1$) does that thinking and change the "sort" of selection technique the "I" ("I"$_0$) derives to choose that thinking.

As the "I's" different positions refer to each other, the reflexive exchanges of different levels of awareness is obvious. Although this exchange of positions of order can sometimes appear to be a construction and move from part to whole, I argue that it is only an appearance. This picture of the reflexive "I" as a "superset-in-variations" is of a lattice structure, which indicates that *it is the "I" itself that can exchange positions of order.* The "I" emerges at the nexus of its originating and at its conscious viewing of this origination. This nexus is elemental to the structure and functioning of the individual's "I" and self; it brings together the "I" and identity by way of the structures of logic and figurative thinking. I now briefly state my proposal to set in place the relation of the "I," its logical form, and the dynamics of figurative thought. The paragraph's last two lines are the punch line:

> The "I" is reflexive. It is a self-realization of the boundaries of the individual as a closed system. In its reflexivity, the "I" *is* and can see itself *as* a trope and *as* a logical synecdoche in relation to itself (the self) and to the whole of the individual person. Accordingly the circle inherent in the autoregulatory organization of the "I" and in its capabilities takes us to its closed system dynamics. Thus, *the trope and its relation to synecdoche are part and parcel of the very structure of the "I."*

In the next sections, I discuss this proposal in terms of the following issues: relating political structure to the individual's identity, examining the creativity of socially shared product creation with that of individual origination, and reexamining the definitions of primitivity and pathology in relation to these issues.

Equivalence and Identity, Part-Whole Patterns, Logical Order, and Originating

Inclusiveness and the Direction of Thought

Empirical-minded persons look at what people say—their sentences, for example. Rationalists look to find a logical structure that accommodates thought and thinking. Hence, they focus on *terms* and *properties* and on their relationships in propositions. "Properties" appears as a logical term. Assigning properties helps to apply logic to sentences about existing things and actions. Capturing the logic of terms by way of properties, on one hand, and nailing down the meanings by way of the terms of sentences, on the other, form a good two-pronged way to get an integrated picture of language and thought.

What are the relationships of logical propositions and their terms to sentences and their subjects and objects? When I think about "properties," they are "of" something. In relation to a person, S, I think about S's property of calm or, more generally, that "mood" is a property "belonging to" S. However, "mood" can also belong to all individuals. Can one focus the "property" as a discrete entity or state of affairs that can be assigned specifics such that it can "become" a unique system? The logical idea of property as a variable would be assigned such unique valuations that it can belong only to an "individual constant." When a person, S, is treated as an individual constant, the constant is usually assigned a lowercase symbol, such as s, in comparison to the property itself, like that of "mood," which can be symbolized as an uppercase character, M. This distinction is meant to indicate that in the expression Ms, the universal property of mood is attributed to the person S. In all this, the problem of the "One and the Many" (or that of the "whole" and the "parts") can be conceptualized by the relationship of the properties to the individual constants.

However, in relating "property" to "constant," *the direction of inclusion is reversed when the person's (S's) agency is in focus.*[1] Consider anger as a mood when S's agency is *not* in focus:

"If you want to see what anger looks like, look at Sid."

Anger as a mood can be a property, and "Sid" a constant. Thus the property-constant direction is indicated, Ms. But consider the direction when agency *is* in focus:

"Sid made a show of anger."

"Sid" is the property; "anger" the constant. Hence, Sm. Now consider the two sayings:

"Time heals all wounds," and

"I am making time"

differ in regard to the agency accorded "Time" (in the first) vis-à-vis the "I" as agent (in the second); and the properties of "Time" differ in each of these instances. But the emphasis I follow is on "what includes what." Where T symbolizes time and S an individual, T$_s$ represents the first aphorism, and St the second.

The form of this logical issue seems that of genus and species, the exchangeability of which depends on the *direction of thought*. If you begin by focusing on M as an object, then when you specify s as a feature, M is the genus. When you begin by focusing on s as an individual and you think about M as belonging to s, then s is the genus.

I have pictured the "I" as superordinating relative to "its" thoughts. From the natural logic of this point of view, the vicissitudes of *inclusiveness* and the *direction of thought* would depend on the identification of the "I" as an origin point for the thinking. *When the "I" is identified at the origin point, we picture the matters of inclusiveness as having a metaphoric characteristic of shifts of genus and thus, overall, a lattice structure.*

I might be aware that in order to "make time," I decide to go to work via a faster route than usual. In this awareness, the "I" is at an origin point and includes "driving" and "time"—or, we might say, thoughts about driving and time. But to go the faster route, I have to look for a large blue sign with an arrow on it. When I see the sign, "it" tells me in which direction to proceed. At that point, agency switches, and the "I," although not at the immediate origin point, advances its capabilities by incorporating an "automatic" routine, which can later come under its surveillance. There appear *spirals*—movements that not only continue the genus-species exchanges, but also by doing so produce patterns of growth and change in thought and its capacities.

Hofstadter's (2007) strange loop explanation of the "I's" logical trajectories also takes on a geometric shape. The "loop" describes a way of structuring categories and propositions, such that exchanges take place between two different levels of category ordering. As Hofstadter sees it, the meanings of different order categories get bollixed up—just as they would when a person attempts to reason in the case of the "Liar's paradox." However, when dealing with such category exchanges, keep in mind that they occur in tropes, as a matter of course. Tropes are figures, and they do have a meaning-form relationship (Turner 1998). So, Hofstadter is right to point to the peculiarities of meaning that occur as the "I" swings from category to category. However, my choice is to focus on the form of the trope, not only as dynamic logic ordering and re-ordering of the categories to be exchanged, but also as a key to the "I's" functioning to effect that ordering.

In point of fact, Jaynes (1976), by his formulation that the self involves a "metaphier" function, comes closer than Hofstadter to viewing the metaphor—and the trope in general—as a figure encompassing the logic of the self and these exchanges. The exchanges of category ordering are trope-like, and the logical aspect of the figure is understandable when you spell out the relation of the categories or sets in terms of lattices.

When the "I" enters its reflexive mode, the spirals affect (and effect) logical order. "I" can look at myself as a driver of a car and then from the "driver" point of view look at myself more generally—say, in regard to mood. As I do, the "I" exchanges operator and object positions as genus, and there is growth and change in understandings.

Spirals, "I," and Time

How does the "I" look at the way it comes to look at itself? Certainly, growth in the person's thought capabilities can be the product of operator and object exchanges by which the "I" can look at its various levels of awareness and activity. However, although we can picture the spirals that take the "I" in and out of its position as an object and as an operator, it is difficult not to fall into the traps of a category mistake. The "I" as an aware agent can view itself, but when it is an object, it is not aware in the same sense—even if the difference is that it once was and will be again, but isn't at the moment. So it would be worthwhile to account for the logical vicissitudes of the "I" as a penultimate category—with powers of superseding whatever level it finds for itself.

In a sense, the ever-spiraling "I" *can* put itself aside. This shifting of location in different contexts and of place in a logical ordering of awareness doesn't require the extreme of denying the "I's existence on the grounds of its extreme instability of identity. The "I" can feel its jumps to different orders of awareness. Yet despite those differences and the different roles and circumstances, which can be the contextual housing for the "I," its consistency seems solid in its independence from past and future segmentations of time. The feeling can be that the "I" is there, but that it is *time* that is moving or conceived to be moving. So why assume that this feeling is other than an illusion? The answer is that it would be metaphysically extravagant to mush together the evidence and the categories for the existence of the "I" with the evidence and categories for the nonexistence of the "I." Within its mode independent of time, the "I" might get itself all balled up in reflexivity traps if it asserts that it knows it *is* there! But, strikingly, the "I" *can* consider its bounds in terms of when it "was *not* there." So we might say that *its operator function appears independent of time, but, as an object, the "I" is ultimately time bound.*

In all, the "I," who originates its own thoughts and beliefs and monitors them, can contemplate its own origins as a matter of its embodied identity. In this way, the "I" can set (and/or apprehend) logical bounds to identity. This setting amounts to a stance concerning the issues of inclusiveness and the part-whole patterns. Thus, the "I" can assume that its own identity can have been formed in one of two variations of the part-whole resolution ([Parts → Whole] and [Whole → Parts]). I can assume that

I create my own roles.

Or,

My roles create me.

In addition, how the "I's" thought—itself—projects inclusiveness may also be subjected to the same two separate explanations. First, take this sort of explanation: I think that "I create my own roles" because I visualize separate domains of my functioning, divide off influence spheres within those domains, name the roles possible, and set goals to achieve specific positions that legitimate my occupying those roles. I am aware of what I want to achieve, state it, and direct the activities toward that goal. Then compare this explanation to the Willy Loman prototype. The latter calls for the person to be "well liked" in order to achieve role goals, and it visualizes his thought about events responsible for his nonachievement of goals.

Because both the choice of stance and direction and the explanation for adopting the stance are engaged, *recursiveness* (the "I" within the "I" within the "I" ... "I" ... *n*) and *reflexivity* (the exchanges of reference and representation) appear to be bottom-line features of the "I," and any account of its logical nature requires them. Taken this way, the operations and structures of the "I" seem a replicable phenomenon and therefore should put to bed the extravagant idea of ignoring the "I's" existence.

Direction of Inclusiveness, Causal Properties, and Finding a Place for the "I"

Some people assume their thinking is not emanating from the "I," but instead is either a social effect or a behavioral reaction. Does this viewpoint deny the necessity for a start point, which would be something like a capability inherent in the individual? No. If your behavior is a social reaction, you still need to say that you have to have the "machinery" to react. Although there may be more than one way to avoid threat, whether you move away from it or take it on, the structures and capability for the response have to exist. Yet with the assumption that thinking is a social effect, the external is the presumed start point, setting in motion the reactions of the individual, thereby tying the meaning of the individual's actions to social events or external causes. Do those who assume this sort of external start point retain the sense that their own "I" has causal properties? Does the theorist say to himself, "This is a theory, but 'I' personally am outside this theory? After all, I thought it up!"

The case may be made that Skinner (1953) argued the self into a corner by focusing on exteroceptive factors only. His argument was that there was no percentage to a psychology of mind, cognition, or self. So the idea that there was any cognitive machinery or structural capacity was off limits. Yet as Robert Epstein (1997), a psychologist who worked with Skinner and deeply admired him as a person, points out, Skinner himself continued a life of the mind! Of course, if you follow Skinner to the extreme, you can see he would say that the life of the mind is really a life marked by linguistic acts—all of which are shaped by social and linguistic events. (I don't want to enter into a dispute here about Skinner's concept of an "operant." Charles Catania (2000) sums up the concept: "The operant, as a class of behavior selected by its consequences, is a fundamental unit of behavior." Skinner's (1950) idea is that the organism's response does not require a stimulus. We still have the notion of reactivity, but it is in relation to what follows the response.) In any case, Skinner could have lived a life of the mind, yet paid no attention to "it" as a mediating process—let alone an initiating one.

Times changed with the cybernetic revolution of the 1950s. Nevertheless, in most of the second half of the twentieth century, "mind" did not have to be a "thing" or a structural entity. The pragmatic tradition for which Skinner was standard bearer continued as an ethical and metaphysical basis for a psychological science even after the eclipse of behaviorism and the age of information was upon us. Before the current period of probing and observing the action and structures of the brain, a pragmatic approach greatly facilitated what was to come as great leap forward in robotology—a march toward the cyborg! *"Mind" should still be considered a matter of observable actions.* The computer scientist Marvin Minsky continues to argue against the existence of a "thing" corresponding to "consciousness." He claims that scientists who want to understand "consciousness" should list the things "it" does and see if they can "design" a machine to do those things. Then they would have an understanding of what a human mind can do. It "contains perhaps 40 or 50 different mechanisms that are involved in a huge network of intricate interactions" (1998). But you define a machine after the fact of what it does (Minsky 2006). So, in a sense, it too does not exist as a "thing." If its actions occur, then it can be said to have been a stage in the construction of circumstances producing the actions. *That's awfully similar logic to Skinner's focus on outcomes!* Of course, Skinner is building a technology without paying attention to any mediating process. Minsky is turning the technology another notch to build its prior steps as a machine, which albeit a mediating process, is an observable one. However, stay away from the machinery for a moment and focus on just the "things done." The question for Minsky is, How does he explain how *he* formulated this endeavor and what motivated it? Obviously, as he pursues an explanation for the results of a life of the mind, he continues to live a life of the mind to make this pursuit. So he's copying the model, and the causal theory is that the model is a copy—and that he is too.

In sum, *the belief that you are what you do is a synecdoche!* It is a belief that opens you to empirical observation, but what is open is a fragment of "you." These arguments wind up fragmenting the individual's beliefs and self by isolating interoceptive causes. They are shunted off to "Never-Land" in that they are outside any time frame that would count them in as causal events. We see the same phenomenon in the touchy example of the Israeli-Palestinian conflict over "who started this round of violence." The syncretic organization of self and action stops the unfolding of perspectives that take the "I" to points of redefinition of actions and of the nature of self. Exit inner self; enter "things you do"—and their surrogate prior steps.

You can put the shoe on the other foot and do just the same by isolating exteroceptive causes. The explanation in terms of external causes of behavior has a number of problems, including that of the nonreplicability of the exact scene of the behavior-context mix. Therefore, the logic of synecdoche can step in to focus the self and the "I" as the belief giver. But this focus can become myopic, and all other frames of reference can be locked out. This is what happens when megalomania explodes into a promulgated vision and movement that appear to have religious verve (look at or read the text of some of Osama bin Laden's pronouncements). It also occurs when political vacuums give rise to individuals who disregard all causes but their own will to power.

The issue of overexpanded meanings and underreliance on external sources of belief brings up Freud's well-known position that the meaning of memories is a matter of the individual's perceptions and their reconstruction. Meaning itself would be a matter of self-propelled and ego-invested images and thoughts. Did Freud therein obviate any credence to the childhood training thesis? Well, as psychoanalyst Alan Stone (1997) reasons, perhaps he did. But, once again, the process of the growth and change of beliefs in a context of social factors, influence, and interpersonal dependency is really not incompatible with a phenomenological position. How you view your childhood training is neither solely as the training "really" occurred nor merely as your perceptual bias predisposes you to represent or misrepresent that training. Freud did not eliminate the idea that events outside of the individual's phenomenology cause the individual's thoughts and actions. Biological factors can be considered external to thought and subjective phenomena. Freud continued to believe strongly that fate played a fundamental role in assigning biological determinants of the individual's psyche.

When we say that the individual's "I" can assume explanations of its origins and bounds, we include ourselves as individuals. *Therefore, each of us—you and I included—should carefully run through a wringer our assumptions about the origin of our identity and about the originations of (and by) self. For the sake of stability of identity, the "I's" "wringer" should squeeze out fragmentations and self-defeating divisions of thought, language, and phenomenological experience.*

Too many gifted psychoanalysts apply phenomenological analysis of events to the person or persons analyzed. The persons become the psychoanalysts' "objects," and their subjectivity is the focus. What is missing is the psychoanalysts' consideration of how to utilize their own power of argument and its influence on the assumptions that these other persons make about themselves. But the psychoanalysts, like you and me and like their objects, would have to have derived that power of argument and influence by putting their own assumptions about the origin of identity through their own "I's" "wringer"! Leaving out the "wringer" is a lopsided emphasis on phenomenology without taking into account its interactive and interpersonal potential. An example of an upshot of this lopsidedness is the inability to apply dynamic principles to challenge and attempt to change militants' rationalizations and distorted stereotyped thinking.

Too many gifted neuroscientists claim that mind and conscious experience is explained by neural structures and events, so they leave untouched their own "life of the mind." Ask them if their mental experiences and investment of self in thinking about and researching neural structures and events experiences are no more than those events themselves. The lopsidedness here is in the focus on observable structures and in the eclipse of phenomenological experience of agency. An example of the upshot of *this* lopsidedness is the overemphasis on building theory of psychological productions as neural functioning. What's missing is consideration of how dysfunctional persons—say, with Asperger's incapacities for social and emotional articulation—can learn to orchestrate themselves despite and in light of their dysfunctions and/or structural deficits.

Direction of Development and Identity

With these admonitions in place, I return to our discussion of origination patterns and their effects on inclusiveness and the part-whole resolution. And the payoff question: What are their effects on the nature of identity? I have discussed the relation of properties and individual constants, for which there are part-whole considerations and the problem of "what" includes "what." Although that problem clearly involves the issues of logic entailed when tropes are used, the part-whole considerations appear logically as the matter of genus-species inclusiveness. Further, we gain some leverage on tropes when we say "genus and species," instead of merely referring to set-subset relations. Tropes have troublesome logical forms, but their payoff is multileveled perspectives on meaning, reference, and variable coding of reality. With the terminology *genus and species,* we can infer that existential matters and their causal pressures interpenetrate logical form. Existential objects and processes move along and change, and we can talk about a direction to the change that is a function of time. Some things change in the process of growth, some in a process of entropy and decay, but we can look at each of the changes over time as *development.* So far so good for formal requirements for the individual organism as a unique closed system with a stable identity and a capacity for interacting with contexts—social, biological, and informational. This *development* can result in growth, knowledge, and progress. However, matters of change—in time frames, role and behavior requirements, contrary intentions, and contrary expectations—contribute "negations." Matters of conflict and opposition in concepts of self, goals, role, and identity—the effective individual stymied by environmental constraints, the individual's identity contradicted by competing role attributions—contribute to "negations" of meaning, category, and concepts, as well as to the basic negation introduced by the oppositional processes of growth and decay.

In cognitive terms, the negations can be resolved. As Bateson points out, conflicts can be resolved by regrouping categories and submitting the contraries to a next-higher-order categorization. For example, if the Israelis' conflict is over retaliation or submission to terror, these alternatives can be pictured as subcategories of an ethical view that argues that both alternatives are futile foci. But such a resolution of negations can also take a destructive turn, producing massive negation—such as the destruction or death of a form. To stay with our example, it may be more "correct" to focus on broader sociopolitical problems in the case of the

Israelis' conflicting choices. But the time frame may be such that when a more long-term solution is engaged, the whole state of Israel may disappear because of no action reciprocal to the level of action from which others do *not* disengage.

Let's apply the notion of a *direction of "development"* in terms of the whole-parts issue. How does this notion interrelate with the origins of the "I" and the stable identity of the individual? The *direction of "development"* can be seen from the two perspectives of the whole-part resolution (Parts → Whole) and (Whole → Parts). To spell out the issues in relation to a general course of development, each alternative resolution has its own combination of cause and temporal mode and sequence. Thus, a whole is either constructed from its parts or a whole entails its parts. In the latter version, the causal direction—like that of *final cause* is from the whole to its parts.

With this Whole → Part causation, I reason that direction of "development" can be a general determinant of the total logical/psychological formulation of the individual's identity.

By the way the arrow appears, we see the causal role and logical genus status of the whole. We also see the arrow's direction, by which we depict time moving forward. The alternative *direction of "development"* is Part → Whole.

In a Part → Whole causation sequence, too, time moves forward and can be a determinant of a constituent component—perhaps the modular construction—of individual identity.

This process can end up with an unpredictable whole—as might well happen in evolutionary phenomena. The fulfillment of individual identity is correspondingly indeterminate! As the individual continues to "build" to her whole identity, like Zeno's creature moving forever to some position short of the goal, identity remains elusive.

However, because the morphological boundaries of a given organism are highly predictable, we also conceive of a Part ← Whole process in which a causal role is assigned the whole as a pattern to be fulfilled and in which the time line thereby appears to move "backwards."

When we raise this issue of "backwards" movement and thus consider *direction of development,* three issues of classification, causality, and time come to the fore.

Issue 1 is that of metaphysical classification: *direction of "development"* can be derivational when it shows the part-whole sequence.

Issue 2 is that of causality: *direction of "development"* can be causal when it shows the direction of cause and effect.

Issue 3 is temporal: *direction of "development"* can be temporal when it shows the movement of time relative to the evolution and/or unfolding of form.

To make more specific this shift in the direction of time and the effects on the bounds of identity, I express it in symbolic statements. Once again, I use a lowercase letter for an "individual constant"—a given and specific person—and I assign the letter a subscript status to indicate that it modifies the property of personhood. The purpose in doing this is so that the property can be individualized and referred exclusively to a particular—in this case to a person as a unique individual.[2]

Where Ps stands for the "personhood of particular person, S," PIs for the individual's "personal identity," H_s for the individual S's "human form," and GSs for that individual organism's "genetic structure,"

$$P_s \leftarrow PI_s \leftarrow H_s \leftarrow GS_s.$$

The $\leftarrow$ symbol here represents several things. First, there is a causal direction from the more simple form of the individualized property (GS_s) to the more complex (P_s). Second, there is a derivational direction logically from the subunit to the superordinating one. Third, these arrows represent the temporal directions of the evolution of forms and structures. The complication is that the more primitive form (GS_s) is at one and the same time the more general, or universal, and the least particularized; in terms of the outcome relations of the units, it is also at the lower end of aware decision making and agency. For these reasons, *the lattice structure governs genus-species categorizations of the determinants of identity.*

As this formulation relates to "identity," it can be assigned either by the individual himself or by others. The formula is an object that can be defined and subjected to consensual validation. As such, it appears closer to Goffman's (1963) idea of the person's *social and personal identity* than to what Goffman calls *ego identity*. *Social identity* is a matter of *categories into which others put you on the basis of the relationships of attributes.* Attributes may be physical, as in height or other features of physical appearance, or they may be role related, as in "he holds a job." *Personal identity* is a matter of identity "pegs" or "markers." The total set of pegs and markers is the individual's unique history and in toto point *only* to that person.

Both social and personal identities differ from a subjective "felt" identity, which Goffman calls *ego identity*. To focus the Part ← Whole process as one of consensual validation may appear an odd statement when you look at and consider the temporal structure as the "key" dimension of identity. However, even if the whole or general structure precedes individuation, the particularized outcome reads out as a P_s ← PI_s ← H_s ← GS_s view of identity of constituents! This outcome lends to the idea that smaller units or "primitives" construct identity of a whole. This idea includes the concept of the whole as built from parts or portions. The parts or portions begin as "primitives," or simple units, *but their identification as units means that whatever they are and in whatever stage of development or potential for becoming a part or portion of a more complex thing, they are closed systems.* The *but* clause of the last sentence puts a cap on identity and therein is the "key" feature that stabilizes identity. The process I am depicting thus proceeds logically from *species to genus* when you consider properties and individual constants. This case is easy to make when considering the existential elements and the development of complex meanings and structures of identity. We might say loosely that the process also moves from subset to set and that its logical form, when made more abstract, allows all sorts of tropelike latitude for views, stories, and theories of derivation, causation, and temporal direction.

BACKWARDS CONSTRUCTION, STRONG CAUSALITY, AND RESOLUTION OF PATHOLOGIES   I call the constituent view just depicted *backwards construction*. In doing this, I seem to compress Part → Whole and Part ← Whole patterns, and in so doing I appear to favor the idea that the outcome of evolution is predetermined. No such thing! As I conceive the compression and the category "backwards construction," they are based on an after-the-fact observation. You view the whole in a time frame that I will call "the present." You look backward to some "past time" when the project to achieve the whole began with simple parts. You assume that if you deconstruct the whole, piece by piece, you would be in a sense returning to that "past time." Moreover, were you to achieve something like that, you might then "reconstruct" the whole. The implications are that if your "whole" at "the present" has a faulty or missing piece, the reconstruction can produce a better product in terms of any intention toward perfection or nondysfunctional status of the whole. An example would be the "re-do" of the visual experiences in communication that the autistic person is "missing," according to formulations like those of Baron-Cohen (1995, 1997).

The purpose of my theoretical "compression" approach is to select a temporal mode that provides reversibility as a cardinal feature of the possibilities of construction. One of the things I mean by *backwards construction* is that the causal direction, when viewed from the time frame that includes the completion of the whole, is seen as having begun from the primitive to the complex. The *causal direction* of the construction appears logically backwards. But when you consider both the de facto causal direction and the *apprehension of that casual direction*, they appear to move in opposite directions. *From a point of observation at the completion of the causal action—the "present"—the theory that the causal action can move backwards is derived in retrospect.* However, the theory that the causal movement had been from "the past" to the "present" is also retrospective. You apprehend the causal process in the present when you see the whole—the outcome product. The resultant causal theory is clearly based on the idea of assembly, but it also implies that disassembly is possible. Thus, this way of retrospectively theorizing from the present dictates a construction process in which *the causal arrows for disassembly go in a forward direction consonant with the apprehension of that causal process.* Accordingly, the disassembly

$$P_s \rightarrow PI_s \rightarrow H_s \rightarrow GS_s$$

results in a systematic degradation of the "final form." To a certain extent, the process resembles the course of primitivation due to the pathological destruction of cognitive functions. This course might be somewhat like the cognitive ravages of brain cell death in the person who has a stroke.

The point here is more general. The idea that a person's identity is built or deconstructible seems a powerful ontological statement of causal determination. Someone who needs to establish her individuality by markers can formulaically aspire to and reach personhood by putting that portion of self-definition in place. This can be the case even if the person does not achieve what would seem positive at a given marker position. Thus, it is part of C. S. Peirce's identity that he did not achieve a professorship at Harvard. This aspect of his life, along with what we know about his abilities and accomplishments, tells us a great deal about who he was. It's also part of Sigmund Freud's identity that he gave up the search for such an academic marker, which—along with his sensibilities, reactions to the prejudices of his time, and soaring talents—tells us a great deal about who *he* was. Contrastingly, if markers are gone, destroyed, or missing, depersonalization takes place.

Do the Peirce and Freud examples, as I cited them, refer to identity from the point of view of the "I" or from the point of view of the "me"? William James put it that the empirical self is the object we index—and select—as *me*. Who you are includes your possessions. James wrote, "*a man's Self is the sum total of all that he CAN call his,* not only his body and psychic powers, but his clothes and his house, his wife and children, his ancestors and friends, his reputation and works, his land and horses, and yacht and bank account" ([1890] 1952:188, emphasis in the original).

With some missing markers, the "me" is deeply affected. Changing your address and leaving your house is one such case. But with other property or markers, I am not sure that the difference in a person's perception of self lasts much longer than the "TV makeovers" that produce such feelings. The loss of a credit card or a mix-up in your address can throw you into a tizzy. Retiring or getting fired from a job, where there are role markers involved, can require a major adjustment that many people simply do not make. Further, the military's depersonalization of individuality seems to involve an ascendance of the group's goals and to foster thoughts about the enemy in stereotyped terms. This sort of thing is depressingly dramatic when religious leaders stir up vicious stereotypes, as is reported in the case of present-day stimulation of terrorist sentiment. But these depersonalizations are accompanied by some compensatory gain in the ideas of group goals and martyrdom. In contrast, dehumanization of prisoners in concentration camps results in fantastic losses of what we would usually add up to be included in a person's identity. Yet, even so, in some cases survivors rebuild all the lost markers and roles.

STRONG CAUSALITY, NEGATION RESOLUTION, AND IDENTITY RESILIENCE   Thus, as regards an existential condition, *backwards construction* is a *strong* causal view of identity, and the causal arrows of both directions are implied in the ← sequences. This point can be expressed by saying that when the "backwards" direction

$$\text{"I"} = P_s \leftarrow PIs \leftarrow H_s \leftarrow GS_s$$

is the case, if any negation irresolution, which appears as a step in a forward direction, is "compensated," identity is restored. If after a shock an individual has a temporary loss of identity, we have something like this:

$$\sim(PI_s) = \bar{I}\,[P_s \Leftarrow \sim(PI_s) \leftrightarrows H_s \leftrightarrows GS_s].$$

As a function of the temporary loss of memory, the individual's identity $[\sim(PI_s)]$ is as not-herself, $(\bar{I})$. But the remaining aspects of the self ($H_s \leftrightarrows GS_s$) are intact. However, in that remainder ($H_s \leftrightarrows GS_s$), I use the two-way arrows to show that causal movement in the progression toward identity as a unique person (P_s) may either advance or become degraded further. The degradation in this symbolization is $P_s \Leftarrow \sim(PI_s)$.[3] Another point in this example of loss of memory of identity is that you might look at identity as being compromised because of more than the usual uncompensated negations. Although I don't know how valuable this sort of formulation can be for therapeutic purposes, it is theoretically subject to the strong view of identity, which provides that progress, degradation, and restoration of identity are functions of construction and negation compensation.

This strong view in the form of ideas of construction and reconstruction is central to present-day theorizing about therapeutic approaches. There are serious approaches to such disorders as autism in which social-interactive patterns and brain patterns are thought to be interdependent and interpenetrating. The approaches to the remediation of degraded social capacities are based on concepts of reconstruction. It is also important to say that James Mark Baldwin's idea of the individual as a "'socius,' an element in a social network or situation" is central to the whole set of assumptions and enterprise of such remediations (1913:109). For example, Deák, Fasel, and Movellan (2001) consider the concept of "shared attention" a practical lever for socialization development. In brief, the formation of identity and cognitive capacities is by way of the complexities of meaning and social capabilities in a compilation of social actions and interactions, which in turn are interdependent on neurological levels. At these levels, development and compilation also takes place. The various therapeutic approaches make use of schematizations of action patterns as a basis for cognitive, linguistic, and social action.

It would not be out of order to trace the various therapeutic approaches to the cognitive linguistics point of view of Rosch (1978) and Lakoff (1987, 1999). But let's not forget the pragmatic and empirical traditions that make action—in the form of behavior and performance—the start point for change techniques. "Backwards construction" can be described in terms of changing the conditions under which a person's action and complex functioning reach "the present" whole. But the same terms can apply to Skinner's concept of the "operant" and the changing of behavior and of what he would call "meaning." I admit that for building and rebuilding human cognitive functions, the strong causality view can be excessively narrow. Also, Skinner's concepts do *not* clearly involve structures such as schematizations. Yet as a "backwards construction" start point, the "operant" is a *category of responses*. Hence, it's a primitive and general category, which as a structural consideration disappears in a Skinnerian blind spot.

So we return to structures for the construction and reconstruction processes. The type of symbolic formulations I offer,

$$\text{"I"} = P_s \leftarrow PI_s \leftarrow H_s \leftarrow GS_s,$$

can express the "I." When considered in the form

$$\sim(PIs) = \bar{I}\,[P_s \Leftarrow \sim(PI_s) \leftrightarrows H_s \leftrightarrows GS_s],$$

the formulation can give a good idea of what pathologies might look like dynamically and logically when compared to the functions of an "I." It would take one more step to place these terms and their relations in an operator role relative to dreams, tropes, myths, or primitive protological relations of icons. How do you mark *that* step? Although we can resolve such a symbolic presentation with a cumbersome rendition of an operator by putting an entire expression such as

$$[\sim(PI_s) = \bar{I}\,[P_s \leftrightarrows \sim(PI_s) \leftrightarrows H_s \leftrightarrows GS_s]$$

in brackets as itself a function, we might just simplify the expression. One approach would be to work out a series of modifiers and subscripts for the "I" and the "Ī."

Where are we these days in terms of approaches to cognitive pathologies? Myriad approaches laud brain research and theorize from it to models of mind, consciousness, and self. Baron-Cohen (1995) argues vigorously that our ability to apprehend and empathize with others is based on capabilities to "read" the other person's mind. Consonant with what psychoanalysts such as John Bowlby (1969) have argued for a long time, Baron-Cohen points to the critical role of patterns of interaction of infant and mother, but he gets specific in focusing on the patterns of their "gazing" at each other. Although this focus can have its own possibilities in therapeutic reconstruction, as in cases such as autism, Baron-Cohen links the gazing experiences to specific brain-pattern development. As in the formulations given earlier, there is construction going on, and the after-the-fact observation of brain-pattern development would argue that reconstruction is possible. Perhaps restored missing experiences of gazing might augment brain-pattern development, although the intervention might be best the other way around. Early intervention in the development of brain structures might result in more articulated gazing behavior—and its sequelae. The value of the construction view of the *direction of development* would be significantly enhanced if the various brain theories of mind were someday to reach a point where strong identity of mind and brain gives rise to all sorts of remediation of pathologies. Yet sole reliance on this view of identity and of the individual is too reductionistic and can lead to tragic oversimplification—a point I discuss in Fisher 2003b.

Counterpoint in Part-Whole Patterns, Tropes, and Identity

Do you hear the *counterpoint* between the synecdoche in the origination of individual thought and the part-whole relations of tropes? The simultaneous cognitive experience of different orders makes them like musical themes and subthemes. The themes are the individual's patterns of logical forms and their relations. The subthemes are the individual's way of relating the terms and icons in tropes. You hear their exchanges of similarities and differences. The subthemes convert to themes. You rhythmically tap out intimations of unique but logical genus-species rules. They're in cadences striving to announce the formation and deployment of a figure of thought! The procession is logical, a march of unlikely resonance of forms with contrary content.

If the individual's identity has these echoes and variations in structure, it takes some doing to have a model of logical and psychological identity, so I understand some theorists' need to simplify. If one were to settle for a constituent model, it *would* offer a strong view of identity. With it, the self is constructible. Patterns would not be those of complex multilevel counterpoint. There would be a more consistent presentation of the logic of identity, the identity of the individual, and the dynamics of change. Yet the constituent view can run out of control—like the assistant's decision to act portrayed in Goethe's 1779 poem "The Magician's Apprentice"

(Goethe 1882). As the assistant takes on the decision role, his causal direction stays on course, defining purpose and intent by production, its only value. Think of the therapist reconstructing an identity. Who is doing the reconstructing? Is the person being constructed? Construction or reconstruction cannot substitute for the purpose of the construction. Even so, there are obvious advantages to the → causal relations between a state of affairs and its construction or reconstruction. These advantages oddly produce an insight into the structure of the trope.

Trope as a Part-Whole Equivalence in a State of Constant Tension with Identity Status

The strong view of identity (the constituent view) can be described in terms of a whole-as-part position. "Part" is an abbreviation for the fact that the parts are never other than an assemblage falling short of an ideal identity with the whole. For a constituent view, the causal relations have to be seen as interchangeable from part to whole. The constituent view of identity is based on the proposition that the whole is equal to the sum of its parts (Σ Parts). Thus, to mark the interchangeability of the constituent view, it would appear as

Whole → Σ Parts,

and therefore,

Whole = Σ Parts.

Logically, this would be the nature of an identity, and as far as human identity is concerned, it would be compatible with all I have said about construction. Tropes would be *one step down in the construction process.* Instead of the achievement of an identity with the

Whole = Σ Parts

status, the trope takes a form that allows equivalence, but leaves unresolved negations. Thus,

Whole $\leq \Sigma$ Parts.

The primary negation is

$$\sim(\text{Whole} = \Sigma \text{ Parts}).$$

So we have the statement of identity that is the hallmark of a trope:

$$\{[(\text{Whole}) \equiv (\leq \Sigma \text{ Parts})] \cup [(\text{Whole}) \neq (\Sigma \text{ Parts})]\}.$$

This statement marks a situation involving the coexisting of an equivalence, $[(\text{Whole}) \equiv (\leq \Sigma \text{ Parts})]$, and a negation, $[(\text{Whole}) \neq (\Sigma \text{ Parts})]$. *The trope's enigmatic statement of identity is one with a nested unresolved negation. The statement is the structure of the trope as an equivalence in a state of constant tension with identity status.*

Explore how this tense coexistence of an equivalence and a negation works in the hands of the "I" as interpreter—or, if you want, the arbiter of the tension! *The equivalence of the whole to a part or to a sum of less than the parts becomes a symbol for the identity of the whole. A figurative statement can define an individual.* Juliet is the Sun—as far as Romeo is concerned. Epithets can "define" the individual. Tropes are constructed for particular uses by an unstable "I"—namely, an "I" that can be redefined at any frame or schematic of itself and/or the situation in which it is operative. If I define Mohammed Ali as "the Greatest," the "I" who accepts that definition is tentative and would use that epithet to identify Ali only in certain sets of circumstances. Comedian Jackie Gleason used to refer to his "Honeymooner wife," Alice, by declaring, "Baby, you're the greatest!" When *I* (the viewer or audience member) believed that she was the greatest, it was "I" in a different situation than when "I" was in agreement with one of Ali's declarations about himself! But are both "the greatest"?

The person defined by the trope has an identity that depends on a specific array of social schemata and on the beliefs and perceptions of the person constructing and/or using the trope. (You and I as viewers or audience members form a trope to view the person's identity). When we are applying the trope to someone such as Mohammed Ali, our idea of his identity corresponds to Goffman's concepts of personal and social identity. It is *personal,* as is easy to see, when his many iterations of "I am the Greatest!" are reviewed. He offers his markers. It is *social* in that others pick up the trope and hold it as an equivalent marker for Ali—under a particular set of circumstances. "Ali, you are indeed the Greatest," Howard Cosell used to say when he could allow his admiration to overcome his reportorial dialectical challenges. So what about the "I" forming the trope as a belief and evaluating it from schema to schema? Does *this* "I," who floats from schematic to schematic in her perceptions, beliefs, and perspectives, have an identity? Who was Howard Cosell that he could criticize the "I am the Greatest" argument and then convincingly praise the veracity of Mohammed Ali's declaration? The extravagance of "I am the Greatest," Cosell would seem to agree, had reached the point of identity! Cosell was in tune with the part-whole equivalence of Ali's trope!

You might argue that a proliferation of "I" in different frames creates a kind of complexity that is at odds with the assumption of a singular identity. However, uniqueness of identity might be established as a pattern that repeats itself. The pattern might be something that occurs within each schematic, or it might be part and parcel of every schematic of that individual. This latter description would make the uniqueness pattern similar to the progression of the individualized properties of an individual constant, as in the formulation $(P_s \leftarrow PI_s \leftarrow H_s \leftarrow GS_s)$, which accounts for the construction of a "person."

In terms of the constituent view, look at the syntax of—or within—the particular schematics of an individual. The cadences of syntax do not differ from schematic to schematic. The routes and node points of movements can be traced and the contrary subthemes reconstructed with modifications. Modules can be built to function well and quite similarly in specific scenarios. Where the observable language and iteration of beliefs can be specified and instances can be piled up or patterns established, we begin to get a collection of parts so that the equivalence of the parts to a whole identity seems achievable. The "I" can be its collection of modules. Identity would be splintered, yet identifications would construct the self. All the rights, privileges, and downsides of a simulation theory obtain.

The Arrow Goes Forward

Come away from the construction model and the considerable advantages of the → relations of time and causality we have reviewed. These advantages make use of the *constituent* view of the self and the "I."[4] Consider the situation where the view is that the person is *constitutive*, that individuality is a given, and hence that the inclusion relations of the different categories of the individual, the self, the person, and the "I" are formally a matter of the genus to species order and direction of organization. I offer this formula:

$$(s) \ P_{s_1 \dots n} \subset P_S.$$

I introduce this formula with a *function operator* (s). It relates to the properties of a particular person, S. There are two views by which to relate properties to personhood:

The *constituent* view presents S as that person, *iff* her specific attributes or properties add up to her personhood. For this developing personhood, I assign the symbol, $P_{s_1 \dots n}$. The lowercase form "$_{s_1 \dots n}$" subscripted in that symbol indicates an ongoing collection of the person's attributes and properties that develops a picture of her personhood.

In a *constitutive* view, personhood is more of a genus type categorization, which at a species level, formally includes the specific person, S— even if denoted via her attributes and properties. So, to indicate that S is not an ongoing "work in progress," but instead is categorical species to the genus, P, I assign the symbol P_S, in which the subscripted letter "S" is upper case.

In the (s) $P_{s_1 \dots n} \subset$ P formula offered, the constitutive view predominates, and so, not only is the developing picture of the person ($P_{s_1 \dots n}$) taking place within the constitutive nature of the person (P_S), but also both ($P_{s_1 \dots n}$ and P_S) are taking place within a function performing mode, (s). Where we mark as "S" the whole particularity of the unique individual, (s) is our view of S as a function. This functional view is one by which we often see a person in terms of her role. Eleanor Roosevelt was a specific person, S. As she was perceived to be a person, P_S, she was often recognized in terms of specific functions she performed, notably for the effectiveness of her speeches (s). (Here, P_S is the person S's constitutive property of *personhood*. We think of S as P_S. We see her as having a constant identity. But we also need a concept of *any* property belonging to S, which I note as $P_{s_1 \dots n}$ to say that the individual S attains personhood via a host of properties, such as "speech making." The particular property is less than total personhood, which is the genus category that includes all of the particular properties. The function operator (s) governs and exchanges categorical assignments to S and her personhood. Accordingly:

The function (s) is a trope.

This sort of category exchange by a trope can read out something like "Eleanor the speechmaker," "Omar the tent maker," or even "Eric the Red."

The logical transactions that move the elements of personhood and particularities into different inclusion relations are all functions of the thinker's focus. Thus, when we assign moods, properties, and functions, we always have perspectives from which to choose. Hence, "what subsumes what" deeply depends on the *constituent* and *constitutive* viewpoints we have identified.

From a *constitutive* viewpoint, for the function (s), any property of S (which we have noted as $P_{s_1 \ldots n}$) is included in P_S (namely, S's property of "personhood"). P_S is the inclusive category. Thus, S's property of personhood is a property that includes all others that belong to S. This inclusive relation can also be read in terms of a temporal direction, as follows:

$$P_S \rightarrow (s) \, P_{s_1 \ldots n}$$

Read the $\rightarrow$ as a causal direction within a forward temporal mode. To get the flavor of this temporal and causal direction, consider the observation that the embryo, in its progression to the full-grown infant, seems to proceed in a particular pattern of growth, differentiation, and formation of independent structures, later subordinated to a whole system of structures. The system operates in a coordinated way to fulfill a complex function, and what had been the several independent structures can now be seen as "subsystems" coordinated to that complex function. This picture of subsystems coordinated within a system is known as the developmental principle "functional subordination." Although you might argue that all this looks like a progression of construction, the issue I now focus on is the "fulfillment of the form." The embryo may *construct* from simple to complex or from individual properties and functions to coordinated ones.

However, concerning the present view of constitutive identity my point is that the principle of functional subordination and the particular patterns it realizes depend on autoregulation. *What appears to be the case is that such patterns follow a temporal and causal direction and fulfill a function that is of an inclusive nature.* In sum, developmental phenomena, such as functional subordination, work within an autoregulatory dynamic and appear to involve phenomena subject to a proposition of this same *constitutive* form of inclusive relations—namely, (s) $P_{s_1 \ldots n} \subset P_S$.

Even with all this shifting from constituent to constitutive perspectives, I am leading to the idea that a form of inclusive relations can be considered in terms of a logical proposition. Its conclusion is an unfolding of a premise—that is, the conclusion is inherent in the premise. Issues of direction of development involve not only logical forms, but also causal considerations. So the direction of development may go "backward" toward the realization of what appears retrospectively to be the whole originating nature of the individual's uniqueness and identity. Yet the causal direction may go "forward" in the manner described for functional subordination. These issues of conflicting directions of the logical form and of the causal and temporal modes require resolution.

Logical Form in Pursuit of Identity: The Weak Causality of Forward Temporal Direction

The direction of development I have indicated as "forward" in the determination of identity is a *weak* condition for identity. I say "weak" in the sense that the identity is a whole, established only in a forward direction. It is *not* reversible or reconstructible from the parts. An example would be to say, "It ain't over till it's over!" An individual isn't a whole self until he's lived out his life and made every possible contribution to markers of personal and social identity. As far as ego identity, the person "ain't finished till he's finished!" In the patrimony of Erikson's (1950) approach, there is always another phase of the psychosocial stage as long as the individual's life keeps rolling along toward resolution. I therefore also offer the expression

$$P_S \equiv (s)P_s$$

to show that personhood is achieved as the exercise of functions over an array of instances. So at any particular point in time that a $P_S \equiv (s)P_s$ proposition is put forward as a definition of the individual's identity, an *equivalence* relation obtains—not an *identity* relation.

Anthropologist Gregory Bateson (1979) reminds us *that logical identities offer reversible transformations, whereas natural causal patterns do not.* Accordingly, S is a unique person. He can be identified in terms of all his properties, but all his properties do not add up to S. The simulation would *not* become the model, nor would the clone become the person. The genus, although implying and leading to the development of the species, is *not* itself the species. This statement would be another form of the idea that *this forward direction of identity is weak: each term is related by equivalence, not by a logical identity.*

The weak form of the determination of psychological identity is a *one-way* direction. *Time* moves forward. *Time* is a contextual constraint of this form, and we can look at this constraint as marking a difference between a biologically based organism and a mechanically conceived construction. Forward construction appears to involve final—but also efficient—cause. Thus, P_S exists in terms of a sequence governed by efficient cause, imply-ing a linear temporal sequence—with time segmented in discrete particu-lar units. Yet, strictly speaking, there's a temporal context for P_S that can be indicated as another "constant variable," which can be symbolized

$(t_p)P_S$

This is to say that time (t) moving toward a potentiation (p), but *not there* at the moment of efficient causation, is a constant variable condition (t_p) for P_S, the person that is uniquely S. The full unfolding of the person may be *inherent* in the P_S, but *not there until the entire unfolding is done.* This view, by the way, is not unlike Erikson's idea that an identity changes in relation to all sorts of aspects or dynamics, but serves as a point from which dynamics and coordination continue to take place. In all this, we see contrarieties not only within the patterns of causal direction, but also between the direction of development and the logical requirements of identity transformations.

10   The "I," Entropy, and the Trope

Form and Dynamics Irresolution and Direction
Toward the Growth of Forms

Who, if she *is* a tension system, has a stable identity? What objects would attain stable intelligible meaning if tropes or their variants could describe them? The complication is that these forms are also tension systems! Picture the tropes or their logical forms as the tectonic plates supporting continents. The plates have their own inner tension and various external gravitational forces encouraging their movement. They jostle around and, in seeking Archimedian displacement and replacement of form and position, may break their wholeness as support structures. The continents (objects) move as if magnetized in planes above the plates. They are influenced in their own plane, pulled by the great waters, away from their mass coagulation. They come to a resolution as islands that have bounds, and, thus, each becomes an "identifiable" object.

The forms of our thought are tension systems, and they reach toward— and in the hands of the "I," they *covet*—logical and psychological identity. These systems require regulation, even though the powers to regulate are in dynamic forms, on all levels besieged by limits and entropy. Nevertheless, we, with all due tentativeness of our awareness and dubiousness of our causal powers, can originate forms and their perspectives in the face of the inevitable constraints of closed systems. However, this idea of origination attributes some engendering features to the very constraints that can delimit, can cut down potential, and can degrade. So these constraints and irresolutions can result in primitivized, partial, pathological, and immature forms.

What were at one time the myths that guided our thinking about origins and creative powers would now not qualify to offer a place for the individual and a story for the understanding of human development. We can't go back in time and be a mythmaker. We are at a point presumably beyond the beliefs in homeopathic magic. Few of us would incant "Open sesame," and expect doors to open! Sears has the knowledge and technology to make electronic door openers; scientific principles for electronic transmission have superseded the idea that a wave of the hand or a "door-opening chant" will work because of some similarity between the wave and the door's intended motion.

For a present-day artist to comment in his art about the classical mythical Vulcan's powers, his physical infirmity, and his fall from on high—as did fifteenth-century Florentine painter Piero di Cosimo—brings us to images that require suspension of centuries of the art of science and technology. Piero's rendition of primitive existence prior to Vulcan's inventions may be a commentary on what would result with the suspension of scientific and technological thinking! (See art historian Erwin Panofsky's [1939] 1962 analysis.) But even the depiction of a "natural primitive state" is not something we can now tolerate without considering how anthropologists have developed methods to picture it. Are Piero's images hallucinatory? Immature? Pathological? To offer icons of the primitive successfully is to connect them to ancient history, but, as in any effective use of a trope, the icons have to be articulated with enough breadth of reference to relate to the present—and *that* is the viewer's "I."

This form connecting the present to the icons of the past is a clue to the "I's" temporal and causal tension; it accommodates a breadth of reference; and it is an origin point for a trope's logical structure. Our time is an era of the mix of robots and terrorists, with all their sorts of threat as forces that would diminish the "I" and its capacities for originating. Yet this multidirectional and referential form, even in the face of tensions threatening identity, dynamically directs resolution of the great vicissitudes of primitivation and creativity. I should—and I do—summarize the chapter and the book by stating the terms of resolve and resolution.

The life of the individual as an organism is time bound. To picture the person's experience of self as logical transformations can clash with the different ways we capture the flow of time in cause-and-effect

explanations. Obvious contrarieties ensue when relating logical transformations to cause and effect and to the necessities of time flows and
bounds, relative to the organism as a closed system. Bateson concisely
notes the problem in using logical reasoning processes as a model for
causal events to explain biological or mental dynamics. The problem,
as he sees it, is fundamental: "The *if... then* of causality contains *time*,
but the *if... then* of logic is timeless. It follows that logic is an incomplete model of causality" (1979:59).

We run into less of a problem with efficient cause as a sequence in a
future-moving field of events than we do in the case of final cause. We
easily conceive a causal process proceeding from a logical form, forward
in time, to the fulfillment of that form. The fulfillment of the form *is temporally nonexistent at the moment of causation*, but when the form is seen
to "cause" its own fulfillment, a time warp bends the logic of the form. Yet
the form is producing this causation—from a point in the future because
it is not possible to realize without engaging a forward sequence of cause
and effect events. Piero's look back as a look forward is an intriguing way
of thinking. The as-if quality has all the causal power of the dream of a
future outcome as a determinant of that outcome—even if, like Piero's
image, it returns us to a primeval time. Aside from such contraries introduced by the interactions between efficient and final cause, *the very
presence of final cause in a logic model produces an effect → cause sequence for
temporal purposes, but retains a set → subset sequence, logically.*

Invocation: Why Grow Lilacs out of a Dead Land

To the extent that we lose the battle for the internal locus, we enjoin the
Other and the belief givers as other than ourselves. We subscribe to an
external determination of our ideas and beliefs. Call it external locus, or
EL. "I" is not to be my locus of origination. It might be my folks, my
culture, the statements of my leaders, the rules of law. It might be my
unconscious wishes, the bends and gaps in my neurological structures,
my genetic instructions. To subscribe, to join in this alternative, to believe
that our beliefs are determined by EL require that we give up capabilities
for figurative interpretations. What do we gain? We would see the meanings of tropes and their degenerate forms (stereotypes, propaganda, and
so on) as *concrete* and the logic of their application as *clear-cut*.

It *would* be a relief not to worry about ambiguity. A rose would be a rose; pornography would be "you know it when you see it"; and Mary Jo Buttafucco would know "if it walks like a snake, talks like a snake, and acts like a snake—it's a snake!" If logic were merely applied to terms that could not vary in their meaning or denotation, perspective on possible transformations would be minimized. If you adopt the EL as your guide, the terms you would subject to clear-cut logic can be at different orders of reference. They can be numbers or letters as equivalent units. Or the terms can be concretely specific; any equation of the terms would be invariable. What's so bad? The rules for the different orders would be the same, so they would act similarly and mix up orders—people can become numbers—being exchangeable with fully exposed contrarieties and oppositions of meaning and inclusiveness. So, what's the cost?

Look again at what happens in dream logic when one object is remembered as equivalent to another. "You're the top" as a dream image is literal; the equivalence is not debatable by a perspective denying the logic. In fact, it is not that simple. The meaning of "You're the top" can syncretize the figurative and literal meaning; the dream thought can mean both. Yet the opposing figurative meaning in the dream is subordinated to the concrete. The subset *is* the superset! There is a topsy-turvy order; species categorizing prevails as superordinating. When this happens, on the top of the ordering appear the contraries, which otherwise would be resolvable within a genus → species organization.

What I am calling "concrete" can be a status you would apply to an overly abstract notion, such as a symbol that cannot be modified, or it can be a quite specific picture that cannot be other than a symbol, in the sense of a prototype. So for abstract "symbols" and for concrete equations, it comes down to the same thing. When the "I's" perspectives—which would include negations to challenge the EL's perspectives—are unavailable, logical transformations that take them into account cannot take place.

All this inertness is because the believed picture, if bound within the EL, would shut out an internally inspired view as a perspective, which may otherwise be evaluated relative to that of others and the belief givers. This evaluation *cannot be made* when the dependency on an external source is complete or when the individual has no access to conscious thought about the meaning of the phenomena and about his own self and its perspective.

What "I" Always Wanted to Know About Monads, but Was Afraid to Ask

Without resolution to the problems of causality and temporal mode, it would be unbearable to face the "I" as locus of origination. The hornet's nest of contradictions, sending time and causality in contrary, opposing directions, stings like the inability to hold your identity in place. It can feel like knowing who you are by what you can do, but not knowing when you can act and when the effect would exist. The hero of *The Prisoner* (Mc-Goohan and Tomblin 1967), the classic TV intelligence spy game thriller I mentioned in chapter 6, acts like this in episode after episode. Or it can feel like knowing that you did or will act "in" a particular time, even though, by the rules of time and space, you also know that you cannot be there. This is what happens to the hero of the movie *Somewhere in Time* (Matheson 1980). Or it may feel like you cannot grasp the time frame of others while you remain a prisoner of your own constant present. In the movie *Portrait of Jennie* (Osborn and Berneis 1948), the hero-artist Eben Adams feels this as he hears the statements of the character Jennie Appleton. He watches her move, grow, and think at a different speed than that of his own time zone. He tries mightily, but he cannot scale his being with her temporal destinies. What emerges in the trials and tribulations of these selves inwardly facing a dissolute identity is the theme of inevitability and fate as ruthless external forces. Even the fear of these determinants seems preferable to the horror of a dissolving identity that's eating the person away from the inside.

Gombrich's theme is of the disgust that sets in when art gets overly organized and too many contraries are exposed. Insult is added to injury when the picture is patched together by too many concretized niceties. They organize but elude elegance and do no more than dangle access to identity.

The search for the primitive and the loss of the "conscious 'I'" occurs in art as the return to the concrete. This search can appear in the symbolic abstractions that irreverently cross from one order of meaning to another. Or it can be within the concrete, which inheres in the form that cannot rise to mean more than itself. The issue is the artistic vehicle as trope. My point is that the return to the concrete is a violation of the structure of a trope!

Many artists have gotten lost in either form abstractions or color abstractions that presume to denote a quality, but nevertheless fall short of a noesis. As artist Mark Rothko's career advanced, his work became increasing abstract, but the progress was from the abstract as symbolic to the abstract as concrete. Color was depicted without form and without any content save the color (see Mills 1998). However, the degree of concreteness did not constitute a trope. The art as icon fell short—whether in terms of the artist's conception or by way of the engagement of the interpretative function of the viewer's "I." Rothko's view was not even part of a movement to get the viewer to supply the reference points. The involvement of the "I" was in the artist's and the viewer's *immersion in the subject*. Adolph Gottlieb, Mark Rothko, and Barnett Newman make Rothko's point in a 1943 letter to the *New York Times:* "the subject is crucial and only that subject matter is valid which is tragic and timeless. That is why we profess a spiritual kinship with primitive and archaic art" (Rothko and Gottlieb 2006).

So if the color red were portrayed as "the color red," the viewer can immerse himself in the icon and thus illuminate the subject "red" because the icon is the subject. The subject is that of "red"—a perception within the individual's life experience. The icon, though, is art—that which *is* the subject. If Rothko's ultimate comparison was of art and life, and this comparison was made through the concreteness of the subject, the problem still remains. *The distinction between art and life was not made in the terms of the trope.* The "art qua subject" approach amounts to "the subject qua subject." This attack on categorization is an assault on comparison and reference, and is *incomplete synecdoche.*

In contrast, take the case of simplifications and symbolizations, where the primitive form *does* refer to a rich array of references. In art where the form resembles a symbol of a bioform, you have an abstraction—but there is a remnant. Sometimes it's a recognizable structural similarity between a biological form and the abstraction. Sometimes it's a pattern of apparent interactive relations that is reminiscent of the patterns of biological forms. Sometimes it's the transmogrification of the biological form by way of how it appears in a perceptual, conceptual, or dream format. In such a case, origins are in biological objects and in the perceptions of those objects. You, therefore, *can* argue that the abstraction includes an interpretation of origins, which is why artists such as Joan Miró and Paul Klee can be seen to articulate how the imagined form has articulated reference points. A description of Miró's work by the Art and Culture Network shows this articulation between the primitive and the various reference points:

"Since the age of the cave-dwellers, art has done nothing but degenerate."

Miró's statement, aside from revealing his views on the history of art, also says something about his own artistic aims. He wanted to bring art back to its primitive, playful origins, to paint as if he were painting on the wall of a cave. Eschewing every technique of representation that the intervening centuries of art had spawned, Miró's paintings are flat. There is no attempt to create the illusion of depth; his simplified, linear figures float in an ethereal, boundless space. Never interested in representing real places and objects, Miró painted from a land of dreams, a land in which fanciful, sporadic associations determine the distribution of things. (1999–2003)

But to show further that Miró provided a rich array of reference points, a description by Diana C. Dupont is on the mark:

FOLLOWING THE SPARE APPARITIONS of the mid-twenties, Miró moved to develop a unique iconography of signs and pictograms drawn from his imagination, his environment, and his Catalan heritage. Closely linked historically to the Surrealists yet isolated from them by consummate individualism, he delved into the realm of dreams and fantasy, using images that evoked subconscious recognition and universal emotions. The compositions again became more complex, sometimes taking the form of a landscape in which disparate object-images are combined within a single arena, sometimes becoming ambiguous arenas where biomorphic forms float on amorphous backdrops. Miró fused poetry with pictorial concerns, alluding to the literary conjugation of beauty with lyrical titles that provide keys to the symbols depicted. (San Francisco Museum of Modern Art and Du Pont 1985)

Robert Hughes shows how Klee's articulation of the primitive has rich connections to a vast array of reference points:

[Klee offers] not only close but ecstatic observation of the natural world, embracing the Romantic extremes of the near and the far, the close-up detail and the "cosmic" landscape. At one end, the moon and mountains, the stand of jagged dark pines, the flat mirroring seas laid in a mosaic of washes; at the other, a swarm of little graphic inventions, crystalline or squirming, that could only have been made in the age of high-resolution microscopy and the close-up photograph. There was a clear link between some of Klee's plant motifs and the images of plankton, diatoms, seeds, and micro-organisms that German scientific photographers were making at the same time. ([1981] 1991:306)

The case is made for artists who utilize and vary the reference base for bioforms. But in contrast to the incomplete synecdoche of Rothko's depictions of color, Klee's reference points for color follow the same principle of rich comparisons and reference points, as do his "bioforms." (See Januszczak 1993.)

Thus, for Miró and Klee, the forms are *completed synecdoches* in that the part terms (the paintings bioforms) are comparable to the whole (biological potentialities in biological forms): "But there are many instances where the symbol is poorly coordinated with either aspects of design or other levels of meaning. For example, consider composer John Cage's 'silence' to represent 'pauses' between 'sounds'" (L. Solomon 1998:sec. 1, para. 1). The work is supposed to show the importance of pauses, but it winds up being a synecdoche with the absence of a comparison term. The absurdity of the "silence" as something to which the "listener" brings the reference points surfaced recently when the technique was tried again. BBC News reported that

> Musician Mike Batt had paid a six-figure sum to settle a bizarre dispute over who owns copyright to a silent musical work.
>
> Batt, who had a number of hits in the 70s with UK children's characters The Wombles, was accused of plagiarism by the publishers of the late US composer John Cage, after placing a silent track on his latest album, *Classical Graffiti* which was credited to himself and Cage. (2002)

The error of representing something without adequate articulation of the trope occurs in many art forms. There is the "boring" movie to represent "boredom," the disorganized piece of music to represent "disorganization." Formally to have a meaningful synecdoche, you have to have some reference to the whole and some to the part. John Cage's trope falls short simply because there's no reference to music. The "boring" movie does not relate to nonboredom and therefore is too dependent on the viewer's bored state. (An egregious case is the movie *My Dinner with André* [Shawn and Gregory 1981].) The viewer is disengaged from the process of defining the boredom because she's so enmeshed in its experience. Here is the key. *A trope is not a trope if the "I" is disengaged. The disengagement takes place when the synecdoche is so concrete in its involvement with the part that the "I" cannot visualize the whole.*

A Brief Pointer About the "I's" Disengagement When the Trope Is Incomplete

We argue whether a work of art can be a concrete thing or a wall of white coloring, like what artist Robert Rauschenberg might present (*White Painting,* 1951). The idea is that the opposites, the contrasts, the category of everything else (universal complement), the implications for analogues on other levels of abstraction should not be presented by the artist. They should be conjured by the viewer's "I," forced by the absence of the complement of the concrete category to present it. Solomon writes: "It wasn't until 1951 that Cage was inspired to proceed by seeing the white, empty paintings freshly done by his friend, Robert Rauschenberg. 'I responded immediately,' he said, 'not as objects, but as ways of seeing. I've said before that they were airports for shadows and for dust, but you could also say that they were mirrors of the air'" (1998:sec. 2, para. 11).

So if you listen to silence, you can provide the nonsilent aspects of music. This argument is easy to refute because music is not necessarily the opposite of silence. On the part of the "I" as viewer, the contrast is going to be either "nonsilence," which can be considered a "universal" logical complement, or a particularized opposition, which is not necessarily that of the artist. The "catchall" complement simply begs the question that anything in the world can be compared to what you say is not it. But the main point is that for you, as viewer, to come up with your own complement or with an opposite of a concrete category does not require an artist! If the artist's category is in fact coterminous with the subject, then you can experience the subject directly. You don't need the artist for that; you can go to your room for a "time out" and experience silence. Then, if you like, you can get Benjamin Moore paint and paint the whole room white! In a word, you're *entitled to your own perception and categorization,* and it is *that choice* for which a complement is also yours. Ditto for the artist. Here is how I cast this argument about absent complements as a rule of logical form:

Art as a (category ≡ concrete subject) has neither a logical complement nor a specification of subcategories.

Therefore, this brand of art does not present the viewer with what a scientist would call opportunities for falsifiability, and therefore it lies outside the judgment of whether it is art! Its form resembles more of a totalitarian approach to presenting a belief than either a dialectician's argument, which has another side, or some sort of empirical approach to a standard that makes for art.

The search for the primitive and the primitivation of tropes is not merely a phenomenon in art or the arts. *The search for the primitive and the passion to lose the "conscious 'I'" occurs in sociopolitical contexts of logic, reason, rules, laws, and their impact on beliefs. It also occurs in contorted symbolic abstractions such as slogans and absurd categorizations. With concrete forms that do not rise above themselves, ideas are not destroyed by argument; instead, "killing"—the idea—can become "killing"—the acts. The conscious "I" is submerged in a sea of primitivation, and the belief givers are given any control they grab. In fact, with constriction of the "I's" powers, the belief givers are also given any controls that they do not grab.*

Both in Nazi Germany and in the former Soviet Union, the constriction of personal expression and the totalitarian control of sociopolitical events and products included division of art into the acceptable and the nonacceptable. In the heyday of the former Soviet Union, social realism was married to political art and the "mystique" of the state. Paintings, posters, and sculpture became not only repressive, but also "boring." In fact, sociologist Alex Inkeles writes that in the received view, "The actual characteristics of 'good' art, whatever the precise label, tend to be much the same in all modern totalitarian societies ... concrete rather than abstract, directly representational, 'wholesome' rather than dealing with 'unpleasant' subjects, light in color rather than dark in shade and tone, 'social' rather than predominantly 'private' in subject matter, and 'cheerful' rather than somber or 'depressing'" (1968:79–80).

However, Inkeles's view is that totalitarian art also served primarily to show how the mystique of the state would change people's lives and what their future would (and should) be like. He concludes that such art is not merely representational, but also symbolic! This conflation delimits the "kind" of primitivation involved. The primitivation is *not* purely and simply "concrete." Yet it is a form of primitivized icon that constricts the individual and the "I." Art, as trope, is "symbolic" in its logical prototyping organization. Like stereotypes and propaganda with their constrictions, art's meanings are contorted to produce the "good" and the "bad," the "acceptable" and the "unacceptable." And when combined with political power, totalitarian-controlled art, like stereotypes and propaganda, strait-jackets individual agency. The sources of belief are "given over" to the belief givers; the art becomes their tropes. Just as one can learn to build stat-ues of Stalin and to incant adorations of Lenin's actual—but pickled and preserved—body, art can be the metaphor for social correctness that can be uttered "trippingly on the tongue." Inkeles's twentieth-century point is eerily pertinent in terms of current-day emphasis on religious rewards in the future for politically "correct" acts. The totalitarian control of art aims at control that extends to the future and to the individual's conception of it—including of the afterlife and its rewards! The symbolic icon goes all the way. It remains a species of the primitive because it is symbolization with no complementary categories, and any contrary or contradiction is expunged. So not only are *endings*—the future, the afterlife—controlled tropes, but so are *beginnings*. Inkeles writes that "the totalitarian seeks the elimination, the physical destruction of the bad art.... He actually pro-scribes not merely its circulation but its very *creation*" (1968:81–82).

In Soviet realist art, there was no excitement of the tension, conflict, and juggling of reference levels and domains that one can see in the bioforms of Kandinsky and of those following in his patrimony. A long story of the escape hatches from totalitarianism of both the left and the right is needed to account for the survival and continued production of Kandinsky's (and Klee's) work. Kandinsky left Russia in 1921. In 1932, the Bauhaus, at which he taught in Germany, was dissolved and then finally closed in 1933 (Roskill 1992). Kandinsky left Germany in 1933. In 1937, "his paintings would be among those included in the exhibition of con-demned works called 'Degenerate Art'" (Faerna 1996:8). Minus a full re-view, here is a summary formula. It has been my argument all along:

The "I" has to be engaged to enable internal perspective on the icons that encompass and give form to beliefs, attributions, and ways of viewing ourselves and the Other.

But remember the case of Eben Adams—the artist-hero character of the *Portrait of Jennie*. The contraries of cause and time sabotage this character's ability to cause events and to define his identity in relation to objects and external events. So part of the formula is the "I's" immersion in a natural-logical form and organization of forms—a situation involv-ing tension heading toward dynamic resolution. But what a concatenated tension this is! There are the problems of the "I" and its perspective on time frames, of the cause-effect sequences within those time frames, and of the sense of the "I's" agency as causal within its temporal mode. Concatenated and dynamic, all this forms a tension system that should engage—full-time—the powers of autoregulation.

Although I am aware of empirical studies (McGregor et al. 2001), archival and statistical approaches (Sales 1973; Doty, Peterson, and Winter 1991), I do not review them here for reasons I have cited concerning the limits of operational measures of psychodynamics, on one hand, and the differences between measures of external context and explanation of interior processes, on the other.

In this book, I go in the direction of using an analysis of natural logic and its structures to view the tension system and the autoregulative resolutions. Such an analysis can be too thin if not complemented by semiotic latitude. The quest for harmonies and balances in a tension system invite a search for elements and simples before new combinations can be made. Although the "I" is a prime element, analysis of autoregulative resolution needs to be semiotically inclusive of all sorts of levels of description and explanation. It would cheat the "I" if analysis were to present too monocategorical a picture! My analogy is with Klee's (1911) attempt to depict and present "primitive forms" to elucidate the "primitive" status of childhood drawings (see Roskill 1992:fig. 11).

The success of such an enterprise—and of my quest as well—is in the "layering" of the trope with diverse types of category and diverse bits of meaning. When legendary jazz bassist Slam Stewart hums, that is a very different phenomenon compared to his evoking sound from the bass instrument (Byas and Stewart 1945). Stewart's body and soul have a different "form" than that of the "bass." The tone is similar enough that you sometimes think two basses are playing and other times you think one bass is being played by someone who can make it sound like two! Layered onto this "equivalence" of the tones in the face of the nonidentity of the process and of the "forms" is an "equivalence" of melodies. Stewart plays a melody with "legitimate variations" and then dares the listener to catch the moment when his joke is to sidetrack the melody into a foolish caprice—which is converted back to the main melody as soon as you notice it. The meanings with such layers of sounds-as-icons are so diverse that there has to be some sort of protologic to interrelate them.

The formula *"Where the layers touch, there should be equivalences that are not identities"* is meant for any trope—whether the iconic form is music, art, or verbal symbol. In fact, I have already referred to the phenomenon of a *semiotic system of sign equivalences and their transactions*. Art critic Robert Hughes quotes Kandinsky: "Modern art can only be born when signs become symbols" (1981:301). The relationship between a sign and a symbol is best understood here in semiotic terms. A sign is a representation, but a symbol is on the next-high logical order. It represents that representation. You can see why Kandinsky took the path leading to an idealist way of picturing the universe. The objects represented were less important than the icons and their transactions. In Peirce's terms, they would be interpretants. They mediate and organize; they help to order and to regulate. Kandinsky proposes that

> the complete negation of objects is equal to their complete affirmation. In both cases the equals sign is again justified because it leads to the same end: embodiment of the selfsame inner sound...it makes no difference whether the artist uses real or abstract forms.
>
> Both forms are basically internally equal. ([1912] 1974:168)

Thus, for the artist, the sign-as-symbol equivalences can involve the layering I have been describing. The different layers can provide a route to deeper layers and/or to universal significance.

In this sense, the equivalences can be aligned with the idea of the "spiritual," as in Klee's theory of "visual 'equivalents'" (Hughes [1981] 1991:304). In this book, *I propose the idea of a semiotic system of equivalence that permits transactions from species organization of categories to different categorical orders.* From my perspective, those things that Kandinsky sought as the "selfsame inner sound" and cast as the apprehensions of the soul are at least in part related to the proposition that the "I" is the arbiter of perspective and logical organization and is the clearinghouse for meaning. The "I" focuses and perhaps seizes on the "pressure points" for the logical organization of meaning.

What is visible for Klee in some sort of "symbolic particular" (color, form, line) "is merely an isolated case in relation to the universe." Klee "believed that eighteenth-century counterpoint (his favourite form) could be translated quite directly into gradations of colour and value, repetitions and changes of motif; his compositions of stacked forms, fanned out like decks of cards or colour swatches...are attempts to freeze time in a static composition, to give visual motifs the 'unfolding' qualities of aural ones"

(Hughes 1981:304). These equivalences are the potential pressure points for the "I" to feel, grow cognizant of, and emerge with understanding, reorganization, and perspective.

The emergence is sometimes a recognition, yet not one with an urgency to resolve things. Novelist, playwright, and psychologist Arthur Koestler (1964) makes such a case for humor. In the case of Slam Stewart's musical improvisation, the suspension of causal patterns is *like* humor—an intellectual holiday from the problem of deciphering the causes that are presented to the listener as causes from the bass-as-source versus those from the person. The question of how the sounds are made is subservient to the phenomenon of the harmony of what the listener perceives, but usually does not codify, as contraries. Another view is that the listener resolves the problem of the bass-as-cause versus the person-as-cause. The listener becomes the causal node for hearing the center-stage harmony, for resonating to the equivalences, and for placing the contrary sources in the background. The melody mix is also his resolution of the linear progression of a melody that drifts into another time zone in his mind. How does he find the confidence that the artist will "return" from his melodic caprices to the linear temporality of moving the main melody along? In the listener's capacity for design, we get the resolution. It is by way of the entrancing equivalences and their bearable suspension of the contraries of causal sources and any contradiction between these sources and temporal progression.

However, where art, humor, and aesthetics leave off, there is a need to go beyond the enjoyment of origins to the decisions of beliefs about the self and about the world around it. Where there is a need to be fully conscious of the conflicts and contradictions—to make sense of things, to know one's rights, to interpret others correctly, to make products, and so on—we focus on tension. We screw up our courage, and we do not eschew tension when so many contraries are present because a new model of self or science or government—or indeed art—is needed. But because the focus on such needs requires the "I's" feel and conscious sense of tension, I directly focus on *this requirement* as the tension system.

What *is* the relation of the "I" to the problem of organizing cause and time to resolve their contraries? Can this relation be expressed as a resolution—although the resolution itself, like a natural-logical form, would be dynamic?

Back to Leibniz's Future

We go back in time to Gottfried Leibniz, who in his *Monadology* achieved a resolution by attributing final cause to the functions of the *soul*, while arguing that efficient cause was inherent in the bodily functions and events. "Souls act according to the laws of final causes through appetitions, ends, and means. Bodies act according to the laws of efficient causes or motions. And the two realms, that of efficient causes and that of final causes, are in harmony with one another" ([1698] 1999:sec. 79).

Translator Robert Latta defines Leibniz's term *appetition* as "the activity of the internal principle which produces change or passage from one perception to another" (in Leibniz [1698] 2007). The term *appetition* is old, but this definition appears very new—au courant! Talk about old terms. What about *soul, final cause,* and *efficient cause*? All of these terms bring us squarely back to preoccupations with Aristotle. *Is* anything new to be found in Leibniz's resolution? Well, first of all, how or why would we consider it a resolution?

It *is* a resolution rather than a mere state of tension because the issue of "form" will not go away! It provides bounds for the tension; it accommodates its elements, forces, and even the contraries of two separate entities—body and soul. Leibniz considered the *monad* to be a form, which accommodates both soul and body. This division of an individual has a long history and would not have been out of order in the seventeenth century. Yet if you read how the *functions* of each entity work, *they* do seem compatible with contemporary views of the natural functions of an organism that has capacities to perceive, to transact perceptions, and to initiate and regulate motion. In our contemporary terms, the resolution is that *different sets of functions with different sets of rules exist within the same form.*

In Leibniz's way of looking at this, "The soul follows its own laws, and the body likewise follows its own laws; and they agree with each other in virtue of the pre-established harmony between all substances, since they are all representations of one and the same universe" ([1698] 1999:sec. 78).

You can take the status of different sets within the same form as a matter of categorical order and organization. Seen this way, the resolution of dynamic exchanges for stable ordering is remarkably consonant with contemporary views of autoregulation, and it appears enough of a basis for a resolution of the two kinds of causation within the same form—or individual organism. The overall form can achieve its organization and dynamics as a set of rules directed toward "preestablished harmony."

Ergo, final cause. But the internal organization is of separate sets of laws (for the soul and for the body). The empirical quest for these rules is usually by some appeal to a linguistic analysis—hence, the "I" as a "shifter," the objective viewpoint as a "third-person" perspective, the subjective as "first-person" mode.

The soul follows rules of final cause. In contemporary terms, this principle can mean that at any point in time the conscious individual can retrospectively lay out what past determinants account for her "moment" of self-realization. George Washington offered this account: "All I am I owe to my mother. I attribute all my success in life to the moral, intellectual and physical education I received from her." Another president, Abraham Lincoln, went further, accounting for future self-realizations: "All that I am or ever hope to be, I owe to my angel Mother."

The body follows rules of efficient cause because an objective observer can capture a frame within which an act or function sequentially results in an observable outcome. The coordination of movement and the factors in perceiving time as sequence is debatable as to a baseball batter's "keeping his eye on the ball." That's because actually doing so may be physically impossible. Even so, baseball players using such techniques as video frame-by-frame analysis can study the sequences of their own acts in swinging at a ball. This framing is categorized as a "swing," and, of course, golfers make the same sort of attempt.

The soul/body split is fraught with so many difficulties that the resolution I adopt is to assume a governing perspective—namely, that the *form* of the individual sets in motion a logical structuring of categories.

If you want, think of the form as bounds. The dynamics and the particulars of an individual are bound into her form. But the idea I'm focusing on is that these bounds are describable as a natural-logical structure. Overall, as such, the form provides for an inherent logical identity as an individual, but it also has dynamic capacity for categorically structuring its internal logical organization. Thus, within an individual's categorical structure, you can see a category of that which is manifested by her "final cause" propositions. "All I am I owe to ..."; "I believe that it [one's fate] is written in the book"; and so on. And, correspondingly, in the same individual there's a different category—one that constitutes the person's framing and efficient-cause proposals. A president who can attribute everything in his self-realization to his mother can calculate carefully in a forward temporal direction what results his political acts will yield. Bill Clinton comes to mind!

The logical structuring of these modes of conceiving causality, like Leibniz's dichotomy of body and soul, is categorical. The categories have their own sets of rules, which govern causal powers and dynamics. Moreover, the entire organization is rule governed and dynamic—in short, it is self-regulated. Therefore, the Aristotelian ring to the causal powers derivable from the categorical structures and, in general, from form somehow resonates with the functioning and structural nature of natural logic.

There *are* contemporary views of the individual's logical structuring of temporal mode and causal attributions. These views include theories of causal attribution, locus of control, self-efficacy, and the ever-popular cognitive linguistics. They all remain wedded to an analysis of information and communication exchanges, but logical ordering—although in the background—is *not* fundamental. Bateson (1972, 1979) tries straddling a fence on this one. His studies of communication include that of persons, schizophrenics, and dolphins. While focusing on communicative process, he *does* picture logical order as a framing event that can reduce conflict in a way not dissimilar to Aristotle's method of getting beyond contraries and oppositions. He pictures conflict in communication in terms of logical "binds" and conceptualizes resolutions as steps in recategorizing contraries. We can make the Leibniz resolution more specific for our purposes by reviewing Bateson's approach to natural logic and causal events. But you can also see that Bateson has a healthy respect for different forms of life as well as for different stages and modes of thought, including the psychopathological. Therefore, he straddles the fence. There has to be an emphasis on different communicative forms in interaction with the question of establishing a logical ordering that will resolve conflict and handle contraries.

Bateson works out a wonderful resolution utilizing the idea that *codes* can refer to different "types," which I have called "orders" of classification. Particularly interesting is his idea about different types of code. There are *codes for "yes-no"* concerning the presence of differences; and there are *template codes,* such that a sequence of changes follows a previously existing pattern. There is also *part-for-whole coding,* so that changes take place on the basis of a code that assumes—one might say entails—a future presentation of a fulfillment or whole.

These ideas have obvious relevance to category boundaries, decision making, and the capacity for causal attributions. But there is a mix of logic—even if it includes the paralogic—with possibilities of framing and empirical viewings of causal sequences. So I will show some of the implications of Bateson's concept of "codes," but not before putting this approach into the context of this book's concern for logical structure, despite two major sources of aversion to it.

The thesis I offer is that specifying natural-logical structures can make the critically needed distinction of the trope from similar but variant forms. To offer this resolution in the face of a long history of aversion to the mix of logic and empirical-friendly causation requires a recovery from warnings about rationalism (Hume, for example) and from outright contempt for the misuses of reason (Bentham). But even more central to our concerns is a second source of continuous aversion to a categorical resolution to contraries and conflictual elements. That is the "disgust" reaction, which returns the effort at complex organization to a primitive state, where categories are more fluid in their interchangeability. *Should we not also call the warnings and the contempt for rationalism "disgust"—the same reaction that Gombrich describes for art when complexity has gone out of control in the regulation of identity?*

Disgust and Reduction in Art, Science, and the Sociopolitical Arena

How do the considerations of reduction and disgust interact in the three arenas art, science, and the sociopolitical?

In relation to art, identity becomes a matter of stability and recognition of what is depicted. You can then argue that the nature of the icon presented has to stabilize the particular presentation, such that the viewer recognizes its individuality—and hence its internal organization. But with waves of disgust, resultant reductions and their primitivations, and the inevitable realization that reduction "has gone too far," perspectives rebound and are given due respect. The artist multiplies her perspectives, and the viewer does too. Oh boy! Now the tension systems are really complex and challenging to resolve! The more perspective offered, the more the stability has to be rationalized. The outcome is the use of reason and logic to tie together any complexity in meanings, which have been mixed together, irrespective of source, domain of reference, or place in an order of inclusion. When the complexity gets to be too much of a stretch for us to hold together and experience harmony, the logic seems false, and disgust sets in.

The modern art period—typified by the work and rationales of Kandinsky, Klee, and Miró—is then replaced by a primitivized and syncretized presentation, which gladly exchanges logic for the excitement of the shocking and the sensual. There is, as I have waxed poetic about it, a similar sequence in science, social science, and political and social design and technology. In the course of successions of the development of order, of the disgust and inevitable reduction of this order, and of recovery through reorganization, logic and reason can and do rise to the challenge. When there is too much of an assault on the need for structure, logic and reason do their job to explain growing contradictions. For Jeremy Bentham, the use of law as a social and societal regulatory discipline had advanced to its misuse. This misuse of reason advanced to the neutralization of scientific knowledge—as Francis Bacon inveighed. As both Bentham and Bacon complained, things go downhill. The law is misused; science is held in place by vested assumptions; rigid reasoning piles up as evidence to the contrary of accepted ways of thinking; questions do not get answered; and problems do not get solved. So the ability to develop and view perspectives can be kept at bay by distortions of the rational lens. However, the worm turns! If the ravages of disgust are too destructive, the ability to develop and view perspectives is at risk. Eliminating organization and logical ordering can also obfuscate access to perspective.

Epochs in art end and revolutions in art take place when "enough is enough." In the same way, there are epochs in the scientific and in the sociopolitical arenas. They, too, have their expansions and reductions of ideas and assumptions that affect our access to consciousness, put to sleep and awaken the role of the "I," and gate off and then open the door to the attention to inner determinants. Let's focus on a social science example wherein these shifts are dramatic.

The Case of an Idiographic Science as Falsifiable

Various contradictions in psychoanalytic formulations began to seem grotesque when they were rationalized and subjected to "logical" explanation (Hook 1959). A great deal was left by the wayside when it was argued that psychoanalytic insights and formulations were so rich in their array of implications that they were "unfalsifiable." In addition, psychoanalytic explanations were retrospective accounts that, when turned around to be predictive, could be so general and account for such a wide array of results that they were useless in terms of a given phenomenon.

So it was with "disgust" that many threw out so many concepts to explain the internal source of agency, self, and action, and some would say that many "babies were thrown out with the psychoanalytic bathwater." Not only the behaviorist movement, which in its radical form refused any mentalistic or phenomenological construct (e.g., Skinner [1974]), but also the sweeping contextualist challenge, which focused social and cultural sources of human conceptions of self and even agency, made these changes. On one hand, John Dollard and Neal E. Miller (1950) turned psychoanalysis into behavior and reinforcement concepts. On the other hand, psychoanalysis was transformed into a grand anthropological explanation that accounted for societal and cultural dynamics and phenomena. But what about the bread-and-butter process of the "small unit"—the individual and the "analyst" (Wax 1995b)? Was this a special case of "clinical" inference? How would that work into universal concepts, either in terms of human behavior or a grand anthropological explanation?

Working with a concept such as that of sociologist Clifford Geertz's (1973) "signifiers," one might reason that the purpose of clinical inference is to set a theoretical framework for further discourse and understanding. To the extent that symbols and signs are to be studied in terms of the role they play in society (Geertz 1983a, 1983b), the meanings referred to by psychoanalytic concepts are part of a broad ethnography. Within this approach, anthropologist Murray Wax (1983) reasons that falsifiability is the wrong criterion for a grand "model" such as psychoanalysis. The model's broader purpose is to be a framework so encompassing that it includes human complexities. As Geertz puts it, Freud's theory is "powerfully unitive" (1983b:150). It sports Freud's "conviction that the mechanics of human thinking is invariable across time, space, culture, and circumstance" (150). Such a framework, Wax feels, is presumed to be flexible enough to change its assumptions through thought and discourse, but also broad enough in its anthropological scope that it is judged on a "normative" level. His example is the incest taboos that exist in every known society. The effort at falsification on the level of the individual doesn't fit the scope of the theoretical framework.

Nevertheless, the theory *is* applied to the individual person, and, more tellingly, there is some version of phenomenological action critical to the psychoanalytic assumptions and methodology! Wax (1995a) deals with the question of idiographic analysis by fitting the psychotherapeutic methodology to hermeneutics and the evincing of *stories*. The revelation of stories and the analyst's interpretation produce an ongoing exchange that moves the individual ever closer to a "coherent biography."

However, myths and other social phenomena come down to earth. In light of the model's presumed power to propose a universal "plot" (such as the Oedipus myth), there are assumptions about the individual's regulative rules and dynamics. Deep in psychoanalytic theory there is the assumption of intrapsychic causality (conflict within the individual and the mechanics of reorganization). There is also the constraint of the individual organism's bounded life stages, problems, and status. In the light of these constraints, the theory's operational levels have been difficult to formulate. The problems that must be addressed are the framing of temporal units such as the length of analysis, the life issue, the area of adjustment focused, and the relation of individual behavior to biographical coherence. That individual behavior is not only related to a massive set of circles of cultural and temporal contexts, but also found within the "plot structures" of smaller schematics—the job decision, the task in everyday life that has multiple meanings, and, in general, acts of decision within bounded causal/temporal frames.

Most attempts to specify "plots" for operational models—which some would call "hypotheses"—do not seem to provide the "complementary" propositions one can evaluate. Thus, if from psychoanalytic data showing a patient's course of insights one were to try "clinical inference" to evince the plot line of that person's resolutions of conflict—for example, with a specific set of job decisions—the falsification process would be all but nonexistent. So whether the psychoanalysis "worked" or produced insights would be a matter of retrospective rationalizing.

Consequently, in and around the mid–twentieth century, the psychological and philosophical communities had simply "had it" with the combination of logical contradictions in and pseudoheuristic explanations of the concept of the "unconscious" (see Kihlstrom [1994] 2005; [2000] 2006). If something was unconscious, you couldn't know it, so invoking it might explain either the presence or the absence of a phenomenon. They had also "had it" with the way any symptom could be explained by a dynamic emotion and its repression OR by the opposite of that emotion (the opposite of repression) OR by the opposite of their total combination! They were outright disgusted with the dance around falsifiability. A major example of this disgust was physicist and philosopher of science Adolf Grünbaum's (1984, 2006a, 2006b) dismissal of psychoanalysis's status as a science.

Many psychologists *did* want to have a dynamic theory of the individual (see Westen 1998). They did not want to abandon the idea of the individual sources of causality either to the cultural or to the hermeneutics brand of a contextualist point of view. But the ghosts of Hume, Bacon, and Bentham raised warning signals. Those who wanted and needed to recognize the realities of subjective organization and the individual as a source of origins remained terrorized by the caricatures of false reason. So the answer was to "objectify" the mind and its products.

The reductions of thought to language or to text were key. The objectifying of cognitive display and action correlates of "mind" proceeded very rapidly, but it showed the incapacity of logic to hold together the growing information in the cognitive sciences or the model building in ecological psychology, social constructionism, and cognitive linguistics. This case was argued ferociously by Lakoff (1987, 1999), whose view was that classical logic has to give way to the bases of thought, which are actions and outcomes and their explorable spaces. Yet all this together would fragment the individual and make her rationalized identity a nightmare split into dream sequences requiring a dream book of supercode to recognize themes, let alone order or coherence. The real victim was agency.

Toward the Coexistence of Time and Logic

Bateson offers a recovery from the terror that many—in the patrimony of Hume and Bacon—have felt about the misuse of logic in relation to characterizing the human condition of contrariety in causal events. He puts it that "[t]his cycle of *if ... then* relations in the world of cause is disruptive of any cycle of *if ... then* relations in the world of logic unless time is introduced into logic" (1979:126).

An empirical fact can include the observation that in the Israeli-Palestinian conflicts there is a "cycle of violence." We can reason that "violence" (V) reproduces. Therefore, logic would tell us that

If V,

Then V.

If the "cycle of violence" is not a mere logical formula, but instead a causal sequence, then one issue affecting it is time sequence. When two combatants are claiming "You started it!" the rationale for continuing the cycle is at stake. This is not to say naively that what is needed to stop the cycle is merely an agreed upon time frame in which to call a "start" a "start." It is to say that in unraveling cause and effect from logical traps, or misuse, agreement on the time frame can torpedo the logic of the "reproduction" syllogism. Of course, some other confusion of logic and causal framing can be introduced to rationalize the "cycle"!

With this example, though, we can say that time has to be factored in to the logic of the organism's communication system. If I truly believe "the other guy started it," I may be forever at odds with the fellow who has a reciprocal belief. There are obvious cross points between "logic," the belief in a causal sequence, and a definition of a time frame. Each is subject to a different kind of code. The relationship of the different kinds of codes to causality involves vicissitudes of temporal and development direction. Thus, an assault on logic can be introduced by not taking time into account, by deliberately confounding different definitions of time, or by primitivizing the frames. In the Israeli-Palestinian conflict, every act of violence is the beginning of a frame—not the end of a previous one.

Bateson fends off the would-be assault on logic. He assumes a hierarchization of codes, which he uses to disentangle the relationships of different codes, of the differences in development direction, and of both as they interrelate with causality.

Codes, Rules, and Icons

Bateson's focus is communication, whereas in this book I focus on the individual's perceptions and phenomenological experience. Where Bateson talks about *codes*, I invoke the semiotic idea of *icons*. However, I now place *both* in a semiotic context. This placement is critical to the logical forms I describe.

As you know, Peirce defined *icon* in several ways (for a list of these definitions in quotes from Peirce, see Bergman and Paavola 2001). I tend to use his encompassing idea that all signs at a fundamental point are *icons*, "motivated" by the perceptual experience and dynamics of the particular individual who is assigning a representation to an object. (How different is this conception from Kandinsky's idea that all signs are symbols?)

Cliff A. Joslyn, in his definitions of semiotic terms, lists a *code* as synonymous with a *rule*, which he defines as "[a] functional regularity or stability which is conventional, and thus necessary within the system which manifests it, but within a wider universe it is contingent, or arbitrary."

Correspondingly, he defines a *code* as "[t]he establishment of a conventional rule-following relation in a symbol, represented as a deterministic, functional relation between two sets of entities" (1998).

A code involves a symbol system that I regard as basically iconic in origin and motivation, but the symbols are of an order-relating "entities." The entities of concern are in a semiotic context; hence, they are icons. In sum, *codes serve to relate icons.*

Now consider Bateson's idea that there *is* a vantage point for a description of the coordinative powers of the individual organism. *When* are you in that "catbird's seat" so that you can grasp and describe your own capacities to coordinate codes, icons, and meanings? I propose that *the mix of codes and rules that would allow confounding of logic and causality is not going to be unmixed unless the "I" can oversee the divisions of time and events and the application of logical rules.*

Unless the individuals in a conflict like the Israeli-Palestinian "cycle" can—as individuals—coordinate time frames and beliefs in logical sequence, they will remain subject to external belief givers. In the case of the intransigent Middle East struggle, we well know that primitivizing of a time frame and the outright use of stereotypes and propaganda can be and is wedded to complex power rationales and games wielded by individuals in power. Given the interplay of external loci of belief and the control over the educational system and cultural system of the imparting of values, the submerging of the individual is to fathoms "unfathomable"!

In this extreme case of external loci and control, individual identity may be a "never was," and the critical period for it to develop and emerge may be gone. This is an attractive idea in terms of the degree to which the conflict appears irresolvable. Yet other massive suppressions of individual identity have occurred, and we have seen that the absence of the expression of self does not structurally mean that it is gone.

In all, if the "I" is not articulated, the individual's identity and his agency are not grounded in logic, nor is the logic a dynamic force in framing—hence determining—the meaning of causal sequences.

The way I state this principle may call for an account of how things got to be. If this account merely shows what conditions obtained and what things happened in a given case of the "I" as a "never was," constructionist principles might theoretically supply steps for a reconstruction. Then again, if individual identity is a "once had been," the principle still holds—with steps for remediation being more economical. How do we knock out the negatives in that principle of the "not-articulated 'I'"?

In his study of William Wordsworth's attempt to "create" his own identity in the *Prelude,* Francis Steen interprets the poet's approach that "what the mind needs to do ... is to suspend belief in the self and allow the mind to play again, so that the normal reality monitoring can resume. Such a move is likely to be destructive to the delusional circuits of identity—a mild creative breeze, perhaps, or a mighty and disturbing tempest: not a transformation with a fixed and final endpoint, but an ongoing self-creation" (1998).

Imagine an individual who somehow is given the chance to suspend the externally defined self. Sometimes good things do not happen. Richard Corey, the famous character developed by the poet E. A. Robinson ([1869] 1919), did not do well with his own self-evaluation. When the hero retires in the movie *About Schmidt* (Payne and Taylor 2002), he has a hard time because there seems to be "nothing there" other than what he had built on the basis of his externally determined beliefs about his wants, interests, and attitudes. When his wants do come up, he is quite awkward about them.

Consider the argument that you can create capacity for internal determination of beliefs by stimulating "play" and its fantasy products. Perhaps, as psychologist Marcia Johnson theorizes, the emergence of "thought, imagination, fantasy, dreams and other self-generated processes" (1991:6) sets in motion aspects of not only fantasy, but also reality. One might rebut any presumption that fantasy and play are positive goals in themselves. Fantasy and play in the hands of a "never was" might be a disaster! Johnson's admission of this point seems the understatement! She writes that the individual's thinking can produce representations and their relations that "originate from imperfect attributional processes" (1991:3). That these processes may constitute a fundamental part of the self is theoretically interesting, but, practically speaking, they may be fundamentals out of control. A need for self-worth can be basic, but where the evaluation is not coordinated with reality checks, you can have megalomania like Saddam Hussein's.

What about the necessary checks on whether you are dreaming or on whether an interpretation is likely to be consonant with various indicators of reality? There are dangers to the lack of coordination of fantasy when the individual comes close to being a "never was" and when his sense of a social and personal sense—as defined by others—is suspended. Yet we can continue to wonder whether the stimulation of fantasy would be productive and what form an "imperfect attributional process" would take.

Would a person raised on stereotypy and dependency on external origin of belief move toward self-corrective and integrative thought simply on the basis of the forces of autoregulation? Does the calculus of autoregulation include an altruistic ethic? Would this movement cost too much in that to get it going, you would have to get past all sorts of excesses?

I'm sorry I can't go too far down this track here. The purpose of these sociopolitical examples is to show the range affected by the logical forms of tropes and their degenerate forms. In connection with this aim, the question of stoking individual identity has enormous social significance—else I wouldn't have raised the issue of such an awful and intractable problem as Middle East conflict and the concretisms of terror and killing. But I don't want my ruminations here to appear simplistic. If individual fantasy as a route to identity is aroused, one might have a fantastic definition of self based merely on simplistic equivalence. It would be worthwhile to build ideas of remediation and reconstruction, but they would be built on applied levels branching in many directions. I need to rein in to keep to the present purpose, which is to outline basic structures and indicate the range of their implications. For the individual to launch perspective on external determinants of belief requires getting the "I" "up and working." To an extent, you can picture this change from a suppressed or dormant or inadequately developed extension from the self to the "I" as regressing to fantasy forms—dreams and the like. The regressing "I" would be courting degraded logical forms by which objects and terms are equivalent to less than themselves. Obviously, the attributions we would be making to the identity of an object—even if the object is that of the self—are going to come out as faulty. And here's the rub: *a faulty attribution can be a synecdoche taken seriously.*

The logic of synecdoche might free a terrorist from his religiopolitical mantra, but his belief in himself might be so concretely enmeshed in a "liberated" fantasy that his actions would be primitive. We can thus conceive concrete and horrendous acts that are not politically or religiously motivated. Another example arises within the sociopolitical context of totalitarianism. Assume the sons of Saddam Hussein had been free to engage fantasy. Their social exploits were perverse, perhaps directly in line with a concrete interpretation of their own fantasies. We can muse that this concretizing of their thought in their acts might be the terrible outcome of freeing up the "I" as internal locus in a case of inadequate self-development—given the various constraints and restraints on belief in a sociocultural context subordinate to supercontrols over belief from external loci.

On to the easier example of regressed logic via synecdoche and its effect on the "I." Suppose a synecdoche like "I am a success in chess" is such that the person takes the metaphor "chess master" too far in relation to an analogy he makes. He says, "I am a master at chess, and this means I will be a master in all other endeavors." We see synecdoche in his generalizations:

$$\text{"I"}_c \equiv \text{"I"}_{1\ldots n}$$

He equates success at chess, ("I"$_c$), with success at other endeavors ("I"$_{1\ldots n}$).
And another synecdoche seems in order:

$$\text{"I"}_{1\ldots n} = \text{"I."}$$

He defines *himself* as his successes. We have seen that a person can define his personhood as an adding up of his attributes or accomplishments

This is extreme. Yet just think of those individuals who are sold on their capabilities because of test scores—or memberships in Mensa! If you like, we can return for a moment to Middle Eastern problems. For one thing, you can see that in the case of a person with a greatly impacted internal identity, like that of an individual dependent on state-sponsored stereotypes, the stimulation of fantasy may lead to an overblown sense of self—where agency is exaggerated and the individual's determinative capacities are overly inflated instead of deflated. Anyway, with such an "I" as operator, all areas of endeavor ("I"$_{1\ldots n}$) are covered, and any two instances can become equivalent. So, our chess "master" can believe that since he masters chess ("I"$_c$), he also can master business ("I"$_b$). Thus, instances c and b become equal:

$$\text{("I")}[\text{"I"}_c = \text{"I"}_b].$$

Our relation to our own thoughts, fantasies, and dreams needs to transcend the trope's insistent logic. Along with the galloping equivalences of synecdoche, this logic can sweep over the way the "I" relates to itself. So, within the overall perspective of this book, let me frame the need to transcend as a positive and enhancing capability.

In that the "I" is an operator in respect to any trope, thinking can become more realistic, more pathological, and more creative, but this depends on the dynamics of self. These dynamics may include adjustments and regulations due to inadequate development and/or other constraints on the self. Nevertheless, the challenge is to specify these differences in terms of the differences in logical structure. The tropes about the "I" determine the relation of the "I" as operator, but also can be linchpins for the species relations of the "I's" objects.

The self-as-origin in dreams and paralogical forms may be the source of thoughts and even poetic imagination. However, there is more to the self as it expands in relation to self-reflective capability and process. Freud's concepts of all this capability and process were in reference to *secondary process thinking* and *ego functions*. However, modern psychological thought leads us to think about the self in computer terms. Thus, for Johnson (1991)—and for many others concerned with self-regulation—another aspect of the self is revealed by the capacity or operator, which would be the self as "reality monitor." Again, just as in many models of self-regulation, the present-day hope is to link such operators and capabilities to brain mechanisms and structures assumed either to underlie them or else simply to *be* them. Johnson, Kounios, and Nolde (1997) take the idea of *reality monitoring* one step further, showing activity of brain mechanisms that may be engaged during the exercise of such judgments.

For Bateson, the apex at which you can conceptualize and relate your interpretations, assertions, propositions, and, in short, the code relations for your thinking is at the point of your autonomy. There, you can accommodate "messages" at a higher order than inherent in and processed by the simple codes of the organism's perception of differences. The higher-order coding capacities include messages *about* the messages. On a higher level of information, what's coded is the nature of the codes themselves. As communication becomes *metacommunication,* we can talk about what we are saying or observing, about the fact that we are saying things, and about the very codes we use to say or even perceive things. In short, a person can note, find terms to describe, and make propositions concerning his own codes, actions, and decisions. Bateson points out that this process is recursive.

As I say, the *logical forms* are reflexive and recursive. In a spiral that takes the "I" out of its embeddedness within an order of codes and icons, there arise succeeding orders of "codes." New orders of superordination proceed so that the next can be one of the self that notes codes, and a next—yet again—can be one of the self noting that self that is noting. (On the embedded self, see Fisher 2003a.)

My work in this book does not involve the ecology of communication or the learning and transmission of information so well spelled out in

Bateson's essays. Nevertheless, his idea of "codes" is compatible with a semiotic context in which icons and rules of relating are the logical bases for understanding terms and propositions. I am therein certainly indebted to his integration of logical ordering and cause-and-effect dynamics of the organism as a closed system. His concepts are an additional source and support for the idea that it is insufficient for a coherent view of the "I" to rely solely either on the constituent view or on the logical/psychological view of identity. Neither by itself coordinates a view of the "I" in its relation to the logical forms of a unique organism and the facts of it as a closed system.

Limits, Logical Order, Entropy, and the "I"

Construction in Relation to Autoregulation and Epigenesis

The focus of my analysis of logical forms and their variations *is* from the logical/psychological position on identity, but my argument is that this position can *subsume—not replace*—the idea of a constituent process. So, yes, Popeye was right: "I yam what I yam an' tha's all that I yam" (Widener 2001).

In the spirit of the constituent view, there is a review of particular meanings in different semantic and action domains, and this review is part of the logic system's autoregulative dynamics. Ego psychologists would follow some version of Freud's idea that the individual's secondary and primary process thinking have to be balanced so that any regression or primitivation is subject to reality monitoring and reality checks. Reversal of genusspecies relations takes place all the time. Examples are fantasy rumination and the characterization by slogan or shorthand figure of speech. Woody Allen presents a fantasy that he is Miss America! He's in deeper trouble than has been the case if he doesn't come away from this synecdoche and view himself as "the writer of that fantasy." Heaven knows what politicians attribute to the "whole" of their opponent's character when fighting for the nomination in the presidential primaries. But when the mudslinger is not the chosen candidate, he quickly abandons his rhetoric and supports the "fuller picture" of the chosen party candidate.

Logical ordering has to be classical, and any reversal of genus-species relations has to have a "snap back" capability—a potential for the emergence of higher-order levels of awareness or, in Bateson's terms, "codes."

But people have problems with various bouts involving concretisms and species organizations. The bouts can be matters of deficient, defective, immature, or dysfunctional inner development of self or repressive, distorted, or inadequate external sources of values and beliefs. To deal with such problems, the construction model is invoked, and it is *not* incompatible with assumptions of autoregulation and correctibility of logical organization.

For instance, ego psychologists would argue that irresolutions in ego identity *can* be compensated. One of the assumptions in psychoanalysis is that the person can call on a "reader" who might augment a person's ability to see his own meanings—or, as I say, to see them in perspective. There would thus be ways of enhancing ego functions as well as ways to recover from inadequate personal identity formation. The social sources—like an external reader—appear to contribute values and even points of reference to reconfigure organizational structures of the individual. But, even so, these influences are subject to the individual's autoregulative machinery and capabilities.

On one hand, you find a combination of *tasks and construction of solutions* within social contexts and the individual's work, which makes use of external sources of concepts of personal and social identity. On the other hand, as a function of the organism's internal workings, *growth in ways of conceptualizing identity* appears in an *epigenetic unfolding*.

Accordingly, Erikson's ideas of developmental tasks and stages are not incompatible with general trends in the organism's organization and dynamics, such as those toward higher integrations of the self. In terms of our discussion, the emergence of higher-order organization would roll on were it not for various social dampers that either stop the process or send it into hibernation. After some enforced suppression of the individual and of the development and expression of the self, a person with a self, like Rip Van Winkle, can awaken. The bones creak a bit, but the learning equipment can oil itself to get up and get going to build new understandings. I refer not only to the psychosocial suppression of an individual, say, in a family unit, but also to a more comprehensive social, political, and even societal deprivation.

I don't mean that a "wild boy of Aveyron" (Itard [1801] 1962) can be rebuilt or that critical damage was not done to Nicolae Ceausescu's children (SoRelle 1998)! I do mean to say that a constructionist theory has to be articulated, but it also has to work to stoke epigenetic capability—in whatever degraded shape deprivation squeezed it.

The general trends to be achieved by an epigenesis and the resolution of developmental tasks are to reach a different "type level" or logical order. Specific learning scenarios can schematize this or that adjustment, which may be to a role relation, a financial change, or a loss or gain of a given ability or property. Object relations may change, such that behavior toward others is different, and the change can signal a new level of awareness of the self and the "I." Hamlet seems aware of such niceties when he advises his mother, "Assume a virtue, if you have it not" (act 3, scene 4). As his mother's "reader," Hamlet here proceeds to help her to build an awareness that can change object relations with others, change her degree of awareness of such relations, but not go so far as to change her view of the total set of values she can consider changing.

These levels are no "mean" levels at which to specify how the self is determined. Here's where the construction view of the self can keep our hopes alive. But let's not go overboard. A reconstruction by "parts" replacement is *not* an exchange that accounts for the identity of the self. Shakespeare and his audience realize this about Gertrude, whereas Hamlet still has some optimism about what his mother can achieve.

What are the practical limits of construction, and what are the factual limits that would take into account the individual's fundamental "being"? If your memory were replaced by modules that had the exact information as the original, that memory's unique sentient experiences would remain just that—unique and irreplaceable. Yet reconstructions and part replacements can affect the self's evaluation of its capability and even its status of fulfillment. These capacities and routes to resolution as contributors to the self's "achievement" of identity obviously take a few pointers from Erikson's page!

Reconstruction as a way of coping with deficits and degradations due to pathology are very worth a place in a model of personal identity. More dramatic than a hearing aid or a memory adjunct of some sort would be the reconstruction of a cognitive capacity to make judgments. I have mentioned the "modeling of other minds" as a cognitive capacity basic to the social foundations of self and the evaluation of identity. Some hypothesize that if very specific patterns in the mother-child interaction are missing, the result is in the nondevelopment or the faulty development of the ability to judge other minds. Visual contact is one variable hypothesized as essential. On deck is an interesting proposal (Deák, Fasel, and Movellan 2001; Fasel et al. 2002) to capture the specific movements that affect visual interactions between mother and child via robot construction and simulation. Rebuilding skills by way of interactive patterns can take many forms, including simulating the simulations, which would spell out specific algorithms on interactive possibilities.

The Attributing "I"

Although I am aware of and excited by these possibilities of construction and reconstruction, I retain focus on the "I's" originating centers and its function as an operator. This focus is best understood here as a merging of the "I's" conscious capabilities and its role in the *forward direction* of causality and development. The consciousness of the "I" can be pictured in terms of the "I's" effect on objects—including on the I *as object*. Thus, the "I" can proactively become the determinant of its own personal identity. This is to say that it is the "I" who theorizes—irrespective of *social* definitions of roles, markers, and social and personal identity. How can we best look at this "theorizing"? My answer is to keep your eye on the "attributing 'I'"!

In the reflexive capacity to attribute to itself, the "I" is always succeeding its exercises of power by attributing originating power to itself. This capacity is a spiral in the exercise and the attribution of power to the self. Note that the "I" is at some apex in a hierarchical organization of the self as an object of the "I." The "I" does the attributing; the self—as object—is imbued with the attributions.

In summary, the issue we have grappled with is that of the logical form of individual identity. I have viewed it as a question: How do you locate a start point at which cause and logical superordination coincide? To find the start point, the buck stops at the "attributing 'I.'" With this concept, we are now in a position to revisit the apparent disparities between theorists' own life of the mind and their scientific pronouncements about mind and self. Turn back some pages to reconsider Skinner and Freud as theorists. Were their public theories also their personal theories of self? Each theorist put forth a scientific theory that clashed with the way he lived his own "life of the mind." Each could, if pressed, explain how his theory could be fit to the facts of his own life—a de jure status for the theory. My next point is a variation of the idea that a person's own de jure theory of the "I" should be consonant with his de facto theory—actually *living* a "life of the mind": Following the assumption of an "attributing 'I,'" *the spiral of power attribution is a matter of logical hierarchization and set-subset reorganization of the operator-object modes of the "I."*

So in the "living" of their lives of mind and of self and the conscious moments of the "I," Freud and Skinner might have adopted a code that marks the meanings within this de facto mode. Such a code does not have to be one utilizing the same verbal vocabulary, nor does it have to utilize a vocabulary. It might be in the emotional, visceral, psychomotor sensibilities or the like. It might be a loop within and to and from one area of the brain. However, this point centering on different codes and different types (or modes) of experience (sentience), as I derive it from Bateson, is not enough. That Freud and Skinner might be "in such a mode" and have the equivalent of a theory of mind and self within the code of that mode tells us that organizational reshufflings have occurred.

These reshufflings can be exchanges of the ordering of the categories of the self. Think of an exchange in terms of the "operator-object" modes of the "I." Keep the differing code modes in mind as succeeding levels of abstraction—the "I," the "I" looking at the "I," the "I" looking at the "I" looking at the "I"—and so on. At some point in this abstraction process, the "I" chooses a "mode." An easy example is a Skinnerian-type claim that he (Skinner) can be observed and that his "operator behavior" would simply be one of the things observable. In this way, the "operator" category is a subset of the "object" category. Thus, the individual can see himself as an "object," with his operator role subsumed. But this organization can be twirled around. As "operator," the "I" can be a category subsuming itself as an "object." Can Skinner have had moments like Popeye's?

My assumption of the "attributing 'I'" does come to an origin point. It takes the logical matter of these set-subset variations and species-type organizations to a hierarchized structure with the "attributing 'I'" at the apex. When each of these theorists is "living" the life of mind, identity, and self, the "I" is an operator—irrespective of the way it is coded. (We have to admit that the term *operator* can be used at two different orders of the hierarchy—the "I" as operator attributes the operator role!)

When the theorist offers his scientific theory of mind, self, and so on, he makes his "I" an "object." I promised this view would come to an "origin point." Here it is. *By virtue of the actions of attribution and the organization of logical order, the person's de facto theory is that a logical/psychological process forms his identity.*

From the logical vantage point of that originating power, the "I" is the genus or set for its products in thought. When these products include thoughts about the self as an object, such thoughts are species or subsets.

With this origin point, we can come down to earth about the limits of such spiraling. The organism is a closed system with life limits, on one hand, and autoregulatory rules and capabilities, on the other. The ordering of a person's thoughts, their categorizations, and the coordination of the person's apprehension of her various and different "types" of code are all part of the organism's autoregulatory system. This system and its characteristics, such as trends toward the dynamics of resolving uncompensated negations, may be seen at work at various levels of whole-part development. Thus, in that the "I" is genus to its products, there is a Whole → Parts direction of development in the "I's" logical relations and ordering. But the same direction is also present in terms of the entire organismic system, which keeps moving toward coordinative capability and its realization in adaptation.

These points about the organism's adaptational movement toward coordination appear to be restatements of Werner and Piaget's developmental principles. The difficult analysis here has been in terms of the interrelations of logical and causal directions, as Bateson pointed out. To see this interrelation within a semiotic context is one route to resolution. But to get the logic, causality, time frames, and coding levels best articulated, do we focus on the whole organism as a closed system, or do we focus on the functioning that moves in the direction from the achievements of subsystem balances to organismic integration?

To tackle these questions, we need the book's main idea. It concerns the form of the trope, its variations, and the capacities for the "I" to swing back and forth in time frames and into and out of paralogical forms. I described the movement as "spiraling." Even with some idea of constant movement in time, the "I" can identify a focus, be in a place, and shift focus. Sometimes the shift is from "night" to "day" thinking; sometimes it's a shift of regression; sometimes it's a shift from devastation—a submerged, a fragmented "I"—to the reconstruction of logical hierarchization from bouts with primitivation. Along with Bateson, we can take matters into our hands. We can become aware of the code level and gain perspective on causality by noticing and defining time frames. With all focus-shifting capability in the hands of the "I," the logical organization can be fit to the natural phenomena of thinking and to the thinking organism.

You and I have focused on the spiraling of the "I," its powers of attribution, and its operator status. From within this focus, the Whole → Parts direction of development appears as a determinant of the individual's identity. The individual can effect this focus—this zooming in, this descent from an apex—to find synecdoches to view the "I." Zooming out after zooming in also affects the person's own theory of locus of the "I's" origins. The more a person can be "on a perch" viewing the "I" and its capabilities, putting the synecdoches into perspective, the more he moves to the psychological/logical view of identity.

However, we have also seen the role of construction and of constituent elements. This role is a viewing point with a significant influence on identity. We can use it to see that negations remain uncompensated and that in the case of pathologies there can be decompensation. The person who feels a positive change in the evaluation of self after plastic surgery can reevaluate the effects on self and the feelings about self. There may be comparisons of feelings about the presurgery and postsurgery appearances. Although the person may feel good about there being less negatives in what she attributes to others' judgments of her, there may also appear negatives between what the person hoped to accomplish and what her own judgment is regarding the surgery outcomes. The latter can lead to her dismantling of various defenses temporarily erected against the feelings of getting old. Decompensation?

It is important to talk about limits to the "I" and to the resolution of uncompensated negations. I rely again on Bateson's idea of the hierarchization of logical types. The idea applies to the question of the total organismic regulation and then also to intrasystemic dynamics. Codes are important in our being able to interpret what the body tells us. For example, a person who doesn't like to admit that he is frightened should be able to recognize what his sweating and his focus on the sound of a heartbeat may be saying. Remember Poe's hero in "The Tell-Tale Heart." Our hero mixes codes; it's as if his intracommunication system speaks in tongues! But what I mean by intrasystemic can include the sort of focus one can have by way of the "I" noticing the powers of the "I."

The powers of origin of growth and of change and integration of the "I" reach both their zenith and their ultimate limits in relation to the autoregulation/entropy rules and codes. So these rules and codes are limits. Yet to cast them as "governing rules" is to shortchange the "I." Thus, they remain accessible to logical forms with their logical and dynamic resolution of uncompensated negations. But it's the "I" who oversees these forms, makes decisions about them, steers their organization through the vicissitudes of the "spiral." Do we return to Plato here? Does the "I" find its way to the sympathetic imitation of the universal rhythms—the code limits of autoregulation and entropy? Is the process of the "I" a "remembering"? Is this "remembering" an appropriate trope when we say that the "I" spirals in its searches for what is to be found in the propositions of the individual organism, which we are casting here as an encapsulated closed system?

Logical Form and the Dynamics of a Closed System

What is a logical form? In relation to a form's identity, there is no hesitation in my commitment to the logical/psychological view, yet I have pointed out the usefulness of the constituent view of identity and the possibilities of construction when there is dysfunction or lack of structural integrity. With the constituent view and its possibilities, however, the very nature of "form" becomes a dilemma. If there is anything to this view, then identity is simply a result of what acts, actions, and building take place. These events, processes, and phenomena are reactive. The implication is that the identity is "formed." Passive voice! If all the determinants are reactive, then the identification of causality is always in exteroceptive factors. Where is the individual's agency? The general sense of what an agent would be comes down on locus as responsibility for action. Thus, *agent* is defined as "[t]he person who performs an action. Ethical conduct is usually taken to presuppose the possibility that individual human agents are capable of acting responsibly" ("Agent" 2002). Does action determine agency instead of the other way around? This is not all there is to this dilemma. To complete the dilemma of the "logical form," we must also look at the alternate view of identity.

If a form *were* made from within itself, then it would be—itself—an originating factor. But would this mean that change in the form would be independent of external causes? Would it be the case that Faust would know so much and be so in charge of his knowledge that if the Sun stood still and the moon did not rise, it would not affect his beliefs in a model of the cosmos? So the nature of a logical form as a closed system presents dilemmas because both types of causality—internally originating and externally affecting—are involved. How can both explanations—the logical/psychological and the constituent—coexist? How can a form accommodate construction and the adducing of "parts" if the form is a "closed system"?

To chase down a dilemma, the questions have to be relentless. What does it mean for a logical form to be a closed system? It's closed in the sense that it's an *object* with its own internal dynamics of autoregulation. Now, from whose point of view is it an object? Suppose the point of view is "from within," but the organism is *not* consciously aware that this view is a point of view? This question series yields my argument that *the phenomenon of reflexivity is part and parcel of the organism's capability of self-regulation of different codes.*

How chemical signals coordinate with electrical and motor signals is an example. There are operators and objects to be seen and classified within time and action frames. But whoever is looking, the object status of a logical form is a matter of the system's coordinated dynamics, of its subsystems, and of their intradynamic regulations.

So the payoff to the line of questioning I pursued about a "closed system" is perspective on reflexivity and recursiveness. Irrespective of the "bracketing out" of the "I" and of the possible multiple levels of awareness, the characteristics *reflexivity* and *recursiveness* are necessary for the transportability and translatability of codes within the individual as a closed system.

This is not to minimize the role of the "I." We know the "I" can shunt meanings into different logical positions and balance and rebalance perspective. We know that without the "I's" awareness of logical shuffling and its capacity to determine and coordinate perspective and preferences in ordering, the individual may be captured by his own concrete or syncretic readings of tropes. His thought may in fact be in a form that is truncated or primitivized or in a degenerate form of the trope.

The "I" *can* handle the trope form in its accommodation for different orderings of parts and wholes and subsets and sets. What I am saying about autoregulation of multiple levels of code—irrespective of the "I's" role—relates to the degenerate trope and to the self minus the "I's" awareness of it. Thus, although there is the absence of aware and coordinated perspective; there can be coordinated transfers of meaning across codes, which would mean a type of thought that has reflexivity allowing protocategorical exchanges. These exchanges are not by way of full-fledged tropes; instead, the forms are "degenerate" in that they do not shuffle to produce a "whole" picture. If the self "analogizes," equating a particular place to find food with "eating," but is unaware that it is doing this, you can still count on communication among chemical reactions to food possibilities and motor movements. A transfer of meaning and a transmission of message across codes take place.

How does this transfer appear to an observer? It seems to the observer that the individual who keeps going to and seeking food in the same place "believes" that his actions will be followed by the appearance of food. Suppose the individual herself observes that she engages in this behavior. In the absence of examination of the pathways of her own thought, of the "theories she has about factors she can notice, her observation might be of the simple co-occurrence of "place" and the "appearance of food." To the observer, the intracommunication can appear similar in form to homeopathic magic in that the "place" is "believed" to "cause the presence of food." The nature of the degenerate trope is specifiable as a communicative and regulative aspect of the self, at least to the extent that the presence of the self is established merely by the category of its beliefs—within their accessible coding. Thus, the necessary conditions for declaring the presence of the self would obtain—*even if the codes available did not include those on the level of awareness of coding and codes!* This might be the case where the self is defined as a module and the awareness of the self as a governing set of metacognitive functions. But we can have interplay in the terms of the definition of the self and the "part" of the self. The self then is seen *as* a "part." Because the degenerate trope as a form of thought appears truncated when we consider a mature person, we see this form of thought as a reflection of less than the whole person—it is only *part of the person's self.* In the mature person, the metacognitive functions would be coordinated and integrated in the interplay with the self. Let's go a step further with this idea of the regulation of a trope form that has degenerated and the allied idea that the self is still at work, although it is a partial self.

If a brain-damaged individual interprets a trope concretely, we can trace out the codes that *do* interconnect, and, just as Freud made sense of dreams, we can try to see what subsystem draining and prodding are involved. We may even be able to predict that a given person who has an insistent set of needs in a constraining environment will tend toward specific ways of interpreting and reacting to tropes. This can cut two ways. We can figure out how to open persons to view, consider, and place their dreams relative to meaningful contexts, but we can also constrict persons. Thus, there are apparently effective ways to prepare individuals to act in concert with stereotypes and to carry out extreme behavior as a way of fulfilling beliefs and propositions that are themselves primitive, as if in sympathy with the forms of the stereotypes. As anthropologist Scott Atran points out, "apparently extreme behaviors may be elicited and rendered commonplace by particular historical, political, social, and ideological contexts" (2003). False leaders, rabble rousers, and "prophets of deceit" (Lowenthal and Guterman 1949; Jessica Stern 2003) evidently know well how to prepare individuals, construct the stereotypes, predict the behavior—and elicit it! Leo Lowenthal and Norbert Guterman described the behavior and tactics of prophets of deceit for periods before and leading up to the World War II time frame. Jessica Stern describes a prophetess who in 1985 used very similar tactics and principles to create stereotypes and to replace individual identity with group identity.

We can say, too, that those involved in "shock art" also know what reaction different symbols will evoke and that these symbols' capacity to evoke reactions is based not merely on an individual's knowledge of moral codes, but on the interaction of such codes with his various needs, strivings, and control mechanisms. Along with the appeal to such autoregulative capabilities, the "shock artist's" calculation includes her sense of the viewer's "bouncing 'I.'" The "I" can be pushed down, only to bounce up from the depths! For example, shock can suppress the "I" long enough to evince a gut reaction, such as "disgust" and its predictable call for traditionalism mixed with censorship. This reaction can take place in relation to art, as Gombrich points out, but it can also lead to epochs of totalitarianism and then the transformation of totalitarianism into more freedom for members of nations. The stability of the transformation depends on the health of the "I" in its social and political contexts. There is always the danger of replacing one primitivation with another simplification.

In the political arena, terror can produce interpretations that mix rational analysis with primitivized resolutions, such as war. The fact that warlike reactions and extreme aggressive behavior can be a near automatic response to great fear might lead to the idea that if decisions are nonautomatic, war is an indefensibly primitive response. However, the mere fact that automatic responses are primitive does not mean that war is not necessary under certain circumstances.

Although I am *not* arguing that war is not justifiable, I am arguing that as long as the "I" is knocked for a loop, perspectives on solutions that can reach for figurative levels of organization are out of sight. I do not mean to be cryptic. Blind spots are produced by terror and trauma. Stereotypes are one such blind spot in that they are depictions of a whole by the substitution of selected parts. And one person's blind spot seems to find its mirror image in another's. Thus, Atran finds that "individuals tend to misperceive differences between group norms as more extreme than they really are. Resulting misunderstandings encouraged by religious and ideological propaganda lead antagonistic groups to interpret each other's views of events, such as terrorism/freedom-fighting, as wrong, radical, and/or irrational. Mutual demonization and warfare readily ensue. The problem is to stop this spiral from escalating in opposing camps" (2003).

But how to engage in disputes, arguments, contests, and the like may not be merely a matter of tactically offsetting your opponents' blind spots. It simply may not be wise *to regress* or possible *not to regress!*

The twentieth-century defense against probing the psychodynamic unconscious had some good reasons behind it. The psychodynamic views were egregious in their displays of excesses. Personal pathology and reified unconscious determinants were too facile when brought forward as explanations not only in relation to wars, but in relation to multileveled rationalizations of the 1960s radicals, whose "means justify ends" approach justified disregard for individual life and the problem of finding a moral center. Yet even if simple—noncatastrophic—disgust were satisfied, when psychoanalysis and the entire "experiential" point of view were driven from the scene (Koch 1964; MacCleod 1964), the simplification and the primitivation had gone too far.

The more science, technology, and society become rational—without facing the "I" and its justification of the autoregulative nature of its own identity and the facts of coming to terms with its codes of entropy—the more the pressure builds and the more catastrophic is the disgust reaction! How to regress to face the deeper hell in our own nature without becoming a suicide bomber or sinking into a whirlpool of syncretic stew may be the question of our time.

Notes

1. The Problem of Analogous Forms

1. Psychoanalyst Paul Schilder (1950:559) refers to Otto Selz's idea of "anticipatory patterns" and the fit of an object to the schematization of such patterns. Gestalt psychologist Karl Duncker ([1935] 1945:19) evokes Selz's concept, the "excitation of equality."

2. When a subclass becomes a species, it becomes a nodal point from which instances are gathered as members of that species and considered subordinate to it. This is how Lévi-Strauss describes totemic organization. To visualize the point, contrast a genus-species organization in a biological taxonomy with a "family" organization in the Mafia. The "family" functions as a nodal point; the members cluster by various associative rules of contiguity and action patterns.

3. The story as a myth may have been believed in a concrete way that we cannot replicate. And the mythic figure may have had a symbolic value with which we presently cannot fully empathize. An example is the value of prophetic powers.

2. Natural Logic, Categories, and the Individual

1. This analysis brings together Aristotle's view of the formal cause in a natural being with more current phenomenological considerations. See S. Marc Cohen's (2003) review of Aristotle's concept of formal cause; Christopher Shields (2005) on Aristotle's views of cause in relation to "substance as form of a natural body" in *De anima;* and David W. Smith (2003) on opposing strains of phenomenology.

2. Would the citizens of Chelm be helped if transported for a semester to MIT? We would have to straighten out their "wired-in" logic module vis-à-vis their shared beliefs about logic and reasoning.

3. Yad Vashem, a Holocaust memorial in Israel "honors the Righteous among the Nations—rescuers who risked their lives to save the lives of Jews."

4. The Psyche D e-list entered into a paroxysm of attempts to justify a theory of perception that left out the idea of representations (see, e.g., Lehar 2005a, 2005b).

5. I am demonstrating the degree to which processing accounts and a procedure-based epistemology capture the various levels of representation and intracommunicative regulations. If the person has to regulate what she sees and thinks to assess change in an object, its features, or an event, the course of the person's judgment appears one of filling in a schematic. This *is* conceptual organization, but it is also logical categorization.

6. See Jerrold Seigel's (2005) "Locke Versus Decartes" distinction relative to the idea of consciousness and the "I."

3. Shift to Individual Categories, Dynamics, and a Psychological Look at Identity

1. To keep this book's vocabulary fluid, instead of the term *hypostasis,* I use the terms *assignment* and *attribution.*

2. Many of these points concerning "primary and secondary process" thinking and "regression in service of the ego" are refinements of Freud's and psychoanalyst Ernst Kris's (1952) views. Psychiatrist Silvano Arieti (1976) articulates these concepts as they relate to pathology, on the one hand, and to creativity, on the other. Psychiatrist and psychoanalyst Ignacio Matte-Blanco (1988) extends a discussion of the logical rules of pathological and normal thinking.

3. I use the term *field* in the way group μ ([1970] 1981) conceptualizes a "progressive succession of fields." The term *organization* refers to rules—*their* ordering—and to their relation to representation. Also included here are logical ordering and dynamics.

4. Meaning is optimized in two very different ways. The meaning of a specific object is expanded by knowledge of its particulars. Thus, when meaning is conceived as the specific identity and class membership of objects and of categories of objects, it moves upwards to an optimally specific and particularized extension. The height of meaning is then noting the unique.

However, meaning is also realized by regularizing the rules that govern the way objects are related to each other. This sort of meaning is optimized when it is conceived as the presence of *relational* rules that permit formal operations, transaction, and transformation of symbols. In this case, meaning moves downward *away from particulars* toward logical form. Meaning is then abstract and symbolic.

Meaning progresses to its apex by moving from level to level in two different directions. The first is via its *construction.* The second is via *deconstruction.* The latter is the direction of the group μ decomposition of meaning.

5. The progression of meaning toward its apex is bi-directional. See note 4.

6. Meaning may be within the bounds of an internal correspondence of signs, although our knowledge of this meaning can also be subjected to social agreement on sign/referent relations.

4. Form Versus Function

1. S would represent "self as a whole." A variation of S could be a "part" of self. Thus, S could be S at work, while S_2 could be S as sibling. Dr. Jekyll and Mr. Hyde are "part" selves.

2. Dr. Jekyll and Mr. Hyde, as metaphors for part selves that subsume a whole self, are good examples of an "awareness < sentience" balance. The "I" of the author and reader is not lost.

6. What Are the Role and Function of the Self Vis-à-vis Consciousness?

1. Note the difference between the symbol f"I" and the symbol fI. The symbol f"I" represents the "I" reflexively looking at itself. This f"I" is present in relation to another function, fI, which symbolizes the "I's" look at an object, such as a trope or more generally a form. The form can be a matter of class or set structure. In this chapter, the symbol fA is an example of an operator in relation to a set structure. In one sense, there is always an f"I" in front of an fI (at the left side of the symbolic formulation). In another sense, if you regard the form as a primary determinant of thought, the fI is in the front.

2. Symbols such as A and Ā are generic and refer to any set or class. Here, specific concepts (A and B) are referred to a context. They are concepts, but they are nevertheless organized by the rules of class inclusion.

3. The I as object does not have quotation marks around it, but the "not-I" or its symbol "Ī" does have quotation marks around it because it is the complement of the "I."

4. Jean-François Lyotard makes a similar point. He indicates that it is not possible to interpret the dream with the same linguistic and logical constraints that are present when one is awake (1989:33).

5. Saussure presents two basic concepts of the relation of form and meaning (see Peter Ekegren's [1999] description). The first concept, *syntagmatic relations,* involves linear considerations of meaning, as it is captured in words. Oppositions of the meanings of those words arise in their relation to the position of "other" words and *their* meanings in the horizontal sequencing of words, as in consecutive sentences. Thus, other words and their sequential positions are restrictions or constraints on the target word. The second concept, *paradigmatic relations,* signals vertical organization of word meanings. Therein, the meanings of words are constrained not only by other words in their context of sentences and text, but by words not there and by words potentially there.

6. This expression technically should have quotation marks around it to indicate the "I's" reflective functions. Because quotation marks would be cumber-

some here, I ask the reader to assume that the f"I" is at work in front of the expression fPost D_1 as well as in front of what I label "the whole formula."

7. In quoting Freud, Carol Rupprecht, professor of comparative literature, observes that "Brill says hieroglyphics" [1999:sec. 2, para. 7] in reference to pictographic script.

7. Development in the Logic from Immature to Mature Modes

1. *Merriam-Webster's* defines *golem* this way: "1: an artificial human being in Hebrew folklore endowed with life. 2: something or someone resembling a golem: as a: *automaton*..."

2. See note 5, chapter 6.

3. See note 5, chapter 6.

4. The quotation marks around such symbols as f"I" are largely missing in this section. I mean to indicate an "object" status for the various representations of the "I." I amplify the concept of "object status" later, when I describe the objects of the "I" as icons. I focus on object status to indicate the autoregulative nature of the transactions and regulations of the "I"—as if we can emphasize the form aspect of the form-dynamics relationship that characterizes the natural form of the self and the "I."

9. The "I," Identity, and the Part-Whole Resolutions

1. There are several conceptions of "direction." One concerns development, a second causality, and a third, the movement of time. These conceptions also come into play in relation to inclusion and categorization. See the discussion of "Direction of Development and Identity" beginning on page 252.

2. We see later in this chapter that personhood, although particular to the individual S, can be a matter of piecing together different aspects, versions, or features of self over time. When I make that point by the use of symbols, I distinguish two kinds of "mood" by which to mark the particular individual S. This distinction is made by my use of subscripted symbols. One mood is abstract and stable and marked by the subscript symbol uppercase S. The other mood is additive, and its symbol is marked by the subscript lowercase $s_{1 \ldots n}$. However, for the immediate argument here, I use only one symbol for the particular, which I mark simply as subscript lowercase s.

3. Although she has lost her specific identity, she is still a person (P_S).

4. See note 2 for this chapter. It refers to the constituent-constitutive distinction I make here.

References

Abelson, R. P. and M. J. Rosenberg. 1958. Symbolic psycho-logic: A model of attitudinal cognition. *Behavioral Science* 3:1–13.

Agent. 2002. In *Dictionary of Philosophical Terms and Names*. Available at: http://www.philosophypages.com/dy/a2.htm#agent.

Anhalt, E., screenwriter. 1958. *The Young Lions*. Screenplay based on the novel by Irwin Shaw. Century City, Calif.: Twentieth Century Fox.

Arendt, H. 1951. *The Origins of Totalitarianism*. Cleveland: Meridian, World.

Arieti, S. 1976. *Creativity: The Magic Synthesis*. New York: Basic Books.

Art and Culture Network. 1999–2003. Joan Miró. Available at: http://www.artand-culture.com/arts/artist?artistId = 171.

Ash, M. 1998. *Gestalt Psychology in German Culture, 1890–1967: Holism and the Quest for Objectivity*. Cambridge: Cambridge University Press.

Atkinson, R. L., R. C. Atkinson, E. E. Smith, D. J. Bem, and S. Nolen-Hoeksema. 1996. *Hilgard's Introduction to Psychology*. 12th ed. Fort Worth, Tex.: Harcourt Brace.

Atran, S. 2003. Understanding suicide terrorism. *Interdisciplines* (July 1, 2003). Available at: http://www.interdisciplines.org/terrorism/papers/1.

Bacon, J. 1997. Tropes. In *Stanford Encyclopedia of Philosophy*. Available at: http://plato.stanford.edu/archives/sum2002/entries/tropes.

Baldwin, J. M. 1913. *History of Psychology: A Sketch and an Interpretation*. Vol. 2 of *A History of the Sciences*. New York: G. P. Putnam's Sons.

Baron-Cohen, S. 1995. The eye direction detector (EDD) and the shared attention mechanism (SAM): Two cases for evolutionary psychology. In C. Moore and

P. Dunham, eds., *Joint Attention: Its Origins and Role in Development*, 41–59. Hillsdale, N.J.: Lawrence Erlbaum.

——. 1997. *Mindblindness*. Cambridge, Mass.: Bradford, MIT Press.

Barrett, D. 1993. The "Committee of Sleep": A study of dream incubation for problem solving. *Dreaming* 3 (2). Available at: http://www.asdreams.org/journal /articles/barrett3-2.htm.

Barzun, J. 2000. *From Dawn to Decadence*. New York: HarperCollins.

Bateson, G. 1972. *Steps to an Ecology of Mind*. New York: Ballantine.

——. 1979. *Mind and Nature*. New York: Dutton.

BBC News. 2002. Entertainment: Music, September 23. Available at: http://news .bbc.co.uk/1/hi/entertainment/music/2276621.stm.

Bergman, M. and S. Paavola. 2001. Icons. In *The Commens Dictionary of Peirce's Terms: Peirce's Terminology in His Own Words*. Available at: http://www.helsinki .fi/science/commens/dictionary.html.

Bermúdez, J-L. 1998. *The Paradox of Self-Consciousness*. Cambridge, Mass.: Bradford, MIT Press.

——. 2001. Normativity and rationality in delusional psychiatric disorders. *Mind and Language* 16:457–93.

Bickerton, D. 1990. *Language and Species*. Chicago: University of Chicago Press.

——. 1995. *Language and Human Behavior*. Seattle: Washington University Press.

Boden, M.A. 2007. Self-assembly. Review of D.R. Hofstadter's *I Am a Strange Loop* (2007). *American Scientist Online* 95 (3) (May–June). Available at: http:// www.americanscientist.org/template/BookReviewTypeDetail/assetid/55116.

Boring, E.G. 1950. *A History of Experimental Psychology*. 2d ed. Englewood Cliffs, N.J.: Prentice-Hall.

Borofsky, J. 1984. *All Is One*. Installation at the Whitney Museum of Art, New York. Available at: http://www.borofsky.com/installations/whitney1984[inst] /index.html.

Bowlby, J. 1969. *Attachment*. Vol. 1 of *Attachment and Loss*. London: Hogarth.

Brackett, C. and B. Wilder, screenwriters. 1943. *Five Graves to Cairo*. Screenplay adapted from "Hotel Imperial" by Lajos Biro. Hollywood, Calif.: Paramount.

Braga, L.S. 2003. Why there is no crisis of representation according to Peirce. *Semiotica* 143 (1–4): 45–52.

Braine, M.D.S. 1998. Steps toward a mental-predicate logic. In M.D.S. Braine and D.P. O'Brien, eds., *Mental Logic*, 273–332. Hillsdale, N.J.: Lawrence Erlbaum.

Braine, M.D.S. and D.P. O'Brien. 1998a. *Mental Logic*. Hillsdale, N.J: Lawrence Erlbaum.

——. 1998b. A theory of *if*: A lexical entry, reasoning program, and pragmatic principles. In M.D.S. Braine and D.P. O'Brien, eds., *Mental Logic*, 199–244. Hillsdale, N.J.: Lawrence Erlbaum.

Bucci, W. 2000. The need for a "psychoanalytic pychology" in the cognitive science field. *Psychoanalytic Psychology* 17 (2): 203–24. Also available at: http:// www.psychology.sunysb.edu/ewaters/552/PDF_Files/FreudCognition.PDF.

Burke, K. 1952. *A Rhetoric of Motives*. New York: Prentice-Hall.

Byas, D. and S. Stewart. 1945. I got rhythm. Composed by George Gershwin and Ira Gershwin. Town Hall jazz concert, 1945. On *Smithsonian Collection of Classic Jazz*, vol. 3. RD 033-3.

Carmichael, L., H. P. Hogan, and A. A. Walter. 1932. An experimental study of the effect of language on the reproduction of visually perceived form. *Journal of Experimental Psychology* 13:73–86.

Catania, A. C. 2000. From behavior to brain and back again (book review). *Psycoloquy* 11 (027). Available at: http://www.cogsci.ecs.soton.ac.uk/cgi/psyc/newpsy?11.027.

Cervantès, M. de. [1605] 1885. *Don Quixote*. Trans. John Ornsby. Available at: http://www.donquixote.com/paronechap12.html.

Chafe, W. [1970] 1973. *Meaning and the Structure of Language*. Chicago: University of Chicago Press.

Chayefsky, P., director. 1976. *Network*. Los Angeles: MGM.

Chomsky, N. 1972. *Language and Mind*. New York: Harcourt Brace Jovanovich.

Cichowski, R. 2001. Exploring the subjective experience of visual imagery in a case of homonymous hemianopia. *International Journal of Practical Approaches to Disability* 25 (1). Available at: http://216.239.57.100/search?q = cache:cISz75wf-TcC :www.ijdcr.ca/VOL01_02_AUS/articles/articles10–14.pdf+Roland+Cichowski&hl = en&ie = UTF-8.

Coates, P. 2003. Review of *Is The Visual World a Grand Illusion?* edited by Alva Noë. *Human Nature Review* 3 (March 15): 176–82. Available at: http://www .human-nature.com/http://human-nature.com/nibbs/03/noe.html.

Cohan, G. M., lyricist and composer. 1917. Over there. Vintage Audio, updated August 2, 2003. Available at: http://www.firstworldwar.com/audio/overthere.htm.

Cohen, S. M.. 2003. Aristotle's metaphysics. In E. N. Zalta, ed., *The Stanford Encyclopedia of Philosophy*, winter ed. Available at: http://plato.stanford.edu /archives/win2003/entries/aristotle-metaphysics/.

Concise Oxford Dictionary of Linguistics. 1997. New York: Oxford University Press. Available at: http://xrefer.com/entry.jsp?xrefid = 573579&secid = .-&hh = 1.

Coppola, F. F. and E. H. North, screenwriters. 1970. *Patton*. Century City, Calif.: Twentieth Century Fox.

Deák, G. O., I. Fasel, and J. Movellan. 2001. The emergence of shared attention: Using robots to test developmental theories. *Proceedings of the First International Workshop on Epigenetic Robotics, Lund University Cognitive Studies* 85:95–104.

Dollard, J. and N. E. Miller. 1950. *Personality and Psychotherapy*. New York: McGraw-Hill.

Doty, R. M., B. E. Peterson, and D. G. Winter. 1991. Threat and authoritarianism in the United States, 1978–1987. *Journal of Personality and Social Psychology* 61 (4): 629–40.

Dumas, A. 1996. *The Count of Monte Cristo*. New York: Modern Library.

Duncker, K. [1935] 1945. On problem-solving. *Psychological Monographs* 58 (5): i–v, 1–113.

Durkheim, E. and M. Mauss. [1905] 1963. *Primitive Classification*. Chicago: University of Chicago Press.

Earle, B. 2000. From ecological to moral psychology: Morality and the psychology of Egon Brunswick. *Journal of Theoretical and Philosophical Psychology* 20:196–207.

Ekegren, P. 1999. *The Reading of Theoretical Texts: A Critique of Criticism in the Social Sciences.* New York: Routledge.

Ellenberger, H. F. 1970. *The Discovery of the Unconscious.* New York: Basic Books.

Emmeche, C. 2002. The chicken and the Orphean egg: On the function of meaning and the meaning of function. *Sign Systems Studies* 30 (1): 15–32.

Epstein, R. 1997. Skinner as self-manager. *Journal of Applied Behavior Analysis* 30 (3): 545–68.

Erikson, E. 1950. *Childhood and Society.* New York: Norton.

Escher, M. C. 1955. *Liberation.* Available at the M. C. Escher official Web site: http://www.mcescher.com.

——. 1960. *Ascending and Descending.* Available at the M. C. Escher official Web site: http://www.mcescher.com.

Faerna, J. M. 1996. *Kandinsky.* Trans. Alberto Curotto. New York: Harry N. Abrams.

Fasel, I., G. O. Deák, J. Triesch, and J. Movellan. 2002. Combining embodied models and empirical research for understanding the development of shared attention. In *Proceedings of the 2nd International Conference on Development and Learning,* 21–27. Los Alamitos, Calif.: IEEE.

Finley, M. I. [1963] 1977. *The Ancient Greeks.* New York: Pelican.

Fisher, H. 1976. The language and logic of forming an idea. *Journal for the Theory of Social Behaviour* 6: 177–210.

——. 1985. The logical structure of Freud's idea of the unconscious: Toward a psychology of ideas. *Journal of the British Society for Phenomenology* 6 (1): 20–35.

——. 2000a. Consciousness is an island (Commentary 16). *Karl Jaspers Forum* (January 18, posted February 1). Available at: http://www.douglashospital.qc.ca/fdg/kjf/22-C16FI.htm.

——. 2000b. The island of consciousness. Unpublished manuscript.

——. 2001. *The Subjective Self: A Portrait Inside Logical Space.* Lincoln: University of Nebraska Press.

——. 2003a. Categories and embodied objects: The subjective self and the psychologist within natural psychology. *Theory and Psychology* 13 (2): 239–62.

——. 2003b. Metonymy as concept: A metaphor for rhetoric, not for thought. *Semiotica* 147 (1–4): 495–525.

Flanagan, B. 2002. Field. In *Word Association: Quanta and Consciousness.* Available at: http://wordassociation1.net/field.html.

Flavell, J. H. 1963. *The Developmental Psychology of Jean Piaget.* Princeton, N.J.: Van Nostrand.

Frazer, J. G. [1922] 1950. *The Golden Bough.* Abridged ed. New York: Macmillan.

Freud, S. 1900. The secondary elaboration. Chapter 6 in S. Freud, *The Interpretation of Dreams.* Available at: http://eserver.org/books/interpretation-of-dreams/chap06i.html.

——. [1900] 1911. The dream work. Chapter 6 (Book VI) in S. Freud, *The Interpretation of Dreams.* Available: http://www.psywww.com/books/interp/chap06a.htm.

———. [1901] 1971. *On Dreams*. In S. Freud, *The Standard Edition of the Complete Psychological Works*, edited by J. Strachey, 5:633–86. London: Hogarth Press.

———. [1920] 1955 *Beyond the Pleasure Principle*. In S. Freud, *The Standard Edition of the Complete Psychological Works*, edited by J. Strachey, 18:3–66. London: Hogarth Press.

———. [1915] 1957a. *Thoughts for the Times on War and Death*. In S. Freud, *The Standard Edition of the Complete Psychological Works*, edited by J. Strachey, 14:273–302. London: Hogarth Press.

———. [1915] 1957b. *The Unconscious*. In S. Freud, *The Standard Edition of the Complete Psychological Works*, edited by J. Strachey, 14:159–216. London: Hogarth Press.

———. [1900] 1958. *The Interpretation of Dreams*, first part. In S. Freud, *The Standard Edition of the Complete Psychological Works*, edited by J. Strachey, 4:1–338. London: Hogarth Press.

———. [1908] 1959. Creative writers and day-dreaming. In S. Freud, *The Standard Edition of the Complete Psychological Works*, edited by J. Strachey, 9:141–53. London: Hogarth Press.

———. [1933] 1964. *Why War?* (letters of Einstein and Freud). In S. Freud, *The Standard Edition of the Complete Psychological Works*, edited by J. Strachey, 22:197–218. London: Hogarth Press.

———. [1913] 1971. *Totem and Taboo and Other Works*. In S. Freud, *The Standard Edition of the Complete Psychological Works*, vol. 13, edited by James Strachey. London: Hogarth Press.

Frith, C. 1992. *The Cognitive Neuropsychology of Schizophrenia*. East Sussex, England: Lawrence Erlbaum and Taylor & Francis.

Gazzaniga, M. S. 1998. *The Mind's Past*. Berkeley and Los Angeles: University of California Press.

Geertz, C. 1973. Thick description: Toward an interpretive theory of culture. In *The Interpretation of Cultures: Selected Essays*, 3–30. New York: Basic Books.

———. 1983a. Art as a cultural system. In *Local Knowledge: Further Essays in Interpretive Anthropology*, 94–120. New York: Basic Books.

———. 1983b. The way we think now: Toward an ethnography of modern thought. In *Local Knowledge: Further Essays in Interpretive Anthropology*, 147–63. New York: Basic Books.

Gentner, D. and K. J. Holyoak. 1997. Reasoning and learning by analogy. *American Psychologist* 52 (1): 32–34.

Gibson, J. J. 1950. *The Perception of the Visual World*. Boston: Houghton Mifflin.

———. 1966. *The Senses Considered as Perceptual Systems*. Boston: Houghton Mifflin.

———. 1979. *The Ecological Approach to Visual Perception*. Boston: Houghton Mifflin.

Glesner, E. S. [1996] 2000. Johannes Brahms—*Symphony No.3, Op. 90*. Available at the Classical Music Pages Homepage: http://w3.rz-berlin.mpg.de/cmp/brahms_sym3.html.

Glick , J. A. 1983. Piaget, Vygotsky, and Werner. In S. Wapner and B. Kaplan, eds., *Toward a Holistic Developmental Psychology*, 35–52. Hillsdale, N.J.: Lawrence Erlbaum.

Goethe, J. W. von. 1882. The magician's apprentice. In *The Poems of Goethe*, trans. in the original meters, 71–75. New York: Lovell.

Goffman, E. 1963. *Stigma*. Englewood Cliffs, N.J.: Prentice-Hall.

Gombrich, E. H. 2002. *The Preference for the Primitive: Episodes in the History of Western Taste and Art*. London: Phaidon.

Group μ. [1970] 1981. *A General Rhetoric*. Trans. P. B. Burell and E. M. Slotkin. Baltimore: Johns Hopkins University Press.

Grünbaum, A. 1984. *The Foundations of Psychoanalysis: A Philosophical Critique*. Berkeley and Los Angeles: University of California Press.

——. 2006a. A century of psychoanalysis: Critical retrospect and prospect. In *Psychiatry Online Italia*. Available at: http://www.pol-it.org/ital/9grunb-i.htm.

——. 2006b. The hermeneutic versus the scientific conception of psychoanalysis: An unsuccessful effort to chart a via media for the human sciences. In *Psychiatry Online Italia*. Available at: http://www.pol-it.org/ita.

Hanna, R. 2006. *Rationality and Logic*. Cambridge, Mass.: MIT Press.

Harré, R. 1984. *Personal Being*. Cambridge, Mass.: Harvard University Press.

——. 1991. The discursive production of selves. *Theory and Psychology* 1:51–63.

Hayakawa, S. I. 1941. *Language in Action*. New York: Harcourt, Brace and World.

Heider, F. 1958. *The Psychology of Interpersonal Relations*. New York: John Wiley.

Hofstadter, D. R. [1979] 1999. *Gödel, Escher, Bach—20th Anniversary Edition*. New York: Basic Books.

Holyoak, K. J. and P. Thagard. 1997. The analogical mind. *American Psychologist* 52 (1): 35–44.

Hook, S., ed. 1959. *Psychoanalysis, Scientific Method, and Philosophy*. New York: New York University Press.

Hughes, R. [1981] 1991. *The Shock of the New*. New York: Knopf. Available through Artchive at: http://www.artchive.com/artchive/ftptoc/klee_ext.html.

Hull, C. 1943. *Principles of Behavior*. New York: Appleton-Century-Crofts.

Inkeles, A. 1968. *Social Change in Soviet Russia*. Cambridge, Mass.: Harvard University Press.

Itard, J. M. G. [1801] 1962. *The Wild Boy of Aveyron*. Trans. G. Humphrey and M. Humphrey. New York: Appleton-Century-Crofts.

James, W. [1892] 1920. *Psychology: Briefer Course*. New York: Henry Holt.

——. [1890] 1952. *Principles of Psychology*. Chicago: Encyclopaedia Britannica.

Jakobson, R. 1956. Two aspects of language and two types of aphasic disturbances. In R. Jakobson and M. Halle, eds., *Fundamentals of Language*, 55–82. The Hague: Mouton.

Januszczak, W. 1993. *Techniques of the World's Great Painters*. New York: Book Sales. Available at: http://www.artchive.com/artchive/ftptoc/klee_ext.html.

Jaynes, J. 1976. *The Origin of Consciousness in the Breakdown of the Bicameral Mind*. Boston: Houghton Mifflin.

Johansen, J. D. 1993. *Dialogic Semiosis*. Bloomington: Indiana University Press.

Johnson, M. K. 1991. Reflection, reality monitoring, and the self. In R. G. Kunzendorf, ed., *Mental Imagery*, 3–16. New York: Plenum Press.

Johnson, M. K., J. Kounios, and S. F. Nolde. 1997. Electrophysiological brain activity and memory source monitoring. *NeuroReport* 8:1317–320.

Jorna, R. J. and B. P. van Heusden. 2003. Why representation(s) will not go away: Crisis of concept or crisis of theory? *Semiotica* 143 (1–4): 113–34.

Joslyn, C. 1998. Semiotic terms. In F. Heylighen, C. Joslyn, and V. Turchin, eds., *Principia Cybernetica Web* (Brussels), May 8. Available at: http://pespmc1.vub .ac.be/SEMIOTER.html#Icon.

Kandel, E. R. 1999. A new intellectual framework for psychiatry revisited. *American Journal of Psychiatry* 156:505–24. Available at: http://ajp.psychiatryonline.org /cgi/reprint/156/4/505?ijkey = 6a48d33e3018e50e97d756a50a4ab52f55070689.

Kandinsky, W. 1919. On Gray. In M. Roskill, *Klee, Kandinsky, and the Thought of Their Time*, plate 26. Urbana: University of Illinois Press.

——. [1912] 1974. On the question of form. In W. Kandinsky and F. Marc, eds., *The Blaue Reiter Almanac*, new documentary ed., edited by K. Lankheit, 147–86. New York: Da Capo Press.

Keen, T. M. 1999. Schizophrenia: Orthodoxy and heresies. A review of some alternative possibilities. *Journal of Psychiatric and Mental Health Nursing* 6 (6): 415–24.

Kemerling, G.. 2001. Aristotle: Ethics and the virtues. Available through Philosophy Pages at: http://www.philosophypages.com/hy/2t.htm#poetics.

Kennard L. R. 1997. Coleridge, wordplay, and dream. *Dreaming* 7 (2). Available at: http://www.asdreams.org/journal/articles/7-2_kennard.htm.

Kihlstrom, J. F [1994] 2005. Psychodynamics and social cognition: Notes on the fusion of psychoanalysis and psychology. In I. Pitchford and R. M. Young, eds., *Burying Freud: The Human Nature Review*. Available at: http://human-nature .com/freud/kihlstro.html.

——. [2000] 2006. Is Freud still alive? No, not really. In R. Atkinson, R. C. Atkinson, E. E. Smith, D. J. Bem, and S. Nolen-Hoeksema, *Hilgard's Introduction to Psychology*, 13th ed., 481. New York: Harcourt Brace Jovanovich. Available at: http://ist-socrates.berkeley.edu/~kihlstrm/freuddead.htm.

Klee, P. 1917. *Above the Mountain Peaks*. In M. Roskill, *Klee, Kandinsky, and the Thought of Their Time*, plate 23. Urbana: University of Illinois Press.

Koch, S. 1964. Psychology and conceptions of knowledge. In T. W. Wann, ed., *Behaviorism and Phenomenology*, 1–41. Chicago: University of Chicago Press.

Koestler, A. 1964. *The Act of Creation*. London: Hutchinson.

Kris, E. 1952. *Psychoanalytic Explorations in Art*. New York: International Universities Press.

Kugler, P. 1993. The "subject" of dreams. *Dreaming* 3 (2). Available at: http://www .cgjungpage.org/articles/pksubj.html.

Lakoff, G. 1987. *Women, Fire, and Other Dangerous Things*. Chicago: University of Chicago Press.

——. 1999. The third culture: "Philosophy in the flesh." A talk with George Lakoff. Introduction by John Brockman. EDGE 51, March 9. Available at: http: //www.edge.org/documents/archive/edge51.html.

——. 2001. How unconscious metaphorical thought shapes dreams. *PSYART: A Hyperlink Journal for Psychological Study of the Arts*, ISSN: 1088–5870, article no. 000801, Filename:lakoff01.htm. Available at: http://www.clas.ufl.edu/ipsa /journal/2001/lakoff02.htm.

Lakoff, G. and R. E. Núñez. 2000. *Where Mathematics Comes from: How the Embodied Mind Brings Mathematics Into Being*. New York: Basic Books.

Langer, E. 2000. Mindful learning. *Current Directions in Psychological Science* 6: 220–23.

LeDoux, J. E. 2002. *Synaptic Self: How Our Brains Become Who We Are*. New York: Viking Penguin.

LeGrady, G. 2000. Velasquez' *Las Meninas*. Available at George LeGrady's Theory Archive: http://www.merz-akademie.de/projekte/george.legrady/theory/holbein/velasquz.htm.

Lehar, S. 2005a. Direct perception vs. representationalism. PSYCHE Discussion Forum, April 11. Available at: PSYCHE-D@LISTSERV.UH.EDU.

——. 2005b. Direct perception vs. representationalism. PSYCHE Discussion Forum, April 28. Available at: PSYCHE-D@LISTSERV.UH.EDU.

Leibniz, G. W. [1698] 1999. *The Monadology*. Trans. Robert Latta. Available at: http://www.rbjones.com/rbjpub/philos/classics/leibniz/monad.htm.

——. [1698] 2007. *The Monadology Glossary*. Comp. Robert Latta. Available at: http://www.rbjones.com/rbjpub/philos/classics/leibniz/monglos.htm.

Lévi-Strauss, C. 1963. *Structural Anthropology*. Trans.C. Jacobson and B. Grundefest Schoepf. New York: Basic Books.

——. 1966. *The Savage Mind*. Chicago: University of Chicago Press.

Lewin, K. 1935. *A Dynamic Theory of Personality: Selected Papers*. New York: McGraw-Hill.

Lightman, A. 2005. *The Sense of the Mysterious*. New York: Pantheon.

Livio, M. 2003. The golden number. *Natural History* 112 (2): 65–69.

Lowe, E. J. 1995. Class. In *The Oxford Companion to Philosophy*. Oxford: Oxford University Press. Available at: http://www.xrefer.com/entry/551604.

Lowenthal, L. and N. Guterman. 1949. *Prophets of Deceit*. New York: Harper and Brothers.

Lyotard, J. 1989. The dream-work does not think. In A. Benjamin, ed., *The Lyotard Reader*, 19–55. New York: Blackwell.

Luther, S. 1997. Coleridge, creative (day)dreaming, and "The Picture." *Dreaming* 7 (1). Available at: http://www.asdreams.org/journal/articles/7–1_luther.htm.

Machamer, P. 2005. Galileo Galilei. In E. N. Zalta, ed., *The Stanford Encyclopedia of Philosophy*. Available at: http://plato.stanford.edu/archives/spr2005/entries /galileo/.

MacLoed, R. B. 1964. Phenomenology: A challenge to experimental psychology. In T. W. Wann, ed., *Behaviorism and Phenomenology*, 47–73. Chicago: University of Chicago Press.

Marcus, S. 2000. Crisis as sign: From change of representation to representation of change. Paper presented at the German-Italian Semiotic Colloquium "Crisis of Representation," February 18–19, Universität Gesamthochschule Kassel. Paper abstract available at: http://www.uni-kassel.de/iag-kulturforschung /archiv2/krise.htm.

Matheson, R., screenwriter. 1980. *Somewhere in Time*. Universal City, Calif.: Universal.

Matte Blanco, I. 1975. *The Unconscious as Infinite Sets*. London: Duckworth.

——. 1988. *Thinking, Feeling, and Being: Clinical Reflections on the Fundamental Antinomy of Human Beings and World*. New Library of Psychoanalysis no. 5. New York: Routledge.

Maugham, W. S. [1921] 2003. Rain. In W. S. Maugham, *Trembling of a Leaf*. Available at ClassicAuthors.net Great Literature Online: http://www.mostweb.cc /Classics/Maugham/Rain/Rain1.html.

McGoohan, P. and D. Tomblin, executive producers. 1967. *The Prisoner*. Borehamwood, England: Elstree Film Studios, Everyman Films, ITC Entertainment.

McGregor, I., M. T., Gailliot, N. A. Vasquez, and K. A. Nash. 2001. Compensatory conviction in the face of personal uncertainty going to extremes and being oneself. *Journal of Personality and Social Psychology* 80 (3): 472–88.

McGuire, W. J. 1960. Cognitive consistency and attitude change. *Journal of Abnormal Psychology* 60:345–53.

——. 1969. The nature of attitudes and attitude change. In G. Lindzey and E. Aronson, eds., *The Handbook of Social Psychology*, 2d ed., 3:136–314. Reading, Mass.: Addison-Wesley.

Mead, G. H. 1934. *Mind, Self, and Society: From the Standpoint of a Social Behaviorist*. Chicago: University of Chicago Press.

Melloni, L., C. Molina, M. Pena, D. Torres, W. Singer, and E. Rodriguez. 2007. Synchronization of neural activity across cortical areas correlates with conscious perception. *Journal of Neuroscience* 27 (11): 2858–65. Available at: http://www.jneurosci.org/cgi/content/full/27/11/2858.

Miller, A. 1949. *Death of a Salesman, Certain Private Conversations in Two Acts and a Requiem*. New York: Viking.

Mills, U., ed. 1998. *Mark Rothko*. Curator, Jeffrey Weiss. Exhibition organized by the National Gallery of Art, May 3 to August 16. Available at: http://www.nga .gov/feature/rothko/abstraction1.html.

Minsky, M. L. 1985–86. *The Society of Mind*. New York: Simon and Schuster.

——. 1998. Consciousness is a big suitcase: A talk with Marvin Minsky (February 27, 1998). *Edge: The Third Culture*. Available at: http://www.edge.org/features _archive/features_archive_1998.html and http://www.edge.org/3rd_culture /minsky/index.html.

——. 2006. Consciousness. In *The Emotion Machine: Commonsense Thinking, Artificial Intelligence, and the Future of the Human Mind*, 94–128. New York: Simon and Schuster.

Mueller, E. T. 1990. *Daydreaming in Humans and Machines: A Computer Model of the Stream of Thought.*. Norwood, N.J.: Ablex.

Mujica-Parodi, L. R., T. Greenberg, R. M. Bilder, and D. Malaspina. 2001. Emotional impact on logic deficits may underlie psychotic delusions in schizophrenia. In J. D. Moore and K. Stenning, eds., *Proceedings of the 23rd Annual*

Meeting of the Cognitive Science Society, 669–74. Hillsdale, N.J.: Lawrence Erlbaum.

Mumford, L. [1964] 1970. *The Pentagon of Power.* New York: Harcourt Brace Jovanovich.

Munch, E. [1893] 1998. *The Scream.* In *Edvard Munch and Symbolism.* Available through the Munch-museet/Munch-Ellingsen-gruppen at: http://www.museumsnett.no/nasjonalgalleriet/munch/eng/innhold/ngm00939.html.

Murray, H. A. 1938. *Explorations in Personality.* New York: Oxford.

Musil, R. [1918] 1990a. Sketch of what the writer knows. In B. Pike and D. S. Luft, eds., *Precision and Soul,* 61–65. Chicago: University of Chicago Press.

——. [1925] 1990b. Toward a new aesthetic. In B. Pike and D. S. Luft, eds., *Precision and Soul,* 193–208. Chicago: University of Chicago Press.

Mussolini, B. 1932. What is fascism? Available in the *Internet Modern History Sourcebook:* http://www.fordham.edu/halsall/mod/mussolini-fascism.html.

Nirenberg, R. 1996. The birth of modern science: Galileo and Descartes. Project Renaissance: Team A Lecture, week of December 2, 1996. Available at: http//web.archive.org/web/19990429195657/http://www.albany.edu/projren/9697/teama/science2.html.

Nöth, W. 2003. Crisis of representation? *Semiotica* 143 (1–4): 9–15.

Nöth, W. and C. Ljungberg. 2003. Introduction. In *The Crisis of Representation: Semiotic Foundations and Manifestations in Culture and the Media,* special issue of *Semiotica* 143 (1–4): 3–7.

Noveck, I. A. and G. Politzer. 1998. Leveling the playing field: Investigating competing claims concerning the relative difficulty of propositional logic inferences. In M. D. S. Braine and D. P. O'Brien, eds., *Mental Logic,* 367–84. Hillsdale, N.J.: Lawrence Erlbaum. Also available through the Lyon Institute for Cognitive Science at: http://l2c2.isc.cnrs.fr/RDP/novmenuen.htm and http://www.isc.cnrs.fr/nov/NovPltz.html.

NYC ordered to restore funding for 'Sensation'; millions of dollars at issue. 1999. ABCNEWS.com, New York, November 1. Available at: http://abcnews.go.com/sections/us/DailyNews/sensation991101.html.

O'Brien, D. P. 1998. Mental logic and irrationality: We can put a man on the moon, so why can't we solve those logical reasoning problems? In M. D. S. Braine and D. P. O'Brien, *Mental Logic,* 23–44. Hillsdale, N.J.: Lawrence Erlbaum.

Ohnuki-Tierney, E. 1991. Embedding and transforming polytrope. In James W. Fernandez, ed., *Beyond Metaphor,* 159–89. Stanford, Calif.: Stanford University Press.

O'Neill, E. 1956. *Long Day's Journey into Night.* New Haven, Conn.: Yale University Press.

Orwell, G. [1949] 1969. *1984.* New York: Signet Classic.

Osborn, P. and P. Berneis, screenwriters. 1948. *Portrait of Jennie.* Los Angeles: Vanguard Films Production, Selznick Releasing Organization.

Pace, D. 1983. *Claude Lévi-Strauss: The Bearer of Ashes.* London: Routledge and Kegan Paul.

Packwood, D. L. 2002. Module Two: Week Ten. Tutors: John Costello and David Packwood. *Art-History-Online.Info*. Available at: http://www.art-history-online .info/pages/ahoteachmo2w10.htm.

Panofsky, E. [1939] 1962. *Studies in Iconology: Humanistic Themes in the Art of the Renaissance*. New York: Harper and Row.

Pavlov, I. P. [1927] 2001. *Conditioned Reflexes: An Investigation of the Physiological Activity of the Cerebral Cortex*. Trans. G. V. Anrep. Available through Classics in the History of Psychology, an Internet resource developed by Christopher D. Green: http://psychclassics.yorku.ca/Pavlov/lecture16.htm.

Payne, A. and J. Taylor, screenwriters. 2002. *About Schmidt*. New York: New Line Cinema.

Peirce, C. S. [1902] 1950. Logic as semiotic: The theory of signs. In J. Buchler, ed., *The Philosophy of Peirce*, 98–119. New York: Harcourt Brace.

——. [1885] 1993. On the algebra of logic. In C. S. Peirce, *Writings of Charles S. Peirce*, 5:162–90. Bloomington: Indiana University Press.

Penrose, Roland, Sir. 1981. *Picasso, His Life and Work*. Berkeley and Los Angeles: University of California Press.

Pepper, S. C. 1942. *World Hypotheses, a Study in Evidence*. Berkeley and Los Angeles: University of California Press.

Pereira, A. 2007. The difference of unconscious and conscious experience: A physicalist perspective. Commentary no. 10. Target Article no. 95. *Karl Jaspers Forum* 25 (August). Available at: http://www.kjf.ca.

Phillips, J. 2001. Kimura Bin on schizophrenia. *Philosophy, Psychiatry, and Psychology* 8 (4): 343–46.

Piaget, J. 1930. *The Child's Conception of Physical Causality*. New York: Harcourt Brace.

——. [1975] 1977. *The Development of Thought*. New York: Viking.

Piaget, J. and B. Inhelder. [1966] 1969. *The Psychology of the Child*. New York: Basic Books.

Picasso, P. 1937. *Guernica*. Painting at Reina Sofia National Museum Art Centre. Available at: http://www.spanisharts.com/reinasofia/picasso.htm and http://www .spanisharts.com/reinasofia/picasso/e_guernica.htm.

——. [1957] 2005. *Picasso's Las Meninas*: Picasso's studies of *Las Meninas*, August 17 to December 30, 1957. In *A World History of Art*. Available at: http://www .all-art.org/baroque/velazquez11.html

Pinker, S. 2002. *The Blank Slate*. New York: Viking.

Plato. [1892] 1996. *The Apology of Socrates*. In *The Dialogues of Plato*, 3rd ed., trans. Benjamin Jowett, 2:109–35. Oxford: Oxford University Press. Available at: http://www.wsu.edu/~dee/GREECE/APOLOGY.HTM.

Poe, E. A. [2000] 2008a. The cask of amontillado Available through the Literature Network at: http://www.online-literature.com/poe/25/.

——. [2000] 2008b. The tell-tale heart. Available through the Literature Network at: http://www.online-literature.com/poe/44/.

Polanyi, M. 1966. *The Tacit Dimension*. Gloucester, Mass: Peter Smith.

Popper, K. R. 1981. Part 1. In K. R. Popper and J. C. Eccles, *The Self and Its Brain*, 3–424. Berlin: Springer International.

Porteus, S. D. 1931. *The Psychology of a Primitive People*. New York: Longmans, Green.

Queiroz, J., C. Emmeche, and C. N. El-Hani. 2005. Information and semiosis in living systems: A semiotic approach. *S.E.E.D. Journal* (Semiotics, Evolution, Energy, and Development) 5 (1): 60–90. Available at: http://www.dca.fee.unicamp .br/projects/artcog/files/Queiroz-Emmeche-El-Hani_SEED-3.pdf.

Rauschenberg, R. 1951. *White Painting*. In *Singular Forms (Sometimes Repreated), Art from 1951 to the Present*, exhibition at the Guggenheim Museum from March 5 to May 19, 2004. Available at: http://www.guggenheim.org/exhibitions/singular _forms/highlights_1a.html.

Rensink, R. A. 2000. The dynamic representation of scenes. *Visual Cognition* 7 (1–3): 17–42. Available at: http://www.psych.ubc.ca/~rensink/publications/index .html and http://www.psych.ubc.ca/~rensink/publications/download/VisCog _00.02.pdf.

Rich, B. and E. Fitzgerald. [1947] 2001. Blue skies. On *Buddy Rich and His Orchestra: The Golden Essentials 1945–1948*, track 13. Stardust Records.

Robinson, E. A. [1869] 1919. Richard Corey. In Louis Untermeyer, ed., *Modern American Poetry*, 45. Available at: http://www.bartleby.com/104/45.html.

Rokeach, M. 1964. *The Three Christs of Ypsilanti; a Psychological Study*. New York: Knopf.

Rollins, M. 2001. The strategic eye: Kosslyn's theory of imagery and perception. *Minds and Machines* 11:267–86.

Rookmaaker, H. R. 1994. *Modern Art and the Death of a Culture*. Wheaton, Ill: Crossway Books.

Rosch, E. 1978. Principles of categorization. In E. Rosch and B. Lloyd, eds., *Cognition and Categorization*, 28–72. Hillsdale, N.J.: Lawrence Erlbaum.

Roskill, M. 1992. *Klee, Kandinsky, and the Thought of Their Time*. Urbana: University of Illinois Press.

Ross, G., director. 1998. *Pleasantville*. New York City: New Line Cinema.

Ross, J. A. 2003. The self: From soul to brain. *Journal of Consciousness Studies* 10 (2). Available at: http://www.imprint.co.uk/pdf/NYAS.pdf.

Rothko, M. and A. Gottlieb. 2006. Rothko and Gottlieb's letter to the editor, 1943. In M. Rothko and M. López-Remiro. *Writings on Art*, 35–36. New Haven, Conn.: Yale University Press.

Rothstein, E. 1995. *Emblems of Mind: The Inner Life of Music and Mathematics*. New York: New York Times Books.

Rupprecht, C. S. 1999. Dreaming and the impossible art of translation. *Dreaming* 9 (1): 71–99.

Rychlak, J. F. 1977. *The Psychology of Rigorous Humanism*. New York: John Wiley.

——. 1997. *In Defense of Human Consciousness*. Washington, D.C.: American Psychological Association.

Sacks, O. 1970. The lost mariner. In *The Man Who Mistook His Wife for a Hat (and Other Clinical Tales)*, 23–42. New York: Harper Collins.

Sales, S. M. 1973. Threat as a factor in authoritarianism: An analysis of archival data. *Journal of Personality and Social Psychology* 28:44–57.

San Francisco Museum of Modern Art and D. C. Du Pont. 1985. *San Francisco Museum of Modern Art: The Painting and Sculpture Collection.* New York: Hudson Hills Press in association with the San Francisco Museum of Modern Art.

Scambler, G. and N. Britten. 2001. System, lifeworld, and doctor patient interaction. In G. Scambler, ed., *Habermas, Critical Theory, and Health,* 45–67. London: Routledge.

Schank, R. and R. P. Abelson. 1977. *Scripts, Plans, Goals, and Understanding: An Inquiry into Human Knowledge Structures.* Hillsdale, N.J.: Lawrence Erlbaum.

Schilder, P. 1950. On the development of thoughts. In David Rapaport, ed., *Organization and Pathology of Thought,* 497–518. New York: Columbia University Press.

Schor, D. 2001. *The Wisdom of Chelm.* Available at: http://www.tsimmes.com /chelm/.

Searle, J. R. 1980. *Las Meninas* and the paradoxes of pictorial representation. In W. J. T. Mitchell, ed., *The Language of Images,* 247–58. Chicago: University of Chicago Press.

——. [1980] 1981. Minds, brains, and programs. In J. Haugeland, ed., *Mind Design,* 282–306. Cambridge, Mass.: MIT Press.

——. 1992. *The Rediscovery of the Mind.* Cambridge, Mass.: MIT Press.

Seigel, J. E. 2005. *The Idea of the Self.* Cambridge: Cambridge University Press.

Set theory. 2001. In *Macmillan Encyclopedia.* Available at: http://www.xrefer.com /entry.jsp?xrefid = 515429&secid = .-&hh = 1.

Shawn, W. and A. Gregory, screenwriters. 1981. *My Dinner with André.* Directed by Louis Malle. New York: New Yorker Studio; Andre Production Company.

Sheppard, A. 2001. History and systems of psychology. University of Maine at Presque Isle Lecture no. 11. Available at: http://www.umpi.maine.edu/~sheppard/475 /syll475.html.

Shields, C. 2005. Aristotle's psychology. In E. N. Zalta, ed., *The Stanford Encyclopedia of Philosophy.* Available at: http://plato.stanford.edu/archives/sum2005 /entries/aristotle-psychology/.

Skinner, B. F. 1953. *Science and Human Behavior.* New York: Macmillan.

——. 1950. Are theories of learning necessary? *Psychological Review* 57:193–216.

Smith, D. W. 2003. Phenomenology. In E. N. Zalta, ed., *The Stanford Encyclopedia of Philosophy.* Available at: http://plato.stanford.edu/archives/win2003/entries /phenomenology/.

Solomon, L. J. 1998. The sounds of silence: John Cage and 4'33". Available at: http: //solomonsmusic.net/4min33se.htm.

SoRelle, R. 1998. Romania's forgotten children. Story by Ruth SoRelle. Photos by Smiley N. Pool. *International Association of Physicians in AIDS Care Journal* 4 (3). (A version of this story was published in the *Houston Chronicle.*) Available at: http://www.aegis.com/pubs/iapac/1998/IA980305.html.

Sowa, J. F. [1984] 2007. Lattices. In *Conceptual Structures: Information Processing in Mind and Machine.* Reading, Mass.: Addison-Wesley. Available at: http:

//www.jfsowa.com/logic/math.htm and http://www.jfsowa.com/logic/math
.htm#Lattice.

Stern, Jessica. 2003. *Terror in the Name of God*. New York: Ecco, HarperCollins.

Stern, Josef. 2000. *Metaphor in Context*. Cambridge, Mass.: Bradford Books, MIT
Press.

Stone, A. A. 1997. What remains of Freudianism when its scientific center crum-
bles? *Harvard Magazine* (January–February). Available at: http://www.harvard-
magazine.com/issues/jf97/freud.html.

Sullivan, H. S. 1944. The language of schizophrenia. In J. S. Kasanin, ed., *Language and Thought in Schizophrenia*, 4–16. Berkeley and Los Angeles: University of California Press.

Taylor, C. 1989. *Sources of the Self*. Cambridge, Mass.: Harvard University Press.

Tejera, V. 1971. *Modes of Greek Thought*. New York: Appleton-Century-Crofts.

Thagard, P. 2006. *Hot Thought: Mechanisms and Applications of Emotional Coherence*. Cambridge, Mass.: MIT Press.

Thagard, P. and C. Shelley. 1997. Abductive reasoning: Logic, visual thinking,
and coherence. In M. L. Dalla Chiara, K. Doets, D. Mundici, and J. Van Ben-
them, eds., *Logic and Scientific Methods*, 413–27. Dordrecht: Kluwer. Available
at: http://cogprints.ecs.soton.ac.uk/archive/00000671/00/FAbductive.html.

Thomson, J. and A. Craighead. 1998. *Trigger Happy*. Installation. Available at:
http://www.thomson-craighead.net/docs/thap.html.

Thornton, S. P. 2001. Sigmund Freud (1856–1939). In J. Fieser, gen. ed., *The Internet Encyclopedia of Philosophy*. Available at: http://www.utm.edu/research
/iep/f/freud.htm.

Tinbergen, N. 1951. *The Study of Instinct*. London: Clarendon Press.

Tribe, M. 2001. *Mark Tribe: Game Show: Net Games Now*. Curatorial project, exhibition
of online games. Massachusetts Museum of Contemporary Art, North Adams,
Mass., May 2001 to June 2002. Available at: http://nothing.org/gameshow/index
.htm.

Turk, D. J., T. F. Heatherton, W. M. Kelley, M. G. Funnell, M. S. Gazzaniga, and C. N.
Macrae. 2002. Mike or me? Self-recognition in a split-brain patient. *Nature Neuroscience* 5 (9): 841–42. Available at: doi:10.1038/nn907, http://www.nature.com
/cgi-taf/DynaPage.taf?file = /neuro/journal/v5/n9/abs/nn907.html&dynoptions
= doi1058547984.

Turner, M. 1998. Figure. In A. Katz, M. Turner, R. W. Gibbs Jr., and C. Cacciari,
eds., *Figurative Language and Thought*, 44–87. New York: Oxford University
Press.

Van Strien, P. J. and E. Faas. 2004. How Otto Selz became a forerunner of the
cognitve revolution. In T. C. Dalton and R. B. Evans, eds., *The Life Cycle of
Psychological Ideas: Understanding Prominence and the Dynamics of Intellectual
Change*, 175–201. PATH in Psychology. New York: Kluwer Academic; Plenum.

Vogeley, K., M. Kurthen, P. Falkai, and W. Maier. 1999. The human self construct
and prefrontal cortex in schizophrenia. Target paper for the Association for
the Scientific Study of Consciousness Electronic Seminar, March 15 to April 9.
Available at: http://www.phil.vt.edu/assc/esem.html.

Voltaire. 1759. *Candide*. Ed. and trans. Theodore Besterman. Available through the Voltaire Foundation, University of Oxford, Electronic Resources: http://www .voltaire.ox.ac.uk/x_voltfnd/etc/e-texts/www_etexts/index_prose_candi.html.

Von Domarus, E. 1944. The specific laws of logic in schizophrenia. In J. S. Kasanin, ed., *Language and Thought in Schizophrenia*, 115–23. Berkeley and Los Angeles: University of California Press.

Vygotsky, L. 1934. Thought in schizophrenia. *Archives of Neurology and Psychiatry* 31:1063–77.

———. [1939] 1975. *Thought and Language*. Trans. Eugenia Hanfmann and Gertrude Vakar. Cambridge, Mass.: MIT Press.

Wax, M. L. 1983. How Oedipus falsifies Popper: Psychoanalysis as a normative science. *Psychiatry* 46:95–105.

———. 1995a. How secure are Grünbaum's foundations? *International Journal of Psycho-analysis* 76:547–56.

———. 1995b. Method as madness: Science, hermeneutics, and art in psychoanalysis. *Journal of the American Academy of Psychoanalysis* 23 (4): 525–43.

Werner, H. 1948. *The Comparative Psychology of Mental Development*. New York: International Universities Press.

Werner, H. and B. Kaplan. 1963. *Symbol Formation*. New York: John Wiley.

Westen, D. 1998. The scientific legacy of Sigmund Freud: Toward a psychodynamically informed psychological science. *Psychological Bulletin* 124: 333–71.

White, H. 1999. *Figural Realism: Studies in the Mimesis Effect*. Baltimore: Johns Hopkins University Press.

Widener, L. 2001. *Thimble Theatre*, by Elzie Segar. In *The NeverEndingWonder.com Library of Cultural Arts. The Greatest Comic Strips Ever, Part One*. Available at: http://www.neverendingwonder.com/comics.htm.

Winnicott, D. W. [1971] 1988. *Human Nature*. Schocken Books.

Wirth, U. 2003. Derrida and Peirce on indeterminancy, iteration, and replication. *Semiotica* 143 (1–4): 35–43.

Woodworth, R. S. 1938. *Experimental Psychology*. New York: H. Holt.

Zaillian, S., screenwriter. 1993. *Schindler's List*. Directed by Steven Spielberg from the book by Thomas Keneally. Universal City, Calif.: Amblin Entertainment, Universal Pictures.

Acknowledgments

The best way to thank those who helped make this book possible is to think like Frank Capra. With angel Clarence's help, Capra's hero, George Bailey, in the fabled film *It's a Wonderful Life*, experiences his diminished self in a world of people who never knew he existed.

What if the people who deeply affected me did not exist? Without their love and influence, I would be less than myself. This book would be an apparition of a front and back cover—no more. Capra is right. The pages of my book do not open to declare my ideas unless I recognize and celebrate those closest to me; unless I acknowledge my indebtedness to those who gave me intellectual impetus; and unless I express my profound gratitude to those, who were closest to the writing tasks.

My wife, Helene, is the other side of my soul. Her love is the heart of her soul and the heartbeat of her intellect and wisdom. If this book ever could have been written without these parts of me, it would be without imperatives or values. My sons, Marc and Saul, embody my ideals, but discard my over-commitment to the inner sources of transcendental dialectic. I burrow in, searching strings for symmetries of thought. Marc and Saul listen to dialogue for harmonies and counterpoint that widen the communication net. Their superb accomplishments challenge me to reach out for the ideals they transmute and extend to their children—who renew the challenge in yet undiscovered dimensions.

In 1962, I interned under David Wechsler, a giant whose tiny 4-by-6 photo of Alfred Binet was displayed in a thin black frame on a barren cracked wall in his City-issue hospital office at Bellevue. Binet's image in stark black and white was imposing. He was in the room as I presented my idea for a structural account of the psychopathology of thought. Wechsler immediately grasped my intention to extend von Domarus' analysis of the logic of schizophrenic thinking. He seized on my ambition by reflecting the challenge back to me. Logic, motivation, and the value of intense inquiry were coiled up in his insistent manner. Coil and spring; the idea could be realized—it was up to me. The project, a rationalist's dream, was not to be, because of the untimely death of my then thesis adviser, Irving Lorge, who had approved the project. Lorge grasped the idea's potential; Wechsler deeply motivated the quest; Binet's image obliged the idea to stay alive. Three great empiricists had invested their excitement.

The project hibernated in an age and atmosphere not overly friendly to structuralist models of subjective processes. But the idea surfaced in various forms. In 1973 I specified the logic of defense mechanisms and defensive thinking. I thank Paul Secord and Rom Harré for their editorial encouragement and considerable effort. In 1975, for the same editors and journal, I formulated the logic of forming a new idea. Along with my 1985 article, "The Logical Structure of Freud's Idea of the Unconscious," I set my continuing query in motion. How wide was the range of cognitive phenomena, which squeeze into a template of logical forms; even as their meanings unfold in a wild array of configurations?

My dear colleague and friend, Harold B. Pepinsky, whose intellectual and emotional support was deep, put me in touch with Rogelio Diaz Guerrero in 1971. Rogelio could uniquely synthesize philosophical thinking with psychometrics, operational analysis, and empirical demonstration. With his invaluable encouragement and with Luis Natalicio's understanding and foresight, I published on the language and logic of repression. With "Pep's" support through Ohio State University, I presented two papers in December 1972 at the Seminar on Language Instruction and Research, Merida, Yucatan: "The Natural Language of Repression" and "Psycho-Analysis and Language Analysis."

In 1978, due to Rogelio's efforts and the request of Samuel Hayakawa, the University of Kansai invited my paper, "Image Grammar and Idea Formation," and sponsored my presentation at the Symposium on Intercultural Communication at the University of Hawaii at Hilo. My attention to linguistic form was foreshadowed by a paper I wrote in 1977, after I had an opportunity to meet with David McNeill. His encouragement and comments were of considerable influence in my work on the transformation of meaning and form from a percept of an object to its representation. McNeill's breadth in his approach to cognition and linguistics was so great, he avoided the excesses of empirically based epistemology, while being an exceptional empirical scientist. The topic in that 1977 paper springs the coil. The project expands beyond the issue of different levels of psychopathology. I relate logical form to a range of poetics as well as to primitivized and pathological thinking.

In the 1980s both the phenomenological and the structuralist viewpoints had been sidelined, and the role of language was emphasized to the detriment of an understanding of the forms of thought. In 1995, I published a conception of self and agency as a refutation of radical constructionism in psychology. My foray into agency and self became a gauntlet to throw to several respondents, including Rom Harré, who properly and challengingly argued the other side of things. Wouldn't you know this opposition of ideas stayed with me and encouraged me to flesh out my view of the subjective self and its logical form?

The rationalist in me persists; I dig deeper to view the primacy of logical form in its relation to thought and psyche. I am particularly proud of my 1998 *Semiotica* article arguing the dynamic qualities of logical form and differentiating the logical form of several different tropes or types of figural thought. When I began my journey to find the role of the self in relation to logic and its range of forms in thought, my then editor, Gary Dunning, suggested I look at James W. Fernandez' work on metaphor. The combination of depth and breadth of Fernandez' understanding of a variety of cultures and his attention to the structure of metaphor emboldened me to pursue the structure of figurative thought. *The completion of my book on the subjective self in 2001 was the proximal force by which I could return to the project I envisioned for David Wechsler, Irving Lorge, and Binet's ghost.*

I am pleased with Frank Nuessel's invitation for and positive reaction to my 2006 article on the nature of metaphor relative to the self and the person's phenomenological representations of an audience. In all, the string of *Semiotica* articles set the direction for my probe into figurative thought, its logic, its dynamics, and ultimately its relation to literal thinking.

To unfold the analysis of tropes, I turned to Hayden White's brilliant writing to gain from his grand insight into a hierarchy of figuration and its linguistic expression in figures of speech. Special thanks and my utmost gratitude not only for Hayden White's incisive grasp of the structure of my effort but also for his read of my manuscript, which he did with his characteristic breadth of understanding and uplifting commentary. I am also hugely grateful to Luis Radford. His scholarly work offers concepts, deep understanding, and magnificent resolutions of the different language yet similar concepts of Peirce and Piaget. In the travail of sifting ideas from different thinkers, with different disciplines and configurations of sources for their thought and definitions, Radford's cut to the elements is elegant. I appreciate his crystal-clear suggestions after a generous read of my manuscript.

For this present book, I am super-fortunate to have Wendy Lochner as Senior Executive Editor at Columbia University Press. She is exceptionally knowledgeable in a variety of disciplines—an absolute boon to the writer, who is inter- and cross-disciplinary. Her gift is the flexibility to navigate ideas without getting stuck in the excesses of either momentary fads or over-dichotomous divisions. If this were not enough, her responsiveness and sense of responsibility were extraordinary. Annie Barva's work as copy editor is outstanding; she is a relentlessly logical reader. Her talent is invaluable for the manuscript. Senior Manuscript Editor Roy Thomas's clarity is awesome. He is the model for thoughtful, precise, and devotedly expeditious review. Indefatigable in his masterful handling of the details of the book's manuscript stages, he skillfully navigated my interminable questions and emendations, and he steadfastly steered the book through to its production. In all, thanks for an impressively helpful, competent, and pleasant experience to all at the Columbia University Press. Praise is particularly due to Christine Mortlock and Marisa Pagano for their fine efforts.

Ah, Clarence has his wings. The book has opened to its pages.

Index